MW01035894

THE FRONTIERS COLLECTION

THE FRONTIERS COLLECTION

The books in this collection are devoted to challenging and open problems at the forefront of modern science, including related philosophical debates. In contrast to typical research monographs, however, they strive to present their topics in a manner accessible also to scientifically literate non-specialists wishing to gain insight into the deeper implications and fascinating questions involved. Taken as a whole, the series reflects the need for a fundamental and interdisciplinary approach to modern science. Furthermore, it is intended to encourage active scientists in all areas to ponder over important and perhaps controversial issues beyond their own speciality. Extending from quantum physics and relativity to entropy, consciousness and complex systems – the Frontiers Collection will inspire readers to push back the frontiers of their own knowledge.

Information and Its Role in Nature
By J. G. Roederer

Relativity and the Nature of Spacetime
By V. Petkov

Quo Vadis Quantum Mechanics?
Edited by A. C. Elitzur, S. Dolev,
N. Kolenda

Life – As a Matter of Fat
The Emerging Science of Lipidomics
By O. G. Mouritsen

Quantum–Classical Analogies
By D. Dragoman and M. Dragoman

Knowledge and the World
Challenges Beyond the Science Wars
Edited by M. Carrier, J. Roggenhofer,
G. Küppers, P. Blanchard

Quantum–Classical Correspondence
By A. O. Bolivar

Mind, Matter and Quantum Mechanics
By H. Stapp

Quantum Mechanics and Gravity
By M. Sachs

Extreme Events in Nature and Society
Edited by S. Albeverio, V. Jentsch,
H. Kantz

**The Thermodynamic
Machinery of Life**
By M. Kurzynski

**The Emerging Physics
of Consciousness**
Edited by J. A. Tuszynski

Weak Links
Stabilizers of Complex Systems
from Proteins to Social Networks
By P. Csermely

Peter Csermely

WEAK LINKS

Stabilizers of Complex Systems from Proteins to Social Networks

With 52 Figures and 12 Tables

Prof. Dr. Peter Csermely
Semmelweis University
Department of Medical Chemistry
P.O.Box 260
1444 Budapest 8
Hungary
e-mail: csermely@puskin.sote.hu

Series Editors:

Cover figure: The cover image shows a detail from "Shock-Induced Vorticity" by J.O. Langseth

Library of Congress Control Number: 2006921156

ISSN 1612-3018
ISBN-10 3-540-31151-3 Springer Berlin Heidelberg New York
ISBN-13 978-3-540-31151-5 Springer Berlin Heidelberg New York

Springer is a part of Springer Science+Business Media
springer.com

Typesetting by Stephen Lyle using a Springer TEX macro package
Production and final processing by LE-TEX Jelonek, Schmidt & Vöckler GbR, Leipzig
Cover design by KünkelLopka, Werbeagentur GmbH, Heidelberg

Printed on acid-free paper SPIN: 12187966 57/3180/YL - 5 4 3 2 1

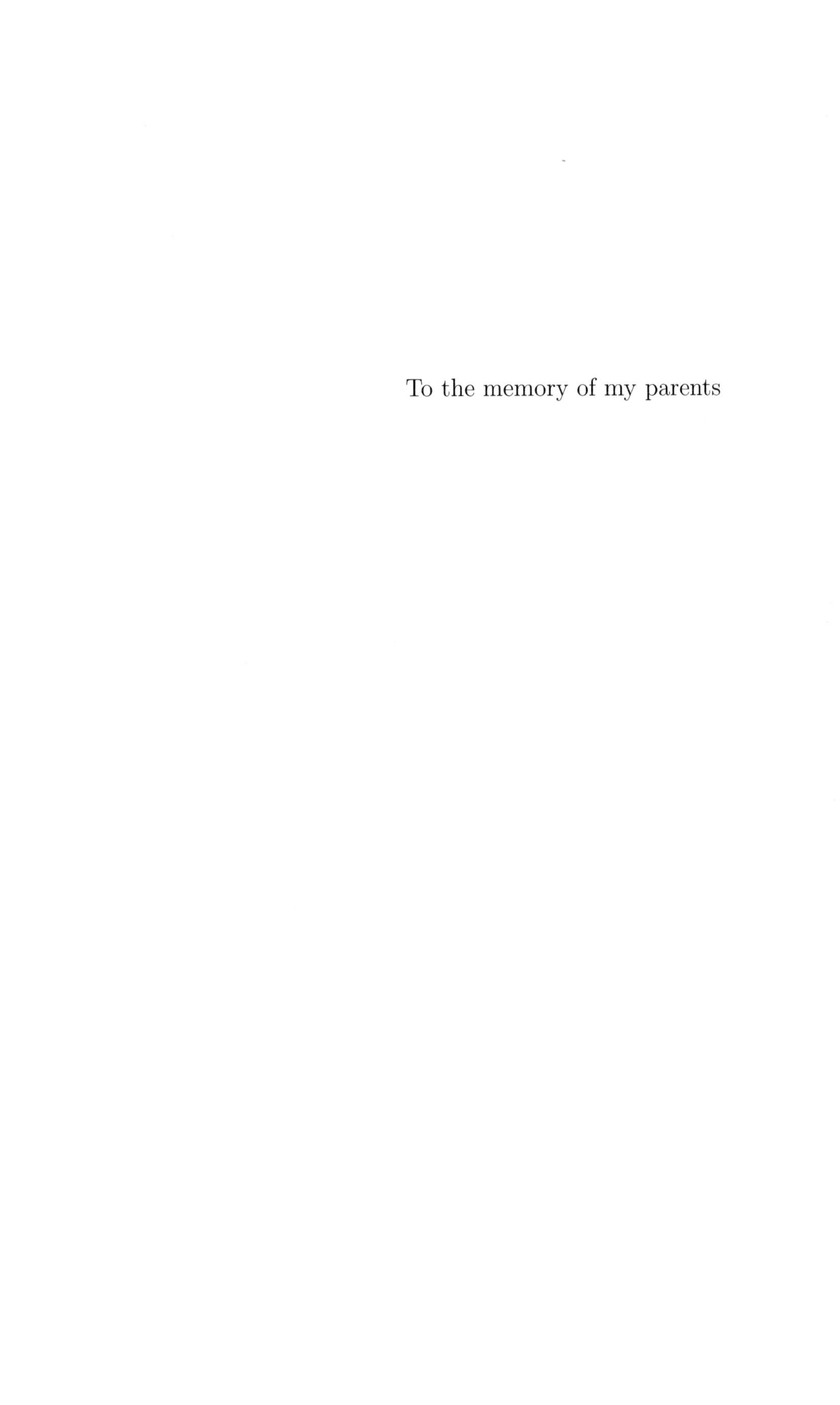

To the memory of my parents

Preface

In 1990 I started to work with molecular chaperones as an ordinary biochemist. Chaperones are the proteins that form our cells' most ancient defense system. I found them fascinating molecules. They protect other proteins and, consequently, help our cells to survive. If we quarrel, if we are anxious, or just run our daily marathon to catch the morning bus, our proteins become damaged. And damaged proteins are sticky. They aggregate, which is toxic to the cell. Chaperones protect these damaged proteins against unspecific, unplanned aggregation, like their eponyms, the ladies at the grand ball, who would protect young girls against unspecific, unplanned aggregation with the boys at the ball. Chaperones are everywhere. They are needed for protein folding and refolding, for proteolysis and transport. Chaperones are highly conserved and form a part of the essential gene set (Koonin and Galperin, 2002). Without them, no life could be imagined on Earth.

Chaperones are truly altruistic. They help, wherever they can. But how do they help? This was my first question. For five years I tried almost everything an ordinary biochemist could do. I purified them,[1] I cut them into pieces, cooked them and soaked them in an arsenal of chemicals and radioisotopes. By the middle of the 1990s, I realized that chaperones are different. They stick. They bind to their target proteins, their modulator proteins, the cytoskeleton, the whole world. If chaperones glue the whole cell together, how can it change? How do cells divide and how do they move?

The secret is affinity. Chaperones make *low* affinity interactions with their partners. Now they bind it, now they don't. They are dynamic. For their omnipresent help, they form weak links which change often. What makes life easy for the cell is a headache for the researcher.

[1]Footnotes will refer to additional information which is not needed to understand the main text. Therefore the reader may skip them. The first footnote is about the word 'purification'. We biochemists use this word for the procedure whereby we cut all original contacts of the protein and extract it from natural conditions, hoping and believing that it will remain unchanged.

Most of the chaperone complexes change if you start to examine them. Now you see them, now you don't. Chaperones give the ordinary biochemist nightmares. It is better to change subject if you want to use your usual assays in a sensible manner. I did not change subject – so the subject changed me. In 1990, I was looking for a well-defined question, and instead, I found a whole world with an astonishing complexity.

In 1998 a seminal paper by Suzanne L. Rutherford and Susan Lindquist appeared in Nature. The take-home message was that one of the chaperones, Hsp90, helps developmental stability. If this chaperone works, almost all the *Drosophilas* look alike. Each of these fruit flies gazes at the world with two complex red eyes, uses six small legs to balance her fragile body, and has two wings, which buzz in unison. If the Hsp90 was damaged, the newly born fruit flies went crazy. Fortunately, not all ten thousand of them did get damaged. Most of the flies still gazed, balanced and buzzed alike. However, *some* of them (exactly 174) became frightening monsters. These poor flies did not have correct eyes, their wings got distorted, their legs were deformed and a number of other malformations also occurred. Indeed they looked pretty miserable. But, miserable in *different* ways. An astonishing diversity appeared, and what is more, this diversity was inheritable. This was due to a variety of preserved silent modifications in the genome of the *Drosophila* population. Normally, Hsp90 buffered these changes and stabilized the appearance of the fruit flies, the phenotype. When Hsp90 was inhibited or damaged, the buffer diminished, and a burst of diversity suddenly appeared.

I got the feeling that something truly new had happened. Chaperones help the proteins around them. I could not figure out exactly how they do this, but at least I had an idea: chaperones bind to their target proteins and stabilize them or change their shape. But how can they stabilize a fruit fly, which is much bigger than them?

The explanation I offered myself was still quite standard, saying that chaperones repair mutant proteins which cannot exert their effects on the phenotype. Hsp90, the chaperone in the Rutherford and Lindquist (1998) experiments has hundreds of client proteins which always require its presence for their activation. Most of these clients participate in various steps of signal transduction. Let us suppose that the gene of one of these client proteins suffers a mutation, and that the mutation changes a critical amino acid and cripples the shape of the protein. Let us also suppose that Hsp90 is able to repair this damage, and that, if Hsp90 operates at full strength, the effect of the mutation is not seen. Finally, let us suppose that the client protein was critical in a signaling pathway of the morphological development. If Hsp90 is damaged, the mutation will impair the client. The missing client causes a collapse in morphological signaling and the fly will become a monster. The explanation seemed to be rather easy. (Well, it was easy for us, but not for the fly.)

There were disturbing signs though. Chaperones were not the only things that could hide the monsters within some of the normal *Drosophilas*, who seemingly gaze, balance and buzz just like their peers with a normal genome. There were numerous other proteins, which provided the same buffering, either in this or in other experimental systems (Aranda-Anzaldo and Dent, 2003; Gibson and Wagner, 2000; Scharloo, 1991; True and Lindquist, 2000). Moreover, in 2003 it was proposed that an astonishingly large number of proteins could regulate developmental stability (Bergman and Siegal, 2003). I became puzzled. I found chaperones fascinating. I loved them, and love always carries us to extremes. We see our beloved everywhere. Everything reminds us of her, she is everywhere. But wait a moment! Most of the proposed proteins had nothing to do with chaperones. Chaperones turn up here and there. But the whole cell cannot be a chaperone! The old explanation was clearly not adequate.

I think I am lucky. When one meets the unexpected, a fresh mind is needed, which finds an immense joy in each playful new thought. There are exceptional people, who have this even in their eighties. There are others, who are lucky enough to be stimulated by others. I started a project in 1996 giving research opportunities for high school students (`http://www.kutdiak.hu`). This movement changed the life of many students, and changed my life too. The students in my lab helped me to take a new look at the world. They were the seeds of the LINK group, who helped to write this book.

Let me put things together again. Rutherford and Lindquist (1998) showed that chaperones buffer the morphological diversity induced by the silent mutations of fruit flies. The inhibition of numerous other proteins can also lead to morphological diversity. However, there was something else here. Rutherford and Lindquist (1998) also demonstrated that stress induces a broader morphological diversity. In fact, the stress-induced, prolonged increase in morphological diversity was first shown by Schmalhausen and Waddington much earlier (Schmalhausen, 1949; Waddington, 1942; 1953; 1959). At first, I did not take much note here. It all seemed rather easy: perfect flies are alike, while stressed flies become damaged, *differentially* damaged. Diversity is revealed in the damage. At the molecular level, stress means more damaged proteins. Chaperones try to repair them, and so become occupied, which is just another form of inhibition.

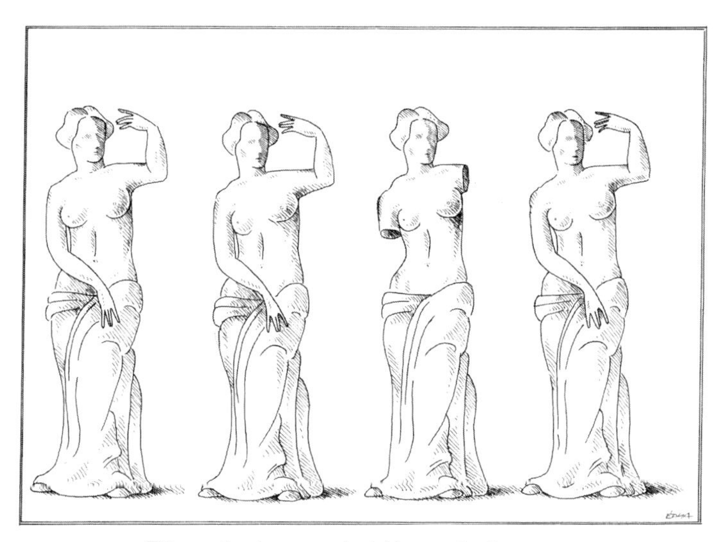

Diversity is revealed through damage

In May 2003, I happened to read the review by Rao et al. (2002). This opened a new world to me. Stress not only induces morphological diversity, but also a thousand other types of diversity. Each bacterium normally swims towards its food. But not when stressed! Here, some of them got really distressed, and either swam in the opposite direction, moved round in circles, or just didn't go anywhere. Here we have diversity again. *Bacillus subtilis* responds to environmental stress with an arsenal of probabilistically invoked survival strategies. Stem cell differentiation or the appearance of various types of cancers can all be a source of similar diversity. Can all these forms of diversity be buffered by chaperones?

Putting this together, we are bound to ask: do we have here a large number of mysterious proteins which stabilize practically everything? This is the time when one goes for a vacation or asks around. As I was too excited to go for a vacation, I wrote to some of my best friends (I can imagine their faces as they stared at their laptops: "Peter went really crazy this time ...") – and got back some great ideas. Tamas Vicsek suggested that I read the recent book by Laszlo Barabasi on networks (*The Linked*, Barabasi, 2002). In parallel, I started to read *Investigations* by Stuart Kauffman (2000). These were the best books I had read for quite a while.

Then I had to sit down. Practically every complex system can be imagined as a network. Atoms form a network making macromolecules. Proteins form a network making cells. Cells form a network making organs and bodies. We form a network making our societies, and so on. Most of these networks are a result of self-organization. In fact, self-organization seems to be an inherent property of matter in our Universe. The resulting networks have a lot of common features, from their topology to their dynamism.

These systems are far too similar. The protein net, where chaperones work, should behave in the same way as every other network. *All* networks must have a component which stabilizes them, like chaperones and the mysterious proteins stabilizing the cells. But what is the common feature of all these elements? Why do they stabilize the whole network? At the beginning I had only one idea, and even this was negative. The common feature *cannot* be anything related to chaperone function. Chaperones protect other proteins by helping to refold them. People cannot protect their friends by helping to refold them! A more general approach is needed here.

Although I did not know it, the solution was already in my hands. Chaperones should give us a clue. Which of their features can be generalized to *all* networks? Chaperones stick. They make links to a number of other proteins. They are hubs. Do hubs stabilize their networks? Well, not really. Hubs are needed to *form* their networks. If we attack hubs, the network collapses (Albert et al., 2000). When we attack chaperones, the network, e.g., the cell survives. It becomes destabilized, but survives. Another chaperone feature must be more important.

What else? Affinity! Yes, affinity. Here was an idea: *The components which stabilize the various systems must all have weak links to the others.* It is not the component that counts, but the type of link it builds to the others. By the end of 2003, the basic idea of this book was born: weak links stabilize all complex systems. Weak links give us a universal key to understanding network diversity and stability, and they are the major actors in this book.

Months of tedious, systematic reading followed. I read dozens of books, collected approximately 600 Mbytes of pdf files, which made a pile of printed hard copies three meters high. I realized that my 'new' idea (weak links stabilize all complex systems) has been an obvious feature in the social sciences for decades (Granovetter, 1973). The same idea had been proven in ecosystems in 1998 (Berlow, 1999; McCann et al., 1998). As I browsed page after page, many other examples appeared, and they will be detailed in the following chapters. This made me rather confident that I had found something genuinely important and general. Interestingly, many authors (like Mark Buchanan in his book, *Nexus*, 2003) had come to the conclusion that weak links stabilize complex systems in their own discipline, but none of them had generalized it to all networks.[2] It seems that it was the chaperones, which stick but form only *weak* links, that had made the important link here.

While I was reading one book after the next and paper after paper, I got more and more surprises:

- I realized that each of the disciplines has a completely different vocabulary for the very same message. (Appendix B is a glossary intended to guide the reader through this jungle of terminology.)
- It was a frightening moment when the LINK group realized that we had completely run out of words and had no way of talking about something so truly simple and beautiful. But let me reassure you: the book is not full of newly constructed pseudo-words. We always managed to get around the problem ourselves and find a novel use for some existing word. However, on many occasions, it took us some time. When one has to use words in a completely different context, one's mind seldom obeys at first.
- The readings gave me a great and sincere respect for the social sciences. In network studies they are a whole lifetime ahead! Jacob Moreno started network studies on friendship patterns and Alfred Lotka published his famous law on scientific productivity in 1926, when my father was born. Anatol Rapoport stressed the general importance of the topology of friendship networks in 1957, one year before my own birth (Newman, 2003).

The book is structured as follows. In Chap. 1, I describe the beginnings of the weak link concept, the Granovetter study, and define weak links. Chapters 2 and 3 summarize the description and dynamics of networks. Chapter 4 introduces the concept of weak links as universal stabilizers, while Chaps. 5 through 11 invite the reader on a journey through Netland, presenting a ladder of exciting examples starting from macromolecules and ending at our own planet. Finally, Chap. 12 summarizes and reformulates the stabilizing role of weak links, bridging it with the

[2]The following remark by Siljak (1978) counts as another predecessor of these thoughts: "A dynamic system composed of interconnected subsystems is reliable if all subsystems are self-sufficient and [...] the magnitude of the interactions does not exceed a certain limiting value." This statement may be regarded as a forerunner of the main thesis of this book, but Siljak's stability criterion can be formulated much more easily using the concept of networks as it is presented here.

concept of stability landscapes and game theory. If you dislike physics or the biochemistry of small molecules, feel free to start your journey in Netland at Chaps. 7 or 8, which describe the networks of our own body and our societies.

When the first draft of the book was finished in January 2004, I realized that I had probably filled a niche. According to Newman (2003): "Studies of the effects of structure on system behavior are still in their infancy." Cross-disciplinary thinking on network properties is also largely non-existent. The moral is that we should use this enormous resource more often, always examining what we have proved in one of the disciplines when it is transferred to all the others. I have done my best. However, I am aware that analogies provide a very fruitful but extremely dangerous field. Therefore I will separate the analogies from the established facts by quoting the original source of information after each fact and by putting most of the analogies into a box in the following manner:

Caution! Hypothesis! As you proceed in the book, wild ideas will appear along the way. I expect a good deal of red ink from the referees: "Speculative!" But I have an excuse: I have *marked* all these hypotheses using one, two or three of the smiley figures on the left. The figure has big hands as a reference to the great Hungarian magician, Rodolfo, who always said: "Watch my hands! Caution! I am cheating!" One smiley means that I do not have enough evidence to formulate the statement as unequivocal truth. Two smileys warn you that, though the statement is logical, its background is largely missing. And three smileys? Well, three smileys will make you either smile or run to the phone to call the doctors in white coats. Three smileys are mostly fiction, rather than science. So why did I put them in this book? They have dared to appear here because they constitute fascinating, mind-boggling ideas. Smiley comments will always be exciting, but I am not quite sure that they will turn out to be true.

Additional information. Those parts where you find the wise head on the left will most probably turn out to be true tomorrow, and even the day after tomorrow, but they are details that will not necessarily interest all readers. Start reading, but if you do not find it interesting, skip it.

Important questions. When you begin to study a new territory, you always have more questions than answers. (In fact, a good scientist *always* has more questions than answers.) So we wondered why we should keep these questions to ourselves, and we decided to share. If you have a good idea for an answer, we would be more than pleased to read and discuss it. Join the LINKs! The email address is at the end of this Preface.

My master's voice: Spite. Sometimes you will see a remark in the text like: *"Peter, you made the typography of this book rather confusing for me. First of all, I cannot read your small letters in the remarks. Moreover, the font you selected for me is the ugliest one I have ever seen."* Spite! Welcome! Spite is my best friend. When you try to write a book, your best friend is the most critical person around you. I am lucky enough to have quite a few such fierce critics among my students.

Some of the sentences in the above remarks were written in the plural. What has happened? Does the author think he has found such a good idea that he may start to speak in the plural, as if to say: "We, the founders of this new science, declare ..."? Not so! The more 'we' know, the more humble 'we' grow. 'We' refers to the members of the LINK group. The LINKs are young people (at least in mind!), who work in different institutions but are strongly linked to each other by their love of weak links and decided to form a virtual lab. Members of the LINK group helped to shape this book. They sent great ideas to each other by emails, by SMS messages and even on slips of paper. Questions arose sometimes during the day, sometimes during the night. The LINKs attacked the sloppy sentences of this text and tried to make the content of the book more understandable. We all hope that this joint effort has brought at least a little improvement. If not, please send all your comments to us. Here are some of the key people in the LINK group:

Péter Csermely left the János Apáczai Csere high school in Budapest, Hungary in 1976. He won several awards in national and international chemistry contests. He is currently professor of biochemistry at the Semmelweis University in Budapest, Hungary and a fellow of Ashoka International. He has published nine books and almost two hundred research papers. He started a project in 1996 which provides research opportunities for more than seven thousand high school students in the best research teams (`www.kutdiak.hu`).

István Kovács left the János Berze Nagy high school in Gyöngyös, Hungary in 2003. He won awards in more than a dozen national physics and mathematics contests. He is currently a physics undergraduate at the Eötvös University of Sciences in Budapest, Hungary. He published his first scientific paper at the age of 19.

Balázs Papp left the high school of the Debrecen University, Hungary in 1996. He received his MSc degree in genetics at the Debrecen University and his PhD from the Eötvös University in Hungary. He is currently a postdoctoral fellow of the University of Manchester. He has published four papers in Nature, one in Nature Genetics, won several honors and awards including two Marie Curie Fellowships and a Pro Scientia Medal.

Csaba Pál left the István Dobó high school in Eger, Hungary in 1993. He received his MSc and PhD degrees from the Eötvös University in Hungary. He is in the Theoretical Biology Group of the Hungarian Academy of Sciences. He has published four papers in Nature, three in Nature Genetics, one in Science, as well as six papers in various Trends journals. He was a Royal Society Postdoctoral Fellow and won the Talentum Award of the Hungarian Academy of Sciences in 2005.

Máté Szalay left the László Lovassy high school in Veszprém, Hungary in 2003. He was the recipient of the 2003 Junior Bolyai Award of computer science. Between 2003 and 2005 he was the president and later the managing president of the Hungarian Research Student Association (`www.kutdiak.hu`). He is currently a computer science undergraduate at the Technical University in Budapest, Hungary.

An interdisciplinary subject is always dangerous. One cannot know, and cannot even understand everything. In spite of this, writing about weak links must not mean weak writing. I owe a lot of thanks to the eminent scientists of various disciplines, my friends, who read the summary of this book or its chapters. I am thankful for their comments, ideas and encouragement. The help of Luigi Agnati, Eszter Babarczy, László A. Barabási, Attila Becskei, Eric L. Berlow, Gustav Born, Zoltán Borsodi, Geoffrey Burnstock, György Buzsáki, Vilmos

Csányi, Ken Dill, Gerald M. Edelman, András Falus, Viktor Gaál, Balázs Gulyás, Mária Herskovits, Gergely Hojdák, Roland Iványi-Nagy, Gáspár Jékely, Ferenc Jordán, Márton Kanász-Nagy, Katalin Kapitány, Mária Kopp, Steve LeComber, Leon Lederman, Susan Lindquist, László Mérő, Ágoston Mihalik, István Molnár, Viktor Müller, Zoltan N. Oltvai, Kleopatra Ormos, Bálint Pató, Csaba Pléh, Zoltán Prohászka, Ricard V. Solé, Csaba Sőti, Attila Steták, Steven H. Strogatz, András Szabó, Péter Száraz, Gábor Szegvári, Attila Vértes, Tamás Vicsek, Denise Wolf and Peter Wolynes is gratefully acknowledged.

Without the encouragement of Tamás Vicsek, the help of our librarian, Csilla Szabó, and last but not least the editorial team, Claus Ascheron, Adelheid Duhm, James Fuite, Angela Lahee, Stephen Lyle and Jack Tuszynski, this book could not have been written. I would like to thank to all members of my family and my colleagues for their understanding during the writing process. Finally, let me introduce Édua Szűcs, to whom I am extremely thankful for the excellent artwork in the book.

Édua Szűcs left the Miklós Radnóti high school in Szeged, Hungary in 1977. She received her MSc from the Szeged University, Hungary. Starting her independent art work as a cartoonist in 1986, she has had more than thirty exhibitions in Hungary and abroad. Her published works include Edua cartoons (1997), Edua cartoons 2 (2001), and illustrations for several books. Awards: Szféra special awards (1996, 1999); Karikatórium special award (1997); Foundation for Hungarian Culture Award (1998); Women for the European Union, first prize (2003).

At the end of this preface, let me invite you once again to send us comments and questions. The LINK group can be reached at the following address and website:

`weaklink@puskin.sote.hu`   `www.weaklinks.sote.hu`

Budapest, Hungary
September 2005

Peter Csermely

Contents

1 A Principle is Born: The Granovetter Study 1

2 Why Do We Like Networks? 5
2.1 Small-Worldness 7
2.2 Scale-Freeness 13
2.2.1 Scale-Free Degree Distribution of Networks 14
2.2.2 Underlying Reason for Scale-Freeness: Self-Organisation 18
2.2.3 Scale-Free Distribution in Time: Probabilities 24
2.2.4 Scale-Free Survival Strategies: Levy Flights 27
2.2.5 Scale-Free Pleasures 29
2.3 Nestedness 32
2.4 Weak-Linkness 43

3 Network Stability 47
3.1 Perturbations. Good and Bad Noise 47
3.2 Life as a Relaxation Phenomenon: Dissipate Locally, Connect Globally 56
3.2.1 Confined Relaxation with Global Connection 56
3.2.2 Self-Organised Criticality 59
3.3 Network Failures 68
3.4 Topological Phase Transitions of Networks 74
3.5 Nestedness and Stability: Sync 79
3.6 How Can We Stabilize Networks? Engineers or Tinkerers 90

4 Weak Links as Stabilizers of Complex Systems 95
4.1 An Emerging Synthesis: Weak Links Stabilize Complex Systems 95
4.2 Weak Links: A Starting Definition 100
4.3 Stability: A Starting Definition 103
4.4 Complex Systems 106
4.5 Weak Links and System Degeneracy 108

5 Atoms, Molecules and Macromolecules 111
5.1 Protein Folding Problems 111
5.2 Energy Landscapes 115
5.3 Weak Bonds in Protein and RNA Folding 119

6 Weak Links and Cellular Stability 125
6.1 Cellular Networks 125
6.2 Stability of the Cellular Net 128
6.3 Stress, Diversity and Jumps in Evolution 139
6.4 Cancer, Disease and Aging 150
6.4.1 Cancer 150
6.4.2 Disease 151
6.4.3 Aging 152

7 Weak Links and the Stability of Organisms 157
7.1 Immunological Networks 157
7.2 Transport Systems 162
7.3 Muscle Net 163
7.4 The Neuro-Glial Network 166
7.5 Psycho Net 170

8 Social Nets 181
8.1 Animal Communities 181
8.2 A Novel Explanation of the Menopause 184
8.3 Stability of Human Societies 186
8.4 Firms and Human Organisations 201
8.5 Dark Networks and Terror Nets 207
8.6 Pseudo-Grooming 209

9 Networks of Human Culture 219
9.1 The Language Net 219
9.2 Novels, Plays, and Films as Networks 224
9.3 Our Engineered Space 231
9.4 Software Nets 238
9.5 Engineers and Tinkerers: An Emerging Synthesis 240

10 The Global Web 243
10.1 The World Trade Web 243
10.2 Turning Points in History 247
10.3 Weak Links: A Part of Social Capital 258

11 The Ecoweb ... 263
11.1 Weak Links and the Stability of Ecosystems ... 263
11.2 Omnivory ... 267
11.3 The Weak Links of Gaia ... 270

12 Conclusions and Perspectives ... 275
12.1 The Unity of the Weakly-Linked World: A Summary ... 275
12.2 Revisiting the Definitions: A Synthesis ... 281
12.3 Prospects and Extensions ... 304
12.4 Weak Links and Our Lives ... 307

A Useful Links ... 313

B Glossary ... 317

References ... 329

Index ... 377

1 A Principle is Born: The Granovetter Study

In the late 1960s students had a rather revolutionary life in universities. In the midst of all this, Mark Granovetter, a PhD student at Harvard University, set himself to figure out how people find their jobs. He interviewed about a hundred people and sent out another 200 questionnaires in the Boston area.

The summary of his first results showed that more than half of the people found their jobs through personal contacts. We instinctively agree with these results. We may browse newspapers or web pages for a new job, but the real hints often come from our best friends. Or do they? In fact, this is not quite true. The really surprising result of the study was that, in most cases, the informants were not particularly close to the job seeker. They rarely spoke to each other, and they saw each other only seldom.

Why was this surprising? Granovetter had good reasons for thinking that strong links would be more useful for finding a new job. Close friends will give all their information to the job seeker and will mobi-

Fig. 1.1. In most cases, the best informants were not particularly close to the job seeker

lize all their contacts to help. Moreover, they meet the job seeker more often, and know more about her skills and preferences. And yet weak contacts still proved to be more useful. Were close friends biased? Did they overestimate the abilities of the job seeker?

Fig. 1.2. Weak ties play a role in effecting social cohesion

Granovetter was puzzled and started to analyze earlier data. He considered an earlier hysteria incident, where more and more workers in a textile plant in the deep south of the USA were claiming bites from a mysterious and non-existent 'insect', until eventually the plant had to be closed (Kerckhoff et al., 1965). Although the rumor starters were isolated people, they had numerous weak links in the community. In Granovetter's meta-analysis, weak links also proved to be useful in the famous Milgram experiment (Milgram, 1967; Korte and Milgram, 1970). In this example, people were instructed to send a letter to an unknown person[1] in the USA by asking the help of persons they knew on a first-name basis. If the starter was white, and the target was an Afro-American, the 'chain of friends' worked efficiently only if the critical point, where the chain of white friends was switched to a chain of black friends, was a weak link. Finally, Granovetter showed that the friendship network of Rapoport and Horvath (1961) was best covered if one used weak links to search for the acquaintances of the acquaintances of a given person. In contrast, the 'best-friend' networks did not cover the whole community. It seemed to be a general result that weak links are more useful for information searches than strong ones.

[1] In the Milgram experiment only the postal address of the 'unknown person' (the target) was not revealed to the starter, and she did not know the endpoint personally. However, the starter did know the name and a few personal features of the target, e.g., the target is Rebecca Smith, a catholic Latin teacher in Cleveland, who is a chess champion.

Granovetter went further. He analyzed social networks in a general context, and observed that weak links also link network modules, a concept confirmed in many later studies. Finally, he came to the conclusion: "Weak ties play a role in effecting social cohesion." He published his findings under the title *The Strength of Weak Ties* (Granovetter, 1973). A principle was born. However, more than a quarter of a century was to pass before we started to learn that weak links not only connect, but also stabilize all complex systems. And now, we have reached

THE END.

Indeed we are already at the end. You have now heard the central statement of the book: weak links stabilize all complex systems. I described Mark Granovetter's landmark paper introducing this idea more than 30 years ago. I indicated the path leading to the generalization of this idea in the Preface. What more is there to say? *"How can you ask such a question? You have not even defined what you mean by 'weak links'?"* Thanks a lot, Spite, for the reminder. I will try to give a starting definition now, but if you would like to have a more complete version, please go ahead and check Sect. 4.2.

Weak links are links between network elements, which connect them with a low intensity. Weak links may also connect network elements with a higher intensity, but in this case they are only transient. I will show later that, in real networks, we have a continuous spectrum of link strengths starting with a few strong links and ending with more and more links, which become weaker and weaker. In most cases, it is rather difficult to cut the continuously changing strength parameter somewhere and say: up to here, all the links were strong, but from this point on, we shall say that they are weak. Consequently, in this book I will use the functional definition[2] of weak links given by Berlow (1999).

Definition of Weak Links. A link is defined as weak when its addition or removal does not change the mean value of a target measure in a statistically discernible way.

[2]It is a question of future research how much these 'functional weak links' overlap with the weak links, which are weak due to their low affinity or intensity.

The target measure here is usually an emergent property[3] of the whole network, or a response the network gives to a certain stimulus. The mean value of the target measure is changed if a strong link is deleted from or added to the network.

Will we lose weak links in the future? Please note that, in this functional definition, the discrimination between strong and weak links depends on the desired or available accuracy of our measurements. If the mean value is measured a hundred times more accurately, the 'statistically discernible change' in the mean value will be achieved by changing much weaker links than in the case of a measurement that is a hundred times less accurate. *"Why are you writing this book then? Your weak links will have vanished in a few years, when my generation has learnt how to measure things more accurately than your generation can."* I have bad news for you, Spite. When your generation has learnt how to measure things more exactly than we can measure them now, you will certainly lose a number of weak links according to this definition, since you will have to reclassify them as strong links. However, with the extension of detection limits, you will be able to measure a thousand times more 'new' weak links instead, which are even weaker than the weakest links my generation could detect. At the end of the day, your generation will have to deal with far more weak links than we ever did. As a conclusion, the younger you are, the more important this book is for you.

Having learnt a starting definition of weak links, this book will show that hierarchical networks are governed by the same principles, from molecules to the whole Universe, and that weak links stabilize us in all these levels. To understand all this, we must first learn more about networks. So let us begin.

[3]For the explanation of the meaning of 'emergent property' and other unusual words in the text, please see the glossary in Appendix B.

2 Why Do We Like Networks?

Networks catch hold of you. They are enchanting and contagious. As a first 'proof' of these statements let me give you my own example. Just before starting to write this chapter, I sat on a train and watched a charming mother and her little daughter just opposite me. The baby fell asleep playing with her comforter. As I continued to watch, my mind went to work. What could be the periodicity of her suckling motions? Was it perhaps scale-free, showing sudden bursts of activity separated by longer and longer periods of stasis? Did it show self-organized criticality? Was this a punctuated equilibrium? My thoughts continued: What if I looked outside? Would I see fractals instead of trees and clouds? *"Let me interrupt you here. Why do you assume that we know what 'scale-free', 'self-organized criticality', 'punctuated equilibrium' and 'fractal' are supposed to mean? And anyway, what is a network?"* I am sorry, Spite. Whenever elements are connected with links, we may call them a network. Networks can be formed from atoms, molecules, cells, plants, firms, words, power stations, Internet routers, Web pages, countries, etc. Even your friends, Spite, form a network. The meaning of the other words will be explained later. If you are curious to understand them now, turn to the glossary in Appendix B, at the end of the book.

Returning to the popularity of networks, I am not the only one who has found this field fascinating. Figure 2.1 shows the number of network-related scientific publications in MEDLINE.[1] The arrow points to the publication date of two important network discoveries, the demonstration of the generality of the small-world phenomenon (Watts and Strogatz, 1998) and scale-free behavior (Barabasi and Albert, 1999). Obviously, these data may just reveal a coincidence. They do not directly prove the profound effects of these important discoveries in the network approach. However, Fig. 2.2 shows the number of

[1]Data of in Fig. 2.1 should be treated with caution, since 'network' may also refer to a network of authors, for example. An additional non-specific effect arises from the fact that the number of annual publications covered by MEDLINE also increased over the period covered.

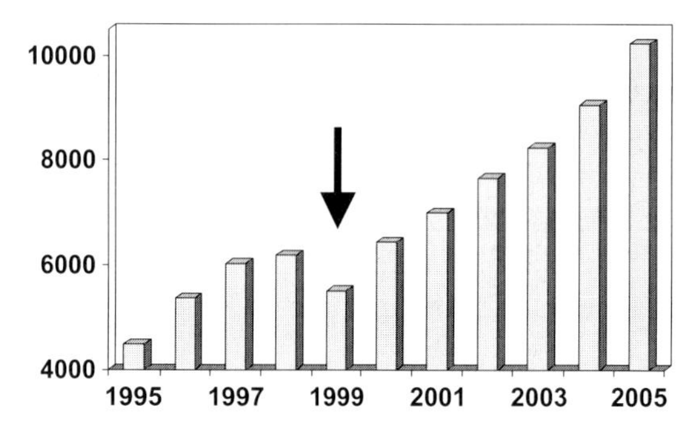

Fig. 2.1. Number of network-related publications in MEDLINE. The number of publications containing the words 'network' or 'networks' in their title or abstract was collected from MEDLINE (`www.pubmed.com`). The 2005 data is an extrapolation. The *arrow* shows the publication date of the two seminal network papers by Watts and Strogatz (1998) and Barabasi and Albert (1999)

annual citations of the above two papers in comparison with the average citations of three randomly selected papers from the same journals having a similar number of total citations. The citations of randomly selected papers peter off after 3 to 4 years. In contrast, citations of the two seminal network papers grow linearly, showing no tendency to decline in this period. No wonder, scientists seem to like networks. But how about the layperson? As a measure of public success, Laszlo Barabasi's book, *The Linked* was translated into 8 languages in the first two years of its existence.

Having these data to hand, I think we may be quite confident in saying that networks really catch hold of people. People do like networks. However, another question arises: Why exactly do people like networks? This chapter attempts to answer this question and uses the elements of the answer to introduce some important features of networks in general.

Small-worldness, scale-freeness, nestedness, weak-linkness: these are the titles of the following sections. All these words refer to properties which are general features of most networks around us, and this is why we have acquired a feeling for them. These properties of the networks we either contain or belong to inherently help us to understand the world around us, being basic, underlying elements of our cognition. Therefore small-worldness, scale-freeness, nestedness and weak-linkness not only mean the actual features of networks (being a small-world network, having scale-free distribution of various properties, containing other networks as its elements as well as belonging to higher order networks and having a large number of weak links,

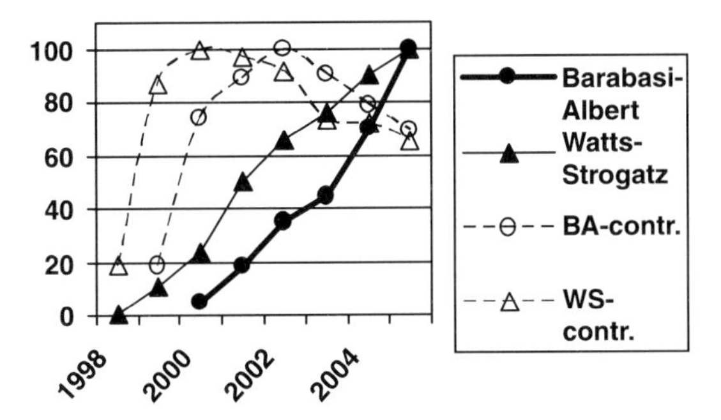

Fig. 2.2. Citations of seminal papers on networks. The numbers of citations for the Watts and Strogatz (1998) and Barabasi and Albert (1999) papers were collected from the Web of Science. Control values show the number of citations of three randomly selected papers from the same journals having a similar total number of citations. Data were normalized to the maximal number of yearly citations. 2005 data is an extrapolation

respectively), but also refer to the help these network properties give us. What is this help? Please continue, if you would like the answer.

2.1 Small-Worldness

Stanley Milgram did many famous experiments. In his small-world experiment he gave letters to starters, persons, who were asked to pass them to acquaintances known on a first-name basis in order to find an unknown, distant target (Milgram, 1967). Imagine that you have the task of sending a letter to the Reverend Lucas Brown, who lives in the capital of Myanmar, Yangon. It is rather easy. I need the address, ZIP code and a few stamps. But not this time! No address is known, and direct mailing is excluded. You may pass the letter only to one of your friends. The important message of Milgram's work, viz., "we live in a small world, and are only six steps apart from each other", became very popular. There is a good chance that your letter to the Reverend Brown will find its target by passing along a chain of around six friends.

A Hungarian prediction of small worlds from 1929. Tibor Braun (2004) quotes the following text from a story by the Hungarian writer, Frigyes Karinthy, in 1929: "To prove that nowadays the population of the Earth is in every aspect much more closely interconnected than it ever has been, one member of our gathering proposed a test. 'Let us pick at will

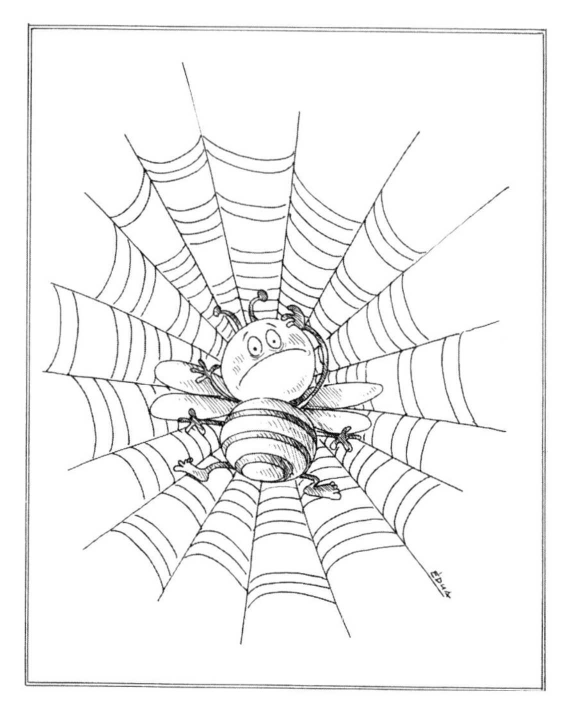

Fig. 2.3. Networks can really catch hold of you

any given existing person from among the one and a half billion inhabitants of the Earth, at any location.' Then our friend bet that he could establish via direct personal links a connection to that person through at most five other persons, one of them being his personal acquaintance. 'As people would say, look, you know X.Y. Please tell him to tell Z.V., who is his acquaintance, and so on.' 'OK,' said a listener, 'then take for example Zelma Lagerlöff' (Nobel Prize for Literature, 1909). Our friend placing the bet remarked that nothing could be easier. He thought for only two seconds. 'Right,' he said, 'so Zelma Lagerlöff, as a Nobel Laureate, obviously knew the Swedish king Gustav, since the king handed her the prize, as required by the ceremony. Gustav, as a passionate tennis player, who also participated at large international contests, evidently played with Kehrling [Béla Kehrling (1891–1937), Hungarian tennis champion and winner at the Göteborg Olympics 1924], whom he knew well and respected.' 'Myself,' our friend said (he was also a good tennis player), 'I know Kehrling directly.' Here was the chain, and only two links were needed out of the stated maximum of five." The amazing foresight of Karinthy (1929) predicting that we are approximately five steps apart from each other on the global scale was proved decades later by Milgram (1967) and Dodds et al. (2003a).

When I gave a lecture on networks to illustrate the smallness of the small world we live in, I asked my audience how many steps they thought they were from the President of the United States. Some of them guessed around a hundred, others were better informed and said: Six! Then I surprised them with the exact number: Three. *"How come? Did you know that someone's parent attended the same school as the President?"* No, Spite, I knew only my own connections. I happen to know the President of my country, Hungary, who did meet the President of the USA. Since the students knew me, this is exactly three steps for them. Our world is really small. However, there is another message from the Milgram experiment: not only do short paths exist between distant network members, but ordinary people are very good at finding them too (Newman, 2003b). How would you kick off your letter to the Reverend Brown in Yangon? *"I happen to have a friend who moved to Kuala Lumpur a year ago. If I recall my geography lessons, it is not far from Yangon. I would ask her to look around. She certainly knows many more people in the region than I do."* Excellent, Spite. If she happens to know a priest in Kuala Lumpur or Yangon, you might even complete the chain in three steps instead of six.

Why was Milgram lucky? Examining the original numbers, I have to conclude that in spite of the seemingly easy navigation shown above, Milgram was lucky. The final conclusion was based on only 18 letters which actually reached the single target of the Milgram experiment in Boston out of the 96 starters at distant locations in Nebraska. In other studies the success rate was even lower (Kleinfield, 2002). It was often hard to define what was causing the numerous drop-outs. However, a later study (Dodds et al., 2003a), using tens of thousands of emails had the same conclusion: we are about six steps apart even in different parts of the world. Experimenters of robust phenomena are lucky. Their instincts often find the right solution even when the actual proof is shaky. However, if you are a young investigator, let me ask you *not* to rely on this. Unsuccessful examples always outnumber the few serendipity stories. We do not hear about the failures: most of them never get published. Moreover, our publication habits mean that we mention only the final success stories and not the very important and frustrating path we had to follow to reach them.

The number of dimensions in our brain. How do we select a direction to send our letter towards the unknown target? In fact, we try to get a match between the character of our acquaintances and the known properties of the target. For this matching task, we categorize our friends into social dimensions, as has been shown in the model by Watts et al. (2002). Spite started the letter to the Reverend Brown by finding a friend in Kuala Lumpur. This was a wise tactic, since geographic information is sufficient to perform global routing in a significant fraction of cases (Liben-Nowell et al., 2005). However, I have no friends in the region, and therefore I would probably double check the priests in my circle of friends instead. The number of social dimensions screened for a search is around 5 to 6 (Dodds et al., 2003a; Killworth and Bernard, 1978). This number is actually quite close to our average cognitive dimensionality, which is measured as the number of persons whose intentions towards each other I may still follow (Dunbar, 2005). What should we do if we want to be even more sophisticated? Should we use more social dimensions? I might have bad news here. The dimensionality of our neural network may prevent this. Even if our world grows hopelessly complex, we will still restrict ourselves to half a dozen, or even fewer, social dimensions to describe it, or start to develop more complex neural nets in our brain. Watch out for contact gurus! The evolution of superbrains may actually be happening around you now as you read!

Our world is a small world. However, it is not only the expanding circles of friends, the social net, which is a small world. Many other networks, like power nets, the networks of neural cells, etc., are also small-world networks. We live in small-worldness. *"This is obvious. If I take a hundred people and everyone knows everyone, their world – let me call it Spiteland for short – is really small. Not six steps, but one, separating any one of them from any other."* Spite, I appreciate your logic, but real life is not Spiteland. We cannot know everyone. Do you have six billion friends, increasing by dozens every second? I doubt it. However, you have hit upon a good point here. Small worlds are not only small in the sense that their members may reach each other easily. Random graphs, where the connections are made between the elements in a random fashion, are equally good at this (see Fig. 2.4). In small worlds, your neighbors also know each other. To use a scientific term, this 'my friend's friend is my friend' effect is called clustering. The clustering of small-world networks is high. These networks are lucky mixtures of 100% clustered regular lattices and highly-connected random graphs (see Fig. 2.4) (Watts and Strogatz, 1998). Small-worldness requires both a dense array of local contacts, which is reflected by the high clustering of small worlds, and a good enough number of long-range contacts. The simultaneous presence of both ensures that the small-world network becomes really small, providing easy conditions for finding any of its members. However, this cannot be achieved by extensive cross-linking

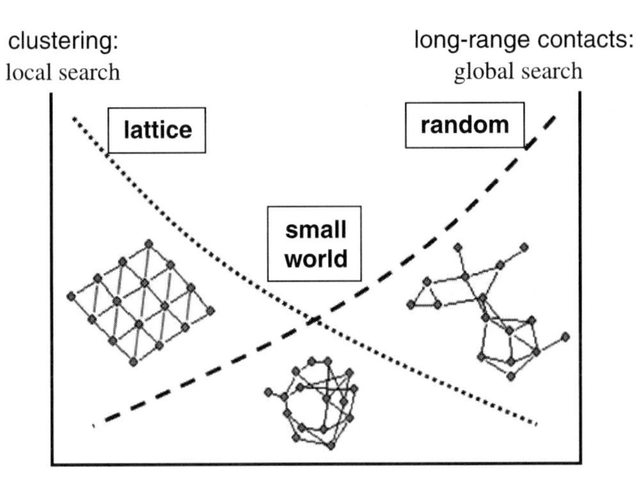

Fig. 2.4. The small-worldness of networks. The figure shows that small-world networks are in-between lattice-type networks and random graphs, having much longer range contacts than the former and much higher clustering than the latter. Note that the measures of both clustering and long-range contacts are purely illustrative

of the network, since building and maintaining links is costly. Natural small worlds are economical (Latora and Marchiori, 2003). In fact, small worlds are much more economical than either random networks or regular lattices.

Some worlds are not so small. Small-worldness depends on what we consider as a member of the network. As an example, the extent of the small-world status of metabolic networks may vary, if we include relatively simple molecules like water, ATP, use directed links, or restrict the network to conserved residues of participating molecules (Arita, 2004; Ma and Zeng, 2003).

How many friends do you need to send your message to anyone? In 2000, Jon Kleinberg published an interesting model for message transmission on a two-dimensional lattice, where lattice elements were linked with random shortcuts. The interesting result was that an optimal condition can be defined for the fastest search. If the shortcuts are neither fully random (where lots of short paths exist, but it is extremely time-consuming to find them), nor restricted to short-range contacts (where no short paths exist at all), an optimal condition can be found where the system transmits the messages most efficiently. Under these conditions, you have exactly the same number of friends in your neighborhood, in the rest of your city, in the rest of your country, in the rest of your continent and in the rest of the world. In other words, you only have to worry about how to send your message to someone in the right neighborhood. Once the message has reached the right region, the fine-tuned targeting will rely on the increasingly denser local contacts as the message homes in on the actual target. This makes the search highly efficient and the system behaves like a small world (Kleinberg, 2000). The Kleinberg condition can be reformulated: for an optimal two-dimensional search, if you go to a higher region (neighborhood, city, country, continent, world), the chances of finding a friend after a random selection become an order of magnitude smaller. Kleinberg's model behaved optimally if the number of connections was the same on all scales, i.e., it was scale-free. Scale-freeness is another important feature of our everyday networks besides small-worldness, and will be discussed in detail in the next section.

Small worlds are easy to navigate. Lattice-type connections with their high clustering ensure the success of the finely-tuned final steps of a target search. Long-range contacts ensure the success of initial steps zooming in on the region of final interest. "The key to generating the small-world phenomenon is the presence of a small fraction of [...] edges, which contact otherwise distant parts of the graph" (Watts, 1999).

Navigation in small-world networks is helped by weak links (Granovetter, 1973; Lin et al., 1978). In many networks such as social networks, most of the long-range links typical of small worlds are weak links (Onnela et al., 2005). As I mentioned above, Dodds et al. (2003a) repeated the Milgram (1967) experiment using more than 60 000 emails. They found that a successful social search was conducted primarily through intermediate-to-weak strength links and did not require highly connected hubs. Moreover, Skvoretz and Fararo (1989) showed that the more weak links there are in a population, the closer a randomly chosen starter is to all others.

Interestingly, groups with lower or higher socioeconomic status, as well as groups under stress, tend to use strong links instead of weak ones (Granovetter, 1983; Killworth and Bernard, 1978). As a possible consequence of this, people under stress and either on the top or at the bottom of society may belong to a more closed world than those living in relative rather than extreme prosperity.

Why do we like small-worldness? Why do we need it? Humans are cooperative animals (Ridley, 1998). Consequently, our brain has developed to keep an inventory of our contacts (Dunbar, 1998). Connection rules have become essential for our survival. Social surveys show that we get used to circles of 5, 15, 35, 80 and 150 people.[2] However, we tend to meet more and more people and our cognitive abilities are not prepared for this. In a modern megalopolis, we feel lost. The expanding world has become alienating. Our only escape is to continually redefine and segregate a small world from the contact wilderness outside. Not only does small-worldness provide a key for efficient network search, as well as respecting our cognitive limits, but it is probably also a prerequisite for preserving our safety in an alienating modern world.

2.2 Scale-Freeness

Scale-freeness is popular. We like it, because it resembles the most important aspects of our life. What is scale-freeness? Why is it so general? Scale-freeness refers to a type of distribution. Here, a distribution is a statistic of any property: how many do I find from its various values in my system. Let us take an example from the last chapter. If I have 100 acquaintances in my village (let's call it Budapest, because that is its name),[3] I can plot their distribution as a function of the strength of our friendship. The histogram of Fig. 2.5A shows that I have one very best friend, two close friends, three friends from the local pub, eight others whom I only see every other week at the bowling club, nineteen from the congregation at my church, and 67 occasional acquaintances from the fitness center, the concert hall and the shopping mall. If the total number of elements in the system is not determined, the actual number may be replaced by probabilities. Taking the above distribution again, from any number of acquaintances, there is a 1% chance that I pick one of my best friends, and there is a 19% proba-

[2]These circles correspond to (1) our family and best friends, (2) our close friends, (3) our colleagues and acquaintances, (4) our fellow club members, and (5) our 'village', respectively, (Dunbar, 1998; Hill and Dunbar, 2003). The number of circles is five once again. This is the same number of circles that we saw before in the Kleinberg condition (2000): neighborhood, city, country, continent and world.

[3]The original saying: "Let's call him Gilberto, because that was his name", is from Leon Lederman's wonderful book *The God Particle* (1993). I am not only extremely privileged to have Leon as a supporter of my scientific research training program for high school students (`www.kutdiak.hu/en`), but I also treasure a dedicated copy of his book, which became my stylistic template when I started mine.

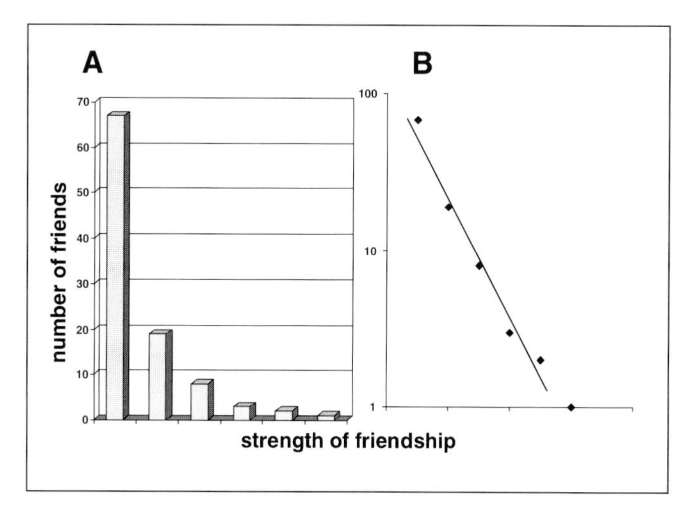

Fig. 2.5. The scale-freeness of networks. The figure shows the approximately scale-free distribution of my hypothetical friends as a function of the strength of our friendship. (**A**) Histogram. (**B**) Log–log representation. Note the arbitrary assumption that the strength of friendship increases by an order of magnitude between each of my friendship circles. Moreover, the distribution range is very limited here, which makes the assumption of scale-freeness very inaccurate

bility that I will find a distant acquaintance, whom I see only once a month.

The distribution of scale-free systems follows a power law and can be written as $P = cD^{-\alpha}$, where P is the probability, c a constant, D the distance of our friendship, and α a scaling exponent. Scale-free distributions are best visualized by taking the logarithm of the above equation to get $\log P = \log c - \alpha \log D$, which shows that the logarithm of the probability is a linear function of the distance of our friendship. If we plot the data of the above paragraph in a double-logarithmical way (see Fig. 2.5B), we get a straight line showing that the distribution of my friends as a function of the strength of our friendship follows a power law. It therefore displays the same type of distribution at any scale, i.e., it is scale-free.

2.2.1 Scale-Free Degree Distribution of Networks

The scale-free distribution pattern has been most studied on the degree distribution of networks. What is a degree? The degree of a network element is the number of connections it has. A scale-free degree distribution means that the network has a large number of elements with very few neighbors, but it has a non-zero number of elements with an extraordinarily large number of neighbors. These connection-rich elements are called hubs. If an element has just a few connections, it is often called a node.[4]

Scale-free behavior was long used as an empirical description of experimental data. This model was first developed by Kohlrausch (1854) to explain the discharge in Leiden jars. The first network with scale-free degree distribution was reported a hundred years later by de Solla Price (1965) analyzing the citations between scientific papers. Since then scale-free networks have been reported in all areas of biology, human relations and constructs appearing in every moment of our everyday lives (Barabasi, 2003).

Table 2.1 summarizes the exponents of degree distribution for a few networks. The aim of this list is not to characterize networks. This would be very inappropriate for at least two reasons. On the one hand, networks have a number of important parameters besides their degree distribution characteristics, such as the number of elements, number of links, mean degree, network diameter, clustering coefficient, assortativity, etc. (see Appendix B for definitions). On the other hand, most networks have different exponents for different regions of the degree distribution, or have an exponential cutoff. None of these features are reflected in the data of Table 2.1, which only give a feeling for the multitude and variability of scale-free networks. However, the data of Table 2.1 nicely demonstrate that networks have surprisingly common features in spite of their vastly different constituents and linkage systems.

Not everything is scale-free that seems to be scale-free. Linear regressions are rather tricky. With a little goodwill, a line can be fitted to practically any number of scattered points. Rather careful consideration is needed to judge whether the result has any real meaning. Here are a few hints to double-check your data, if you believe they may show a scale-free distribution:

- **Data should cover many scales.** A large interval – a scale of several orders of magnitude – is needed to demonstrate scale-free behavior convincingly (Avnir et al., 1998; Eke et al., 2002; Malcai et al., 1997). Scale-freeness proves to be especially difficult to judge in ecosystems, where the

[4] If the degree distribution of a network is scale free, the network is 'hub rich', which means that it has more hubs than a similar random network. Likewise, scale-free networks have no typical degrees.

Table 2.1. Exponents of some scale-free degree distributions. The exponent refers to the reported range of the exponent α from the equation $P = cD^{-\alpha}$, where P is the probability, c a constant, and D the degree of the network elements. The examples here were assembled from the excellent books and reviews by Albert and Barabasi (2002), Barabasi (2003), Barabasi and Oltvai (2004), Dorogovtsev and Mendes (2002) and Newman (2003). Protein network data are from Bortoluzzi et al. (2003); Chen et al. (2003) and Park et al. (2005a)

Name of network	Exponent
Atomic networks	
Occurrence of protein domains	1.6–2.5
Molecular networks	
Prokaryotic protein–protein interaction networks	2.6
Eukaryotic protein–protein interaction networks	2.1
Human protein–protein interaction network	1.7
Gene functional interactions	1.6–2
Yeast gene expression network	1.4–1.7
Escherichia coli metabolic network	1.7–2.2
Biological networks	
Food webs (ecological networks)	1
Social networks	
Scientific collaboration networks	1.2–2.5
Email messages	1.5–2
Zipf's law for size distribution of cities	2
Phone calls	2.1–2.3
Actors' appearance in various movies	2.3
Pareto's law for wealth distribution	2–3
Human sexual contact networks	3.2–3.4
Lotka's law of scientific productivity	2
Information networks	
WWW (in and out connections)	2.1–2.7
Word co-occurrence	2.7
Citations of scientific papers	3
Technological networks	
Software package parts	1.4–1.6
Internet	2.5
Digital electronic circuits	3
Power grids	4

number of established links is usually very small (Jordan and Scheuring, 2002).

- **Frequency–degree plots can be misleading.** To make a distribution like the one shown in Fig. 2.5A, you need to group individual data values into certain intervals. The selection of these intervals is rather arbitrary and may dramatically change the nature of frequency-based statistics. The use of cumulative, rank–degree plots is therefore suggested (Tanaka et al., 2005).
- **Correct sampling of the original network is needed.** In many cases we do not know, or cannot analyze the entire network. In most assessments, we use a fraction of the Internet or protein–protein interaction networks. Sampling of non-random networks requires great care and good controls. With biased sampling, the archetypal random graph, the Erdős–Rényi random graph (Erdős and Rényi, 1959; 1960) can be shown to have a scale-free degree distribution, which is clearly wrong, since it has a Poissonian, exponential degree distribution (Clauset and Moore, 2005; Egghe, 2005; Lakhina et al., 2003; Stumpf et al., 2005). Incorrect sampling changes many other parameters, such as the average path length, centrality measures, assortativity and clustering coefficient[5] of networks (Lee et al., 2005b).
- **Distinct parts of networks may display different distributions.** Part of the above sampling bias arises from the inhomogeneities of the network. This is especially pronounced if the network has modules, e.g., parts of the network where the intramodular contacts are denser than the connections to other modules. In such cases the final distribution can be multiscaled (Tanaka, 2005).
- **Various distributions may resemble each other.** If the range of data points is limited, many distributions, such as the log-normal, stretched exponential and gamma distributions may give rather similar fits to the scale-free pattern. A careful analysis is required to discriminate between these options (Stumpf and Ingram, 2005).
- **Scale-free distributions may be 'false positives'.** A recent report by Deeds et al. (2005) proposed a model for unspecific protein–protein interactions. Here the interactions between hydrophobic surface residues developed a scale-free degree distribution and showed other complex features of network structure as well. This study warns that unspecific data may often overshadow the tiny fragment of meaningful data and may give 'false positive' scale-free degree distributions.

The scale-free uncertainty principle: A little network quantum mechanics. Degree distributions rely on the fact that networks are constructed on the basis of paired interactions of their elements. Can all interactions be described as a sum of paired interactions in

[5]For definitions of these network measures, see the glossary in Appendix B.

the Universe? Not necessarily. The present-day network description may be analogous to the Newtonian world view in physics, serving well as a first approximation to describe complex systems. But later, when the possibilities of this approximation are exhausted, we may well end up with 'second generation' networks, where unpaired, group interactions or interaction halos between many elements will form the basis of the system, parallelling the change in physics when quantum mechanics and wave functions were introduced by Erwin Schrödinger (1935).[6]

2.2.2 Underlying Reason for Scale-Freeness: Self-Organisation

Why is the scale-free degree distribution so general in such a wide variety of networks? The first explanation of scale-free behavior emanates from the work of Herbert Simon (1955). Simon gave an explanation for the empirical law of Pareto (1897) on the scale-free distribution of wealth. The Pareto law, also called the 80–20 rule, says that a small number of people (20%) own a large amount of property (80%) in a given country, originally in Italy, where Vilfredo Pareto lived. Simon (1955) showed that this unequal distribution is a consequence of the 'the rich get richer' effect.[7] The name reflects the fact that the chances of extending existing wealth are bigger than the chances of accumulating it from scratch. This is also quite clear from existence of so many 'How did I get my first million?' stories, implying that to get the second, and other millions is easier. As I will describe in detail later, networks with scale-free distributions also have the common feature that they are built up from gradual events, which are often elements of self-organization.

In 1999, a mathematical method to generate scale-free networks was devised by Barabasi and Albert (1999). The invention they used was preferential attachment. Instead of starting from an Erdős–Rényi random graph (Erdős and Rényi, 1959; 1960) or from a lattice, and then changing its degree distribution by various rearrangements, they generated a network by attaching each of the new elements preferentially to those that previously had more connections. Here, indeed, the rich node got richer. Barabasi and Albert (1999) also showed the generality of the phenomenon by analyzing three different networks:

[6] I am grateful to the LINK group member, István Kovács for this idea.

[7] The 'rich get richer' effect is also called the Matthew effect in sociology (after the gospel of Matthew in the Bible: "For to every one that hath shall be given [...]"; Matthew 25:29; Merton, 1968). De Solla Price called it the cumulative advantage (de Solla Price, 1965) and Makse et al. (1995) referred to it as correlated percolation.

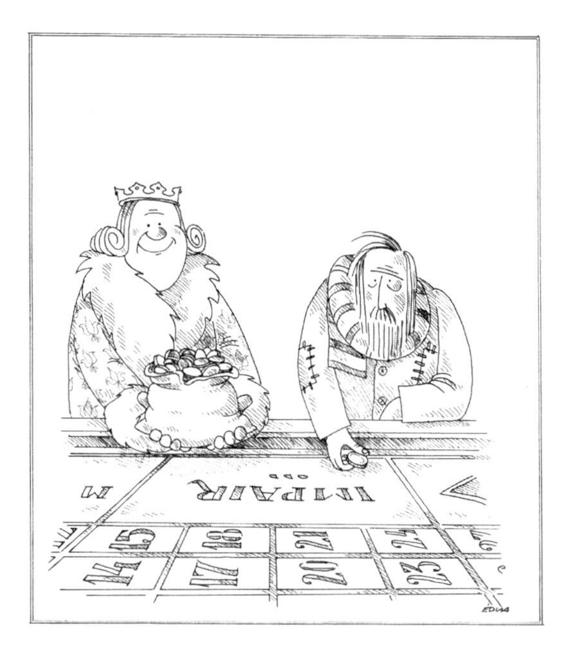

Fig. 2.6. The chances of extending existing wealth are bigger than the chances of accumulating it from scratch

Hollywood actors, the World Wide Web, and the US power grid. Their work gave another aspect to the 'rich get richer' effect: popularity is attractive.

How can one make a scale-free network? Here are a few remarks on the various methods for constructing scale-free networks:

- **Preferential attachment.** After the seminal work of Barabasi and Albert (1999), preferential attachment became a very popular method for constructing scale-free networks. The preference can refer to the degree of existing elements (Barabasi and Albert, 1999) or to their fitness (Bianconi and Barabasi, 2001; Caldarelli et al., 2002). However, preferential attachment does not always result in scale-free distributions. If the fitness difference between the elements is high and there are a few elements which attract the new elements many times more than the others, a 'winner takes all' situation may occur, and a star network develops. In this network, some or one element takes most of the connections (Albert and Barabasi, 2002). On the other hand, 'weak' preferential attachment, aging effects (where nodes and hubs tend to lose connections as the network gets older) and growth constraints may all cause crossovers to exponential

decay (Albert and Barabasi, 2002). Exponential decay means that there are far fewer hubs in the network than in scale-free networks.
- **Duplication and divergence.** Another method to develop a scale-free distribution is the duplication and divergence method. Here the original network is duplicated and then connections are exchanged. Duplication and divergence is an important possibility for the way scale-free distributions may have developed during evolution (Sole et al., 2002; Vazquez et al., 2002).
- **Scale-free degree distribution is not always the optimal solution to the requirement of cost efficiency.** As mentioned before, in small-world networks, building and maintaining links between network elements requires energy. Therefore, network topology is a result of an optimization process. It is optimized with respect to the available resources to ensure optimal communication between different network parts. If the network enjoys unlimited resources, it will have a random distribution with plenty of links. Scale-free degree distribution occurs as a result of optimization in systems with finite resources. If the network experiences even more limited resources, the degree distribution will be steeper than in the case of a scale-free network, and a transition will therefore occur towards a star network (Amaral, 2000; Sole et al., 2003a; Wilhelm and Hanggi, 2003). These phenomena are called topological phase transitions and will be discussed in detail in Sect. 3.4.

What is the advantage of a scale-free degree distribution? As I mentioned in the last section, small-world networks lie somewhere between random graphs and lattice networks. The degree distribution of random graphs (Erdős and Rényi, 1959; 1960) is Poissonian with a maximal characteristic degree and decaying rapidly. In contrast, all nodes in lattice networks have the same degree. In other words, the degree distribution of lattice networks has a single scale only. The scale-free degree distribution lies in-between the two and, similarly to small-world networks, allows easy navigation and travel (Barabasi, 2003; Bollobas, 2001; Watts and Strogatz, 1998).

Scale-free networks give sensitive responses. Scale-free topologies enable more sensitive responses to various changes than those allowed by random networks (Bar-Yam and Epstein, 2004). This can be a very important property for explaining why scale-free networks have been selected and maintained in many systems.

For a wide occurrence, it is often not enough for a network to be economical and help easy travel with a minimum number of connections, as we have already seen for scale-free networks. It is important for the survival of the network to resist damage. Damage usually occurs in the form of random errors which incapacitate one or other network element. Scale-free networks pass this requirement, too. Albert et al. (2000) have shown that scale-free networks show a much better stability in the face of such errors than random graphs.[8]

We already have a clue as to how scale-free networks may have developed and why they have remained with us. In the next part of the section, I will try to answer the question as to why scale-freeness is so popular.

Is the source of our inherent affection for scale-freeness the fact that we intuitively feel the scale-free degree distribution of networks? We are all aware of the connection capital of our class-mates or colleagues in the social networks around us. "Lucky guy, he has hundreds of friends. Just picks up the phone and all his problems are solved." Does this description not fit you? Do not worry if you do not have a thousand friends, it is not your fault. Connection heroes are *by nature* rare. They form the thin tail of the scale-free degree distribution. Somehow, deep down, we all know this. However, our cognitive limits (Dunbar, 2005) make it relatively difficult to keep track of the whole connection network. I might know some friends of my friends but I certainly do not know all of them. Thus our intuitive sense for scale-free distribution does not come from the degree distribution, but emanates from different sources. The solution is space and time.

More on fractals in space. Here I will explain a little more about the term fractal, showing its connection to the idea of fractional dimension and explaining the self-similarity of fractals. For a more detailed explanation, the reader should consult Mandelbrot's seminal book (Mandelbrot, 1977).

- **Fractals, fractal dimension and fractional dimension.** The term 'fractal' comes from the Latin 'fractus' meaning broken or fraction (Mandelbrot, 1977). A fractal object has a fractional dimension. What does this mean? Take a two-dimensional object, for instance, a sphere, and multiply its edge length by 3. You can fit 9 of the old spheres into the new, larger sphere. Taking these two numbers, you find that $9 = 3^2$, which means that the sphere has two dimensions. Writing the above equation in general terms, we get the scale-free distribution of $N = (L/l)^d$, where N is the number of smaller objects fitting into the larger object, L/l is the

[8] The error tolerance of the scale-free size distribution of forms in space (fractal patterns) has been mentioned by West (1990).

ratio of the characteristic measure of the two objects of different sizes, and d is the exponent, called the fractal dimension. In fractal objects, d is not an integer number (see Fig. 2.7), but is always smaller than the geometric dimensionality of the object (e.g., smaller than 2, if the object is a two-dimensional object).
- **Fractals and self-similarity.** Fractals are self-similar objects. However, not every self-similar object is a fractal, with a scale-free form distribution. If we put identical cubes on top of each other, we get a self-similar object. However, this object will not have scale-free statistics: since it has only one measure of rectangular forms, it is single-scaled. We need a growing number of smaller and smaller self-similar objects to satisfy the scale-free distribution.
- **Limits of self-similarity.** A mathematical fractal is generated by an infinitely recursive process, in which the final level of detail is never reached, and never can be reached by increasing the scale at which observations are made. In reality, fractals are generated by finite processes, and exhibit no visible change in detail after a certain resolution limit. This behavior of natural fractal objects is similar to the exponential cutoff, which can be observed in many degree distributions of real networks.

Recent evidence indicates that many scale-free networks can be simplified, renormalized to a self-similar, fractal hierarchy of network motifs. This is related to the underlying tree of the network, which is composed of edges with high traffic between network components. This tree is also called the skeleton of the network (Alessina and Bodini, 2004; Garlaschelli et al., 2003; Goh et al., 2005; Song et al., 2005a). The fractal transport systems of self-organizing networks may also explain the allometric scaling laws, which are probably the most famous of the empirical scale-free laws (West and Brown, 2004).

Allometric scaling laws: the mouse-to-elephant curve. Allometric scaling laws are probably the most famous of the empirical scale-free laws. These laws cover a wide range of empirical scaling relationships, which show the self-similar behavior of various complex systems such as cells, organs, organisms, etc., over a wide range of masses. In the scale-free relationship here, $P = cM^{\alpha}$, P refers to the property, c is a constant, M is the mass of the organism or organelle, and α is a scaling exponent, which varies depending on the nature of P. The expression 'allometric' was first used by Julian Huxley (1932) to show the generality of the concept, which describes the mass dependence of the metabolic rate, lifespan, growth rate, heart rate, DNA nucleotide substitution rate, lengths of aortas and genomes, tree height, mass of cerebral grey matter, density of mitochondria and concentration of

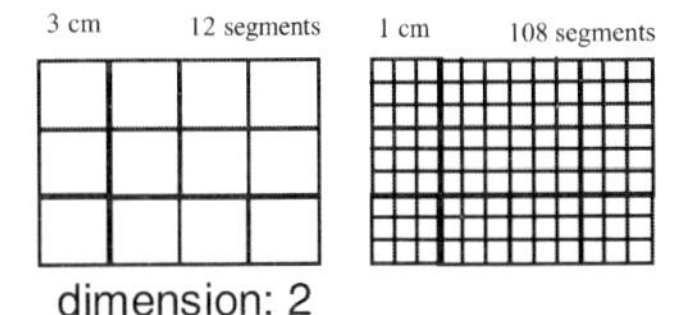

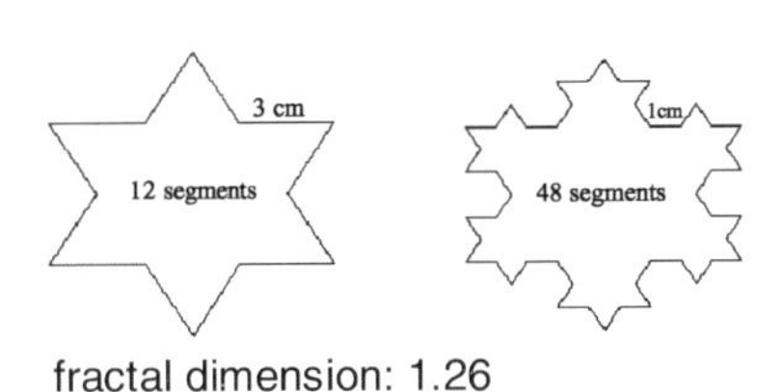

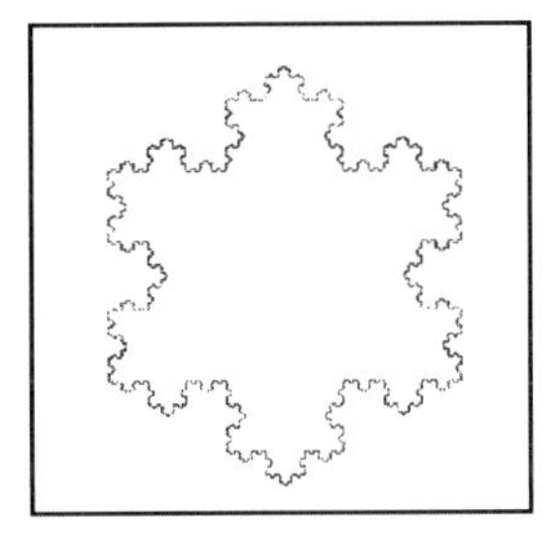

Fig. 2.7. Comparison between non-fractal objects (squares) and a fractal object (Koch's curve). If we decrease the characteristic measure of a square to one third of the original, we can put 108 small squares into the 12 old ones. Taking the equation $N = (L/l)^d$, where N is the number of smaller objects fitting into one of the larger objects, L/l is the ratio of the characteristic measure of the two objects of different sizes, and d is the dimension, we get a value of 2 for d ($108/12 = 3^2$) showing that the square was a two-dimensional object. In contrast, if we do the same with the Koch curve, we obtain $48/12 = 4$ for N, which gives approximately 1.26 for d, showing that Koch's curve has non-integer dimensionality, i.e., it is a fractal. In the *inset*, Koch's curve is shown after a few more recursive steps. With the increase in the number of repetitive steps, the number of scales increases, and the image begins to look like a fractal

RNA, to name but a few (West and Brown, 2004). The most studied of these is the basal metabolic rate, which obeys Kleiber's law (1932). Here P is the basal metabolic rate, i.e., the amount of energy per unit time required by a living organism to remain alive, and the scaling exponent is 3/4. A limited fraction of this empirical scaling law is often called the mouse-to-elephant curve, describing the universality it represents, since it is equally valid for all mammals from mice to elephants. However, recently, the applicability of the formula has been extended even further, from the largest animals down to cells, and individual enzymes (West et al., 2002). The law has become remarkable, since from geometrical considerations (area-dependent metabolism per volume-dependent mass), one would arrive at a value for the α exponent equal to 2/3, rather than 3/4. The deviation from 2/3 to 3/4 has been explained by fractal-type transport systems (West et al., 1997), as will be described in Sect. 7.2. In the other examples, the value of the exponent is different: the mass dependence of the heart rate ($\alpha = -1/4$), lifespan ($\alpha = 1/4$), the radii of aortas and tree trunks ($\alpha = 3/8$), unicellular genome length ($\alpha = 1/4$), and RNA concentration ($\alpha = -1/4$), all have different exponents in their scaling relationship $P = cM^{\alpha}$. However, one has to bear in mind that models always have limitations. Hence the debate concerning a possible over-interpretation of the allometric scaling laws (Dodds et al., 2001).

2.2.3 Scale-Free Distribution in Time: Probabilities

Systems show scale-free behavior in time, too. The probability of the occurrence of a highly connected hub in a network follows similar statistics to the probability of an unusual event. The archetypal example is an earthquake. The Gutenberg–Richter law states that both the occurrence and the magnitude of earthquakes follow a power law (Gutenberg and Richter, 1956).[9] We may know this, although we do not feel it. Fortunately, earthquakes do not happen often enough for us to have any inherent feeling for their statistics, without the meticulous records of decades and centuries. There is a similar scale-free event on a shorter time scale: rain. The dry spells between two rainfalls and the magnitude of the rainfall both follow scale-free statistics (Peters and Christensen, 2002). Our nervous ancestors, when watching out for rain over their drying and dying crops, did acquire a sense for the way this heavenly force behaves. Understanding scale-freeness was crucial for life.

The Noah effect. Not surprisingly, Mandelbrot (1977) called the low, but definite probability of extreme events the Noah effect, referring to the great flood in the Bible. In fact, Noah's case may not be the best example of an extreme rainfall, since the hypothesis of Ryan and Pitman (1998) proposes that the Biblical flood was more probably caused by the sudden burst of the Mediterranean Sea into the empty basin of the Black Sea

[9]The Gutenberg–Richter law states that $N = a10^{-bM}$, where N is the number of earthquakes of magnitude M, a is a constant, and b is the exponent. The value of b seems to vary from area to area, but worldwide it seems to be around unity. Here the magnitude M of an earthquake is gauged on the Richter scale, and it is proportional to the logarithm of the maximum amplitude of the Earth's motion. What this means is that, if the Earth moves one millimeter in a magnitude 2 earthquake, it will move 10 millimeters in a magnitude 3 earthquake, and 10 meters in a magnitude 6 earthquake.

through the Bosporus Strait. *"I have an objection. Recent data suggest that the gigantic spill was not completed in the Biblical 40 days but lasted for 33 years, if not longer (Schiermeier, 2004)."* Well done, Spite! But just imagine that your favorite lake grows 150 meters above your head before your children really grow up. A rather frightening prospect, is it not?

More fractals in time: mono- and multifractals. The functions $f(t)$ typically studied in mathematical analysis are continuous, and have continuous derivatives. Hence, they can be approximated in the vicinity of time t_i by a so-called Taylor series or power series:

$$f(t) = a_0 + a_1(t - t_i) + a_2(t - t_i)^2 + \cdots + a_h(t - t_i)^h + \cdots ,$$

where h is an integer (a whole number). In contrast, most time series found in 'real' experiments cannot be approximated by the above formula. If a non-integer number h is enough to quantify a local singularity in the noisy time series, we call it a fractal series. If we find a single value $h = H$ for all singularities t_i in the signal, then the signal is a monofractal. If we need several distinct values to describe the time series, than the signal is multifractal. (If none of them works, than the signal is not fractal, but has some different behavior.)

Most people in the developed countries do not have to worry about dying crops. World trade provides a safety net against droughts. There should be rain somewhere around the globe. We have a better example to show that scale-freeness is really close to us: the Bernoulli law. No, not those faint high-school memories, if they exist at all, on the fundamental laws of hydrodynamics, which help us to understand and design airplanes that can actually fly, or sails and boats, or evacuation systems that get the sewage safely out of our flat. I am referring to the St. Petersburg paradox, which was also conceived by Daniel Bernoulli, the most talented of the three Bernoullis from 18th century Basel in Switzerland.

While staying as the guest of the czar with his friend, the great mathematician, Leonhard Euler in Saint Petersburg, Bernoulli was thinking about the chances of winning when tossing a coin. *"Now I know why I hated the Bernoulli law in physics. This Bernoulli fellow was a rather dumb guy. Why even think about this? A coin has two sides. The chance of winning is clearly 50%. You do not need to be famous to figure this out."* Spite, you are right, but I am afraid you were not patient enough. Bernoulli was contemplating the cumulative wins and losses. The rule in this game is to play until you have a winning series of consecutive heads (or tails): you flip a new coin each time, and you win all the coins you can flip in a row. And what is the probability that you win in a row? In fact, it is scale-free again. If a billion people play the game, then on the average, half will win one coin, a quarter will win two coins, an eighth will win four coins, etc. In this game we always have a chance of winning an order of magnitude higher, but this chance is an order of magnitude smaller (Bernoulli, 1738; Shlesinger, 1987).

The real St. Petersburg paradox and bet-hedging funds. The well-known form of the St. Petersburg paradox is a bit more complicated than the scale-free distribution of successive wins and losses. It is related to the price one is willing to pay for the game which offers an algebraic payment (2, 4, 8, etc.) for a series of successive heads when tossing a coin. The expected value of the game is $2 \times 1/2 + 4 \times 1/4 + 8 \times 1/8$, etc., which is clearly infinite. However, no one is willing to pay even a modest amount for this game, and this is why it is called the St. Petersburg paradox. In order to solve the paradox, Bernoulli introduced the *relative* value of any amount of money, in relation to one's total wealth. This is now called the principle of marginal utility and has become a central element of economics. Analyzing coin tosses, Bernoulli also introduced bet-hedging as a tactic for spreading risk (Bernoulli, 1738). Bet-hedging is achieved by splitting resources, a tactic which makes better (smoother) use of the winning series than putting all one's money up at once. Diversity of action pays well. Bet-hedging has already had an unbelievably successful career. It has been demonstrated that it has been widely used throughout evolution, and it has proved its worth again since the idea was introduced into modern economics. I will return to bet-hedging as a reason for diversity in Sect. 6.3.

The Joseph effect. Mandelbrot (1977) called the clustering of probabilities the Joseph effect, referring to the seven years of great plenty and the seven years of great famine predicted by Joseph in Biblical Egypt. In fact, the successive high and low flood levels of the Nile and many other rivers are also persistent (Mandelbrot, 1977). The Joseph effect has been observed recently in email communications (Barabasi, 2005), and has been explained by a model based on a decision-based queuing process. However, the underlying cause may be more general, as shown below.

Probabilities give a novel interpretation of preferential attachment (Barabasi and Albert, 1999). To illustrate the link, let me quote the famous saying by Benjamin Franklin:

> A little neglect may breed mischief: for want of a nail, the shoe was lost, for want of the shoe, the horse was lost, for want of the horse, the rider was lost, [which was later continued by] for want of the rider, the battle was lost, for want of the battle, the kingdom was lost, and all for the want of a horseshoe nail!

This shows the extremely strong natural sense for chains of unlikely events. Indeed, the successful (or unsuccessful) completion of many sub-tasks will result in a scale-free probability and clustering for the overall success of the complete task (Montroll and Shlesinger, 1982; Shockley, 1957). This is one of the main reasons behind the emergence of Pareto's law, which shows a scale-free distribution of wealth among citizens (Pareto, 1897).

Scale-freeness is related to the self-organisation of the Universe. The surprisingly general occurrence of scale-free properties in space and time and the scale-free distribution resulting from the completion of successive steps raise the idea that this property is tightly linked to the self-organization of matter in the Universe. The scale-free distribution is related to the emergence and maintenance of life (Kauffman, 2000).

2.2.4 Scale-Free Survival Strategies: Levy Flights

The scale-free probability of the several unlikely events mentioned above gives us the impression that we may be able to predict the unpredictable. This would help us to survive in an increasingly unpredictable modern world. Do we have other scale-free survival strategies? Yes, we do. These are the Levy flights (Levy, 1937). When an albatross, an ant, a bumble bee, a deer, a jackal or a monkey makes a search,[10] the length of individual trips follows a scale-free distribution (Atkinson et al., 2002; Cole, 1995; Ramos-Fernandez et al., 2004; Viswanathan et al., 1998; 1999). In most cases we explore the immediate neighborhood by making a number of small trips, because it is cost-efficient. However, from time to time we make a bigger jump, and on very rare

[10] Levy flight may not require conscious action. As the writer sits here and types this book, and as the reader reads this line, millions and millions of cells are making a Levy flight in our body, hunting for food, infectious intruders, or wounds to be healed.

occasions, we go really far to find our target, whether it be fish, pollen, grass, fruit, or the citation beginning the next paragraph.

Viswanathan et al. (1999) showed that the scale-free pattern of Levy flights has a reason. And the reason is simple (see Fig. 2.8). Scale-free Levy flights are the best strategy to minimize the probability of returning to the same site again (a disadvantage of random search) and also to maximize the number of newly visited sites (a disadvantage of the lattice, Brownian-type search). Once again neither random, nor regular brings us the optimum, only something in-between, which is scale-freeness. Note that this is neither rain, nor gambling, i.e., it is neither something like rain, which you have to observe but cannot influence (with the known exceptions), nor something like gambling, which you may avoid (with the known exceptions). If our ancestors, probably right back at the unicellular stage, had not learnt how to plan a search obeying the scale-free rule of Levy flight, they would not have survived. Non-Levy life died out from Earth early on. Scale-freeness may be embedded in our genes.

Definition and comparison of Brownian and Levy-flight search strategies. Both the Brownian and Levy-flight search strategies have the same probability distribution for the step lengths, obeying the equation $P = cL^{-\alpha}$. Note that this is the same equation as we had in Sect. 2.2, where P is now the probability of the given step, c is a constant, L is the length of the step, and α is a scaling exponent. If the search is in two dimensions and the value of α is three, we speak about a Brownian search, whereas if it is two, a Levy flight is performed. For simple models, Levy flights confer a significant advantage over Brownian search in realistic situations, when the searcher is larger or moves rapidly relative to the target, and when the target density is low (Bartumeus et al., 2002; Viswanathan et al., 1999).

"I do not understand something here. How does a cell make a search strategy? Does she learn it from older cells?" A cell is not complex enough to be conscious. Moreover, Levy flights are characteristic of turbulent vortices (Solomon et al., 1993), anomalous diffusion in polydisperse systems (like Knudsen diffusion; Gheorghiu and Coppens, 2004; Stapf et al., 1995) and electron trajectories (Geisel et al., 1985; Micolich et al., 2001). The common mechanism most probably requires some form of arrest in relaxation processes. This is useful in networks in a low-resource environment since it does not allow dissipation of hard-earned energy. Relaxation-arrest may keep the available resources and also lead to

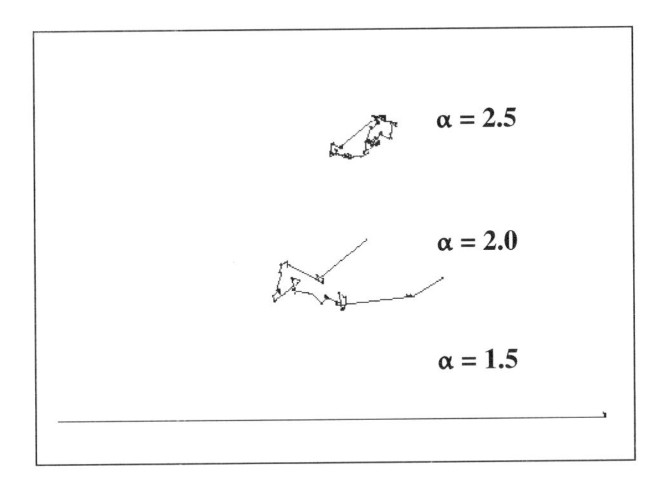

Fig. 2.8. Levy flights are optimal search strategies. In Levy flights, the probability P of the length of a given step obeys the equation $P = cL^{-\alpha}$, where c is a constant, L is the length of the step, and α is a scaling exponent. The search patterns of 1 000 steps are shown for various exponents α. In the model system of Vishnavathan et al. (1999), the case $\alpha = 2.0$, which corresponds to Levy flight, was shown to be optimal. Adapted from Visnawathan et al. (1999) with kind permission

the development of scale-free self-organized criticality (Bak et al., 1987; Bak and Paczuski, 1995; Bak, 1996; Bonn and Kegel, 2003), which may 'automatically' induce scale-free Levy flights. Levy flights may characterize nested networks at many levels.

Levy flights of macro-networks. *"If all these nested networks have Levy flight, then should social groups and ecosystems also make Levy flights, when they try to survive or evolve?"* Spite, your imagination is working well today. This is a very good question for future research and will be addressed in detail in Chap. 12.

2.2.5 Scale-Free Pleasures

If a sense for scale-freeness is so crucial for life, have we figured out any way of practising it?[11] Yes, one form of scale-free training is called

[11]Levy flights may be part of the 'inherent response set' brought with us when we are born. However, any helpful reflex can be overridden by a misdirected mind. We probably need practice to keep our Levy flights alive.

music. Loudness and pitch fluctuations are both scale-free in all classical and folk music from Bach to Mozart, from the pygmies to native Americans. Modern music also follows this trend from Sergeant Pepper by the Beatles to blues and jazz. Here again, random fluctuations produce white noise, the hiss from an unused loudspeaker, which is rather boring. In contrast, synthetic music, obeying regular patterns, produces 'Brownian' music, which is too correlated and becomes boring again. Excitement and beauty come with scale-freeness (Gardner, 1978; Voss and Clarke, 1975).

There are a few exceptions, however. The atonal music of Schoenberg and Stockhausen do not fit this rule (Voss and Clarke, 1975). We may have discovered by the 20th century that no more practice of scale-freeness is needed. Are we right? If anyone reads this book in a few hundred years (if anyone reads anything in a few hundred years), they may be able to give the answer.

Bach abridged. The scale-free structure of music gives us an interesting chance to make a musical sample which is similar to the original. Scale-free systems in space are fractals and are self-similar as mentioned above. Scale-free systems in time, like music, behave likewise. If we reduce a Bach composition to half of its original notes by doubling the scale, the 'half-Bach' and even the 'quarter-Bach' will still sound like original Bach music (composed perhaps when the master was exhausted by the horde of little Bach kids running around, and increasingly sparing with the ornamentation). Interestingly, the final reduction of, e.g., Johann Sebastian Bach's Invention No. 1 in C Major gives the basic three notes on which the whole piece was composed (Hsu and Hsu, 1991).

Is art scale free? After discovering the scale-freeness in music, one is tempted to ask: How general is this? Can we find scale-free elements in the composition of paintings or sculptures? How about the composition of Shakespeare dramas, novels or Hollywood films? Should they not remind us of the Levy flights and other evolutionary lessons? I will return to all of these in Sect. 9.2.

Is play scale free? Going one step further, if exciting music and gambling are both scale-free, how about play? Is it only those parlor games and card games where the probability of winning is scale-free that are popular? Are the chances of placing cards scale-free in the game solitaire? Can a careful series of changes in game rules be discovered in football, basketball, tennis or other games, adjusting the excitement of the game to scale-free behavior? All these are exciting questions for further research.[12]

Good schools are scale-free schools. A good school should have a scale-free probability distribution. Without a few strict rules, the identity of the school cannot be defined and the school lacks the discipline helping the socialization of the students (over-democratic, anarchic schools). Random schools are not optimal. However, the other extreme is probably even worse. An over-disciplined school having almost exclusively strong links efficiently kills all creativity and playfulness (like the Prussian-type schools). Lattice schools are not optimal either. In a good school, not only link strength but also unusual events should follow a scale-free distribution. What does this mean? Most of the time, nothing extraordinary happens. However, from time to time a good school needs an unusual hour or day. The more unusual it is, the more infrequent it can be. But unusual situations should go quite far from the normal, even to extremes at times. And they must occur in a rather random fashion, as in all of our play, as in art, and as in all human activities and institutions, which prepare us for the scale-free probabilities of life itself. A school which either gives too much freedom or is not free enough to create this scale-free behavior serves this key purpose poorly. It is no wonder that students dislike both. Prison schools are disliked instantly, while schools with extreme freedom leave no memories. There is one more important aspect here. Schools socialize us with respect to our society. We learn about the norms of science and social life in schools. Schools tell us what has become acceptable and what has not during the development of human knowledge over the past hundred thousand years. Schools pinpoint certain elements of our knowledge network as reliable and important, and suppress others as superficial, outdated, ridiculous or forbidden. With this labeling, schools strengthen certain links in our knowledge network. If schools do not add many more weak links to the knowledge network by building up creative, playful associations, and providing a rich environment of social interactions, the whole knowledge network will become unbalanced and rather ineffective.[13] Besides teaching the disciplines, a good school also gives ample time for the discussion of the 'non-scientific', 'soft' part of the world. *"I will*

[12]I am grateful to the LINK group member, István Kovács for this idea.
[13]I am grateful to Gergely Hojdák for this idea.

Fig. 2.9. The unpredictable becomes domesticated: the scale-free distribution of unpredictable events gives the impression that they are understandable

print this out and pin it on the wall in my class. My geography teacher will experience a moment of truth, if he understands it."

In conclusion, scale-freeness not only helped our ancestors to survive though the eons of evolution by Levy-flight searches for food and any other goodies we needed, but helps us today to cope with our growing uncertainty about the globalized world. The unpredictable becomes domesticated: the scale-free distribution of otherwise unpredictable events gives the impression that they are understandable. Scale-freeness surrounds us in our environment and appears in all kinds of art to remind us of this task: to be prepared! So there is nothing left but to ask you to be prepared for the next reason why we like networks: nestedness.

2.3 Nestedness

Networks are like Russian dolls. They lie inside each other. My memorable physics teacher, László Holics, used to tell us in class: "Look at this dot. If you examine it closely enough, it is infinite. If you look at it from far enough way, it is a dimensionless point." Networks behave just like this. Most of them can be regarded as structureless elements of a higher order network which I will call the top network in the rest of the book. Similarly, most elements of the top network are themselves whole networks, with a complex structure and stability. These networks, which are elements of a top network, will be called bottom networks in the rest of the book (see Fig. 2.10 and Table 2.2.). In

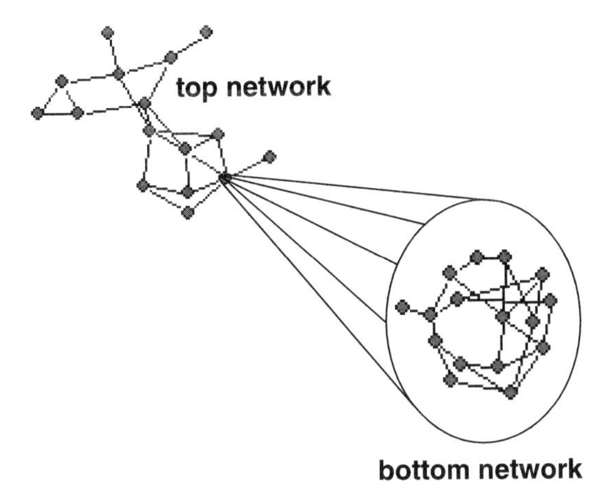

Fig. 2.10. Nested networks. In real, self-organizing networks, each element of the (top) network usually consists of an entire network of elements at a lower (bottom) level. The elements of the lower level form bottom networks. As an example, a top network can be a network of neurons, where the elements are neural cells. Here each top network element, each neural cell, is a network of proteins. Proteins of a single neural cell form a bottom network in this example. Other examples of top–bottom network pairs can be found in Table 2.2. In real, self-organizing networks, both the top and the bottom networks have more elements than the simplified examples in the figure

Table 2.2. Examples of top–bottom network pairs

Top network	Elements of top network	Bottom network	Elements of bottom network
World economy	Countries	Social network	People
Social network	People	Cellular network	Cells
Cellular network	Cells	Protein network	Proteins
Protein network	Proteins	Atomic network	Atoms

other words, bottom networks are nested in their top network. Our world is an onion, which can be peeled and peeled again. Reduction of complex networks to simple elements and the rediscovery of complex networks in these elements are recurrent features of our thinking. From the moment a child disassembles her first watch and encounters the complexity within it, nestedness has been born.

In fact, nestedness is a rather old idea even in the network field. System hierarchy was emphasized by von Bertalanffy as early as 1950. Feibleman (1954) pointed out that the mechanism of any level is found at lower levels, while the purpose of any level is found at levels above. Koestler and Smythies (1969) called nested networks a nested hierarchy of holons. Holons have a Janus behavior. Each higher level is a holon to the lower level. However, the same holon breaks into parts and behaves as an assemblage of them if seen from one level lower. Eldredge (1985) arrives at nestedness (what he calls genealogical, historical and ecological hierarchy) from the genealogy of nested taxa. Nestedness was also described by Oltvai and Barabasi in 2002.

My first encounter with nestedness. For me nestedness did not come in the form of the incapacitation of my father's favorite watch. I had cheese. I still remember the endless minutes I spent at the breakfast table as a three-year old naturalist. I was perplexed. In my hands I had a piece of cheese. However, it was not the cheese but the box which showed me one of the great wonders of the world. On the box there was a bear holding up the same box of cheese. Even on the second box you could see the silhouette of the next bear holding up the next box. Then the resolution gave up and no more bears, no more boxes were seen. My imagination soared. I suddenly saw thousands, millions of bears and boxes inside each other. At the time (after some parental explanations), I thought I had discovered the infinite. Today I know that this was my first encounter with nestedness. Nestedness is an enchanting principle. Once you have felt it, it stays forever.

How was nestedness born? How do networks start to behave as elements and assemble into top networks? Symbiosis-driven nestedness, also called integration by Sole et al. (2003a), occurs if a network enjoys a longer period of relative stability and then, by associating with other networks, all become elements of a top network (Sole et al, 2003a). This self-development is not a purely theoretical assumption. The major transitions in evolution, e.g., formation of the first cells and the first eukaryotes, the first sex, the first multicellular organisms, the first social groups (Margulis, 1998; Maynard-Smith and Szathmary, 1995) may all be regarded as instances of symbiosis-driven nestedness. Life itself requires a higher level of organization, since it is possible only in a macroscopic system. On small, molecular scales, the order required for life would be destroyed by microscopic fluctuations (West and Deering, 1994).

Symbiosis-driven nestedness is the bottom–up, self-organizing, assembling approach to forming a top network. However, there is another way. The top–down approach, the self-structuring, discriminat-

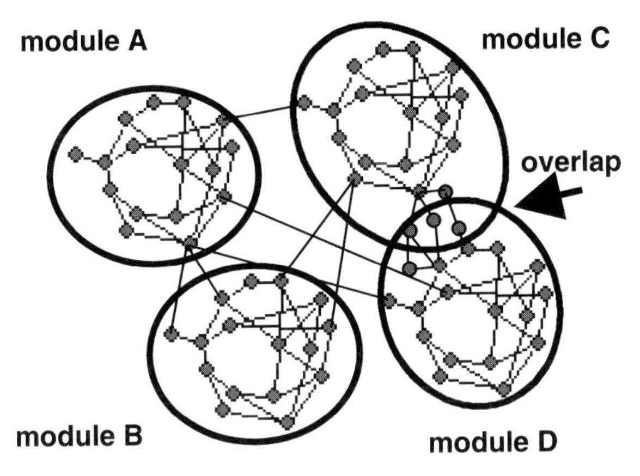

Fig. 2.11. Example of a modular network. The hypothetical network shown here can be divided into 4 modules. Modules A and B, B and D, and A and C do not overlap. However, modules C and D have an overlap region, also called the fringe area. Overlaps play an important role in the defense and regulation of networks, as will be discussed later. In real, self-organizing networks, the modules have more elements than the simplified examples shown here

ing segregation of elements in the top network. This may be called modularization-driven nestedness. This segregative process is referred to as parcellation by Sole et al. (2003a). Here the already-assembled top network first forms modules. Later on, modules can develop into elements of a top network. Module formation and the discrimination between modules and elements of a top network can teach us many lessons. So let us have a detailed look at these processes.

Module formation is a general feature of most networks. Modules are also often called communities (see Fig. 2.11). According to their most widespread definition, modules contain elements which have more links inside the module than outside (Hartwell et al., 1999; Radicchi et al., 2004; Wasserman and Faust, 1994).

More definitions for modules. When we look at a graphical representation of a network, it is often easy to recognize its modules. However, the exact definition of modules is rather difficult. Here I will list some of the available definitions (Radicchi et al., 2004; Wasserman and Faust, 1994). *"I like your approach, Peter, trying to give us as much information as you can in an extremely condensed format, but do you really think we need your survey on the available module determination methods? How about just drawing the network and simply seeing these modules? Would this not be easier?"* If you check the network visualization tools, Spite (a short list of available Web sites is given in Appendix A), you will see that visualizations of networks often inherently contain the modularization protocols. If you want to draw a network 'nicely', you should draw the highly-connected parts close to each other and the outlying elements further from the rest. What is this, if not module formation? We have no other choice, we need to know the module definitions in detail. So here they are:

- **k-cliques.** In the strongest sense, the elements of the module form a clique, meaning that each of them is connected to all the other elements belonging to the same module. As a modification of this definition, the elements of the module form interconnected k-cliques, where k is the number of elements in each clique.
- **LS-sets.** In the strong sense, each element of the module has more connections with other elements of the module than with the outside. In an even stronger sense, this criterion is extended to each subset of the module. Such a module is called an LS-set.
- **Modified LS-sets.** The above criterion can be applied to the sum of the links, meaning that the sum of the outward links of the whole module should be smaller than the sum of the intramodular connections.
- **k-cores.** In a weak sense, the module is a k-core, meaning that the elements of the module are connected to at least k other elements of the same module (Radicchi et al., 2004; Wasserman and Faust, 1994).
- **Co-sets.** A different type of definition arises from the dynamical properties of networks. In this context, co-sets are reaction sets that always occur simultaneously, and these have been defined as network modules (Papin et al., 2004).

How can modules be discriminated from one another?
Here is a brief survey of the methods available for module determination:

- **Clustering methods.** Clustering methods are often called agglomerative since, in one of their representations, all the initial links are deleted and then rebuilt again, starting from the most closely linked communities. In one of these methods, the original network structure is 'renormalized' in such a way that groups of tightly connected elements are considered as one cluster, and this step is then repeated. In this method the network is finally transformed into a dendrogram (Wasserman and Faust, 1994). A very similar procedure has already been mentioned when we talked about the fractal properties of scale-free networks (Alessina and Bodini, 2004; Garlaschelli et al., 2003; Goh et al., 2005; Song et al., 2005a).
- Clustering can be performed by random association of network elements, where the resulting cluster is probed by a modularity parameter derived from a module definition above. Modules emerge as the optimization of this modularity parameter proceeds by successive selection of elements (Newman, 2004).

- A special version of clustering methods uses the Potts model of superparamagnetic clustering, where clusters are identified by assigning a 'spin' to each element. In this method neighboring spins interact with each other, and tend to form clusters, where elements have the same spins (Spirin and Mirny, 2003).
- **Divisive methods.** The gold-standard of these methods utilizes the special feature of intermodular, bridging links, which comes from the definition of modules. If the number of intermodular links is much smaller than the number of intramodular links, then intermodular links should be parts of a much larger number of shortest routes linking two network elements. In other words, the betweenness centrality of intermodular links is larger than that of the intramodular links. The method finds the link with the largest betweenness centrality and removes it. This procedure is repeated until the original network is split into two modules. The method is costly in computation time, since the betweenness centrality changes by the removal of any links, so that it has to be recalculated after each removal (Girvan and Newman, 2002; Newman and Girvan, 2003).
- The edge-clustering method uses the feature that intermodular links are only very seldom parts of link triangles. Deleting links with the lowest amount of edge clustering rather effectively dissects the network at modular borders (Radicchi et al., 2004).
- Another divisive approach starts the procedure with a random split of the original network. Using the module definitions above, the method calculates the 'fitness' of each element after the initial split, and moves the element with the lower fitness to the other module. When an optimal overall fitness is reached, the two modules are ultimately separated and the process starts again with the two modules (Duch and Arenas, 2005). A similar method has been worked out recently by Guimera and Amaral (2005) using simulated annealing to find the maximal number of modules (reasonable splits) in the network.
- Clustering similarities (shared common neighbors) and shortest path similarities have been combined in a recent method to provide a more complex grouping of network elements into appropriate modules (Poyatos and Hurst, 2004).
- **Fuzzy clustering methods.** Neither clustering, nor divisive methods give module overlaps, if executed in their original sense. Module overlaps are also called fringe areas and are important elements of the regulation and evolution of complex networks (Agnati et al., 2004). Reihardt and Bornholdt (2004) worked out a fuzzy version of the Potts model described above. Here, not only is the agglomeration of identical spins rewarded, but the final fitness parameter also has a second term which favors identical spins over the entire network. A proper ratio of these counterbalancing terms identifies overlapping network modules. Fuzziness may be introduced into the network description by adding a random noise to the weights of the network (Gfeller et al., 2005).

- **Fuzzy divisive methods.** Fuzziness can be introduced into the 'traditional' Girvan–Newman method (Girvan and Newman, 2002) by a random selection of the deleted link from those having the highest betweenness centrality. This method is based on the observation that overlapping elements end up in different modules, if we vary the order of deletion of links with a large betweenness (Wilkinson and Huberman, 2004).
- **Topological overlap methods.** These methods utilize the feature that elements of the same module tend to have a much larger overlap of higher order neighbors than elements from different modules (Ravasz et al., 2002; Yip and Horvath, 2005).
- **Network walk methods.** Overlapping modules can be obtained in the strong sense by a k-clique method, where overlapping subnetworks of completely linked k-element subgraphs are identified by a continuous exploration (a network walk) as modules. The method starts with the deletion of a fraction of weak links which usually connect modules, and hence gives overlapping modules in their strongest sense (Palla et al., 2005). Methods using network walks in their weaker sense are currently being developed (Kovacs and Csermely, unpublished work).
- **Dynamical methods.** A new class of approach will take into account the dynamical properties of networks. As an initial attempt, co-sets, i.e., reaction sets that always occur simultaneously, have been determined as modules (Papin et al., 2004).

Modularization is a spontaneously occurring property of networks, where the links are gradually reorganized. Module formation is related to the fractal growth of networks. Modules may also result from the duplication of a network segment and the subsequent divergence of the two arising modules in further evolution. These parallel processes may give rise to modular structures 'automatically', and make them extremely widespread in the organization of life on Earth (Guimera et al., 2004; Sole and Fernandez, 2005; Song et al., 2005b). As an example of this, 74% of known metabolic enzymes of the bacterium *Escherichia coli* are clustered together in modules (von Mering et al., 2003).

Modules as fossils of the past. Symbiotic network modules may be a good source for uncovering the history of network formation. Modules of symbiotic, top networks often conserve an isotemporal cluster of elements which can be identified by detecting the amount of random errors (mutations) in them or knowing their evolutionary history (Kunin et al., 2004; Qin et al., 2003).

Formation of modules seems to be a natural consequence of network development. What is the driving force? What are the benefits of the modular structure?

- Modules localize the effects of deleterious perturbations (Maslov and Sneppen, 2002).
- Modules can evolve rather independently (Hermisson et al., 2003; Kirschner and Gerhart, 1998) and give an advantage, if design specifications (environmental conditions) change from time to time (Alon, 2003).
- Modules allow a separation of various network functions and decrease the cross-talk between them (Maslow and Sneppen, 2002). Modularization usually induces divergence and may be a good source of diversity. All of these things stabilize networks and increase their chances of faster evolution (Kirschner and Gerhart, 1998).
- Modules are not always separated and the module connection is flexible. The intermodular boundary is often not a line but an area, called the fringe area. Fringes can either facilitate or prevent contact between neighboring network elements. The facilitative or occlusive behavior of fringe areas may change from time to time (Agnati et al., 2004).

Modules often have a hierarchical organization, which has been shown to occur in metabolic networks (Ravasz et al., 2002) and in protein folding (Compiani et al., 1998). The organization of the modular hierarchy may follow the scale-free pattern. However, a central module may collect most links to other modules, behaving like a dictator or a black hole. This corresponds to the star structure of networks and reveals an increased level of environmental stress. We shall return to this when describing the topological phase transitions of networks in Sect. 3.4.

Rich clubs and VIP clubs. The hierarchical structure of networks often leads to the development of an inner core. The yeast protein interaction network has a core network of essential proteins, which has an exponential degree distribution (Pereira-Leal et al., 2004). The inner core becomes a rich club if it is formed by the hubs of the network. Rich clubs are formed by Internet routers (Zhou and Mondragon, 2003), large cities in transportation networks (de Montis et al., 2005), etc. Rich clubs are typical of assortative networks, such as social networks, where hubs tend to associate with hubs. However, the rich-club phenomenon extends to disassortative networks, where the non-typical links between hubs become those with the highest traffic. VIP clubs are different from rich clubs. In VIP clubs the most influential members have a low number of connections. However, many of these connections lead to hubs, which can mobilize the rest of the network. Thus VIPs might serve as an interconnected, closed group of the elite, which may even become masterminds of the whole network, as has been demonstrated by Masuda and Konno (2005) for the network of the world's best tennis players.

A rather interesting and hitherto poorly-studied question concerns the nature of the discriminative step between module formation and the behavior of these modules as elements of a top network. *"I guess I have a reason why this interesting question was 'hitherto poorly-studied': it is not appropriate. You said just a few lines above that it is only a question of perception how we define an element of a larger network. If we look at it from a distance, it is a point, but if we go closer, it becomes a whole bottom network. I presume we will have the same inherent difficulties in the discrimination between modules of the bottom network and elements of the top network."* Spite, in a way you are right, discrimination is indeed difficult. However, we have to achieve this discrimination if we want to decide how to describe the network with our one-dimensional information flow, with our words. Before starting to talk about it, we need to know whether it is a module or an element. Let me list a few discriminating features. The higher the values, the more likely it is that the modules of the bottom network will become elements of the top network:

- number of modules,
- network size per average module size,
- structure of modules,
- average number of intramodular per intermodular links,
- weight of intramodular per intermodular links,
- difference of modules from each other,
- availability of module as an independent network,
- transient and unnecessary intermodular links.

Do you know, Spite, if you think over all these steps, what do you imagine? A fantastic step in the self-organization of matter in the Universe: the development of a new layer of nested networks. The gradual enrichment of nestedness is one of the most beautiful events in our past. Without this, neither you Spite, nor me, nor the reader would be sitting here enjoying this book.

??? ?

Are modularization and network nestedness evolutionary tools to restore the stability of overgrown networks? Are the above examples of spontaneous module development (Guimera et al., 2004; Sole and Fernandez, 2005; Song et al., 2005b) general? In other words, do self-organization and growing network complexity automatically destabilize networks? Is modularization and the subsequent development of a top network an evolutionary tool to restore the stability of overgrown networks? *"This is a pseudo-question. You nicely showed that modularization and the subsequent development of top networks are unavoidable steps of self-organization. You even assured us that we could not enjoy your book without this. I beg your pardon, but why are you asking this?"* In spite of the seemingly obvious answer, it is worth asking this question, Spite, since the automatic affirmative has a number of very interesting consequences. If all these processes are spontaneous, the driving force behind cell division might not be just the well-known relationship between cell surface area and cell volume,[14] but cell division may also be triggered by an increase in protein network size and complexity. Similarly, the Parkinsonian modularization of social groups[15] may reflect a stabilization of growing social networks.

??? ?

Is modularization related to the extent of network resources? In Sect. 3.4, I discuss the topological phase transitions of networks described by Tamas Vicsek and coworkers (Derenyi et al., 2004). The preferential random configuration of networks enjoying a resource-rich environment may inherently mean that modularization is triggered by relatively lower resources of network development. However, a formal proof for this is missing.

---

[14]The growth of the cell surface area is proportional to the second power of the number of constituent molecules, while the growth of the cell volume is proportional to the third power. Thus the cell volume occasionally outgrows the maximal possible transport provided by the cell surface.

[15]The Parkinsonian modularization of social groups was first demonstrated by the cyclic growth in the number of members of the British Crown Council, which was accompanied by the development of inner circles of real power (Parkinson, 1957). A possible reason behind this particular example might be that the number of persons able to conduct a meaningful conversation leading to an efficient decision-making process is rather limited (Dunbar, 2004). However, there may be a general explanation behind the same phenomenon.

? ? ? ?

Should we expect the modularization of the Internet, world economy and Gaia?[16] To take this a little further, could we envision the parcellation of other growing networks, like the Internet, world economy, Gaia, and so on? Is there any rule for this? The Internet and the world economy certainly grow. Is it possible that their growth will lead to their modularization, if the resources cannot support the current growth rate? If we think about a sudden modularization of the world economy, it may easily lead to rather violent events, such as an economic crash or war. An even more interesting question concerns the ecosystem of the whole Earth, Gaia. Does this grow? Currently, we cannot even guess the answer. However, a sudden modularization of Gaia may lead to cascading extinctions of several species like ourselves.

? ? ? ?

Is modularization a threat? *"Peter, first you described modularization as a joy, which led to our appearance on Earth. Then, in the last few paragraphs, you described the same modularization as a life-threatening event, which we should avoid. I am confused. Is modularization good or bad for us?"* If we are the top network, modularization of our bottom networks is generally a joy for us, since it shows their growing complexity.[17] However, if we are the bottom network, the modularization of our own network or the top network above us may be a threat, especially if it happens as a fast transition. We need time to adapt to such a change. We need a better understanding of these events to predict and control them. *"Wow, this sounds like something really urgent. What if our understanding lags behind? Do you have any suggestions for NOW?"* My instinctive answer to your question, Spite, is that we should be very cautious when we grow. Until we know more about the sudden phase transitions of network modularization, it is better to avoid them.

In summary, nestedness is a central element in the way complexity evolved on Earth, and probably in the whole Universe. At the same time, nestedness helps us to explain the complexity of the world around us. If I open it, if I go close to it, all of a sudden a miracle happens: it is not a point, but a whole network! Nestedness serves our cognition requirements as well. Nestedness allows the simplification of a whole network to a point, when the complexity of the whole network would

[16] I am grateful to Bálint Pató for these questions.

[17] It is an exciting question in itself, if the numerical measures of complexity which I describe in Sect. 4.4 do indeed grow through growing modularization.

disturb our generalization. Human beings have to enjoy a certain level of isolation to develop their consciousness, and at the same time, they have to be connected to preserve their safety and to enjoy the benefits of labor division. Nestedness is also important to formulate and preserve our own image, our self.

2.4 Weak-Linkness

The task is already completed. We have come to the end of the list of reasons why people like networks. Small-worldness helps to preserve our safety in an alienated world. Scale-freeness helped our ancestors to survive, domesticates the unpredictable, and brings excitement and beauty into our everyday life. Nestedness helps us to explain the complexity of the world around us, and makes us feel at home on Earth. Is there anything left? Weak links, i.e., contacts between network elements which have a low probability, intensity or affinity, do not add very much (yet) to our excitement about networks. However, I hope to show in this book that weak-linkness must also be added to the all-time network favorites, including small-worldness, scale-freeness and nestedness. This section presents a few introductory links between these four properties.

Weak Links and Small-Worldness

Long-range contacts, which are what make small worlds small, are usually formed by weak links. Weak links are necessary for the establishment of small worlds (Dodds et al., 2003a; Granovetter, 1973; 1983; Onnela et al., 2005; Skvoretz and Fararo, 1989).

Weak Links and Scale-Freeness

In real networks, elements are not identical. This leads to the emergence of strong and weak links. Indeed, modeling of Ethernet traffic (Leland et al., 1994), data transport (Goh et al., 2001; Ghim et al., 2004), the metabolism of the bacterium *Escherichia coli* (Almaas et al., 2004), air traffic, scientific collaboration (Barrat et al., 2004a), emails (Caldarelli et al., 2004), and market investments (Garlaschelli et al., 2005a) shows that widely different natural networks develop a scale-free distribution with regard not only to the degrees of various elements in space (self-similarity, fractals) and in time (event probabilities), but also in the distribution of the weights of the links between network elements. Recent modeling of network development has shown that preferential attachment may explain the parallel emergence of scale-free properties of both the degree and link strength distribution of networks (Barrat et al., 2004b; Li and Chen, 2004; Yook et al., 2001). Scale-freeness thus involves the scale-freeness of link strength distribution. What does this mean? The message is rather simple: weak links always accompany strong ones. And in fact, in most networks, we have far more weak links than strong. Somehow networks cannot exist without weak links. What can be the reason of this? This book tries to find the answer.

Weak Links and Nestedness

Network stability may be a key element in the development of multilevel, nested networks. The formation of nested networks obviously requires at least a few contacts between the bottom networks. However, evolutionary selection requires the independence and at least temporary isolation of the bottom networks themselves. Weak links are probably the only tools for solving this apparent paradox. Now you see them, now you don't. Weak links can be broken and reformed again. This feature makes them what they are – weak. Indeed, modules are connected by weak links:

- in proteins (here weak links are provided by water molecules; Csermely, 2001a; Kovacs et al, 2005),
- in cells (here weak links are provided by low affinity protein bridges; Maslow and Sneppen, 2002; Rives and Galitski, 2003; Spirin and Mirny, 2003),
- in societies (here weak links are provided by superficial acquaintances; Degenne and Forsé, 1999; Granovetter, 1973).

Date hubs are probably weak. A recent analysis by Han et al. (2004) showed the existence of two types of hubs in the yeast protein network. One of them had its partner proteins around all the time. This was called the party hub. The other hub, which was called the date hub, kept changing its partners. Date hubs were connecting functional and protein modules together and thus most probably contained weak links. Date hubs have also been identified in transcriptional networks (Luscombe et al., 2004). It will be an exciting task in the future to analyze the stabilizing role of date and weak hubs.

Weak links in cultural evolution. The role of weak-link-induced temporary isolation is not only observed in biological evolution. It also plays an important role in cultural evolution. Cultural innovations often come from a module which is temporarily segregated from the giant component of the society.[18] This relative isolation allows time for the novel idea to develop undisturbed and makes the competition and selection of these ideas by the majority of society possible. If such an idea breaks isolation by weak links, it has a good chance of conquering the society. Blues, pop, the Founding Fathers of the USA were all more or less isolated subcultures originally. 'Incubators' of larger firms also provide an excellent example of isolation-accelerated innovation (Sabel, 2002).

In summary, weak links seem to be necessary for the development of small-worldness, are a consequence of scale-freeness, and make a key contribution to the formation of nestedness. Weak links are also general and important elements of networks, forming most of their contacts. When we talk about the reason why we like networks, we have to talk about weak-linkness.

[18] I am grateful to Viktor Gaál for these ideas.

3 Network Stability

Stability resembles intelligence: easy to imagine and hard to define. Here I will assess network stability at three different levels. First I will describe network perturbations and the concept of local dissipation and global connections. Later a number of different scenarios will be shown, where the perturbation is big or persistent enough to change the stability of the underlying bottom networks, and therefore, the structure of the current network is not preserved. Then stability relationships between the bottom and top networks will be discussed using the example of synchrony. Finally, I will describe the two basic design alternatives: evolution and engineering. Both of them respond to the same question, as does the whole chapter: How can we make a network stable?[1]

If a network has violently changing properties, it is most probably not very stable. How can we measure stability, if a network remains unchanged? The assessment of stability often requires a test, and this test comes in the form of a perturbation to the network. A stable network should try to restore its original status after a perturbation. However, this is not easy. Most networks are open systems and therefore undergo a continuous series of perturbations. In the next section, I will give a survey of such perturbations.

3.1 Perturbations. Good and Bad Noise

Perturbations are often regarded as noise. What is the difference? Noise is usually understood from the point of the experimenter. If we measure it from the outside, noise is the fluctuation of the value we measure. However, from the point of view of the network, noise is a series of perturbations changing its original status. Network perturbations can be called either signals or noise. This dissection is rather artificial and

[1]If you need a definition of stability, let me encourage you to jump to Sect. 4.3. But please come back again!

Fig. 3.1. The understanding of signals depends on the structure of the receiving network

shows our anthropocentric view of the world around us. What is 'good' or 'purposeful' is called a signal, and what is disturbing, undesirable, residual, is called noise.

However, there may be a better way to discriminate between signal and noise. Perturbations which often reach the network and are large enough to disturb the network structure completely, unless the network develops a specific response, provoke a 'learned, adaptive response'. The same adaptive response may arise after perturbations which bring information about important conditions like food or danger, which were experienced by the network in the past to affect the network's integrity in the long run.[2] These perturbations often lead to an unexpectedly large response, which is not proportional to the magnitude of the perturbation, but has been built in to the structure of the responding network as an adaptive response. We usually call these perturbations signals. Other perturbations which are too small, irregular or unimportant to stimulate a learned, adaptive response of the network, are called noise. *"I get it. When a nice girl enters the classroom, she is a signal for my network. When the geography teacher comes in, he is noise."* Spite, I will show later that a better understanding towards the geography teacher is a key point for the stability of your social network. However, your example shows that the understanding of signals depends on the structure of the receiving network. This is the reason why I will neglect the myriad of network-dependent signals here, and I will restrict my-

[2]Preservation of network integrity – or more properly the giant component of the network, network percolation – is necessary for the survival of the organism built by the network. I will describe this network resilience in detail in Sect. 4.3.

self to a description of the perturbation, which is characteristic of all networks: noise. Resistance to noise is an important feature of network stability.

Intrinsic and extrinsic noise. Various parts of the network, like modules, motifs or elements may cause noise to another network segment. This is called intrinsic noise as opposed to extrinsic noise, which comes from the environment of the network (Swain et al., 2002). In the Internet and the microchip, intrinsic noise can be 100 times greater than extrinsic noise. In contrast, in many other systems the levels of the two types of noise are rather similar (Argollo de Menezes and Barabasi, 2004). In gene networks the intrinsic noise of gene expression comes from the low copy number of messenger RNA. An additional component of noise is the transmitted noise from upstream genes and the global, extrinsic noise, which is correlated with the transmitted noise. Here, the total noise of a gene transcription set was found to be dominated by extrinsic fluctuations, and was therefore a function of network interactions (Pedraza and van Oudenaarden, 2005).

Noises are colorful animals. We have white, pink and brown noise to name but a few. White noise is related to fully random fluctuations showing no correlation in time. Brown noise got the name from Brownian motion, since it is typically present in diffusion processes showing no correlation between increments. In contrast, pink noise (which is also called flicker noise, crackling noise, or in a special case $1/f$, $1/t$ or $1/\tau$ noise) has a memory of past events on all time scales, i.e., it is correlated. $1/f$ noise obeys scale-free statistics, meaning that an event that is an order of magnitude larger may always have a non-zero probability of occurrence, but that this probability is exactly an order of magnitude smaller. For a more detailed description of the various types of noise, see the remark below.

A short course on noise. What is the difference between white, pink and brown noise? Scale-freeness is back again. To understand, how noise is related to scale-freeness, we have to do some mathematics again. Noise is usually characterized by a mathematical trick. The seemingly random fluctuation of the signal is regarded as a sum of sinusoidal waves. The components of the million waves giving the final noise structure are characterized by their frequency. To describe noise, we plot the contribution (called spectral density) of the various waves we use to model the noise as a function of their frequency. This transformation is called a Fourier transformation,

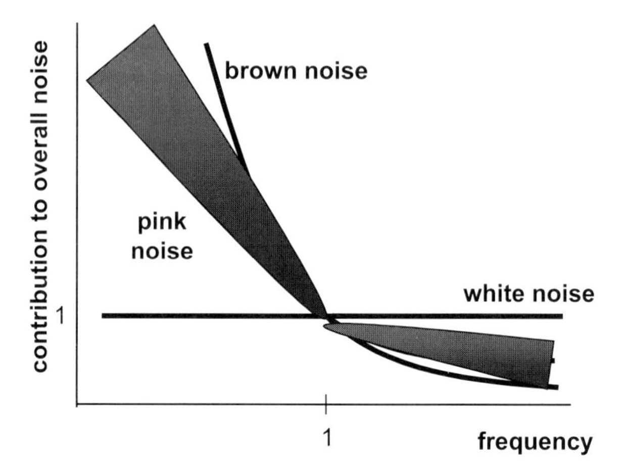

Fig. 3.2. The noise spectrum. Various types of noise (white, brown and pink) are characterized as a sum of sinusoidal waves. This schematic graph shows the contribution of these waves as a function of their frequency. *Shaded areas* around the curve for pink noise are intended to illustrate that pink noise denotes a wide range of noise distributions

and defined by the integral

$$f(\omega) = \int_{-\infty}^{+\infty} f(t)\mathrm{e}^{-\mathrm{i}\omega t}\mathrm{d}t = \int_{-\infty}^{+\infty} f(t)\Big[\cos(\omega t) - \mathrm{i}\sin(\omega t)\Big]\mathrm{d}t \ ,$$

where ω is the frequency and t the time. At the end of this transformation, we receive a distribution of the constituent waves (see Fig. 3.2). Why do we have this complicated mathematics? Part of it is tradition. Noise was first extensively studied in electric amplifiers, and spectral density and frequency were therefore easy choices, since they were the most important properties characterizing the thermionic tube amplifiers (Johnson, 1925). However, this rather complicated representation has proved to be very useful to discriminate between various types of noise.

- **Scale-free behavior.** The contribution of the sinusoidal waves used to compose the noise (the spectral density of the noise) displays the very same scale-free behavior we already observed in the distribution of various network features.[3] The spectral density of the noise obeys the equation $P = cD^{-\alpha}$. Note that this is the same equation we encountered in Sect. 2.2, where P is now the spectral density of the noise, c a constant, D the frequency, and α a scaling exponent. The values of α can vary between zero and (usually) two. If α is zero, we talk about white noise; if

[3]Just as a reminder, our current scale-free list consists of degree distribution, fractal behavior, event probability, and link-weight distribution (see Sects. 2.2 and 2.4).

α is two, we have brown noise; and anything in-between corresponds to pink noise. In some cases we have noise with α larger than two. This is called black noise.

- **White noise.** White noise (when α is zero) has an equal contribution from each wave throughout the whole spectrum. In other words, white noise implies that the value has fully random fluctuations with no correlation in time. White noise has no memory, e.g., it is a Markov process.[4] Compared to other types of noise, white noise shows a strong dependence on short time scale events. This means that high frequencies – short time scales – contribute equally to the final noise structure with short frequencies. This is not true for any other type of noise. There, short time scales give smaller and smaller contributions as we go from pink noise to brown and black noise.
- **Brown noise.** Brown noise (when α is two) is also called Brownian noise, since it resembles a diffusion process with no correlation between increments. However, brown noise is 'better' than white noise, since it 'remembers' the position immediately before the last time step. Here the starting point is defined, but the endpoint of the given step is random. Compared to other types of noise, brown noise shows a strong dependence on long time scale events. This means that small frequencies – long time scales – give a much greater contribution to the final noise structure than in either white or pink noise.
- **Brown and white noise are related.** Brownian motion can be considered as an integral of a white noise process. In other words, if a particle undergoes diffusion (Brownian motion), its position has brown noise, while its velocity shows white noise, in agreement with the above notion that the next position is selected in a random process.
- **Pink noise.** Pink noise (when α lies between zero and two) is the most exciting of all, and therefore, has many other names. It is also called colored noise, flicker noise, crackling noise, Barkhausen noise, $1/f$, $1/t$ or $1/\tau$ noise. The latter three names refer to the situation when α is exactly unity, and the spectral intensity is inversely proportional to the frequency in the equation above. In pink noise, the contribution of low-frequency waves is higher than in white noise. This means that rare events have a greater effect on the noise than frequent events. This is the reason why we call this noise pink. Its spectrum is biased towards the low frequencies, which correspond to red light in the spectral analogy with visible light. The spectrum of pink noise is therefore 'reddened' compared to white noise, i.e., it is pink. Pink noise contains disturbances equally on all time scales, i.e., pink noise is scale-free. In other words, if we speak about a pink-noise process, fluctuations happening once a minute and once a century have the same influence on the present. Pink noise has a memory

[4]In a Markov process, the distribution of future states depends only on the present state and not on how it arrived in the present state.

of past events on all time scales (Halley, 1996; Milotti, 2002; Sethna et al., 2001).

Pink noise is encountered in a wide variety of systems, such as quasar emissions, solar flares, protein dynamics, human cognition, electronic devices, traffic flow, group decision-making, and economics to name but a few, and is suggested to be a characteristic feature of system complexity (Gilden et al, 1995; Gisiger, 2001; Halley, 1996; Lu and Hamilton, 1991; Milotti, 2002; Sethna et al., 2001). However, there are examples of pink noise closer to everyday life. The crackling noise we hear when we crumple a piece of paper also has the structure of pink noise. Additional examples of pink noise will be listed when we consider a specific case of its occurrence, viz., self-organized criticality, in the next section (Bak et al., 1987; Bak, 1996). Let me note here, however, that some of my former examples of scale-freeness in Sect. 2.2, like music, were also various forms of pink noise.

Noise is bad for the network, if high and continuous noise levels disturb all network functions. So far, the take-home message is that we have to stop noise in order to survive. This assumption is wrong. Reducing the noise to zero would mean no interaction of the network with the environment. Isolation is clearly a bad strategy, since such an isolated network will die. However, zero noise is bad for another reason too. Noise can be helpful in many ways. The first documented observations of good noise were sailors' reports on the peculiar phenomenon that disordered raindrops falling on the ocean can calm rough seas (Reynolds, 1900). Another example of the optimal level of noise is opinion formation. A low noise is not enough for modulation of opinion formation, while strong fluctuations prevent the formation of a definitive collective opinion (Kuperman and Zanette, 2001).

A special case of good noise is stochastic resonance (Benzi et al., 1981; Paulsson et al., 2000).[5] Stochastic resonance occurs when the signal-to-noise ratio of a nonlinear device is maximized for a moderate value of noise intensity. The term 'resonance' in the expression 'stochastic resonance' reflects the fact that the weak signal is often a periodic signal. Moreover, a bistable system can be treated as an oscillator, where the rate of switching events gives the typical frequency,

[5] Stochastic resonance has a conceptual resemblance with stochastic focusing. In the case of stochastic focusing, the 'helpful' noise is mostly intrinsic, while in the case of stochastic resonance, it comes mostly from the environment (Paulsson et al., 2000). Since intrinsic and extrinsic noise are often difficult to discriminate, I will use the term stochastic resonance to describe both phenomena throughout the text.

Fig. 3.3. The signal usually has to exceed a threshold to trigger an effect from the adapted network

or eigenfrequency, of the system. This periodic or quasi-periodic signal can be in resonance with the noise, when the eigenfrequency of the nonlinear oscillator and the frequency of the input noise match. A typical case of stochastic resonance occurs in biological systems, when the noise is added to a subthreshold signal and brings it above the threshold, i.e., makes it detectable (see Fig. 3.4). Here, for low noise, the signal will not pass the threshold, and the signal-to-noise ratio is therefore low, due to the undetectable signal. For large noise, the output is dominated by the noise, which leads to a low signal-to-noise ratio again. For moderate intensities, the noise allows the signal to reach threshold, but the noise intensity is not so large as to dominate the output. Hence, a plot of signal-to-noise ratio as a function of noise intensity has a maximum. Pink noise is especially good at helping signals to exceed a threshold, since it is long-range correlated and has a greater chance of satisfying the resonance condition above (Soma et al, 2003).

Stochastic resonance has been invoked to explain climate fluctuations, the sensitivity of fish, cricket, rat mechanoreceptor cells, and the sensitive functioning of ion channels (Bezrukov and Vodyanoy, 1995; Ganopolski and Rahmstorf, 2002; Paulsson et al., 2000; Wiesenfeld and

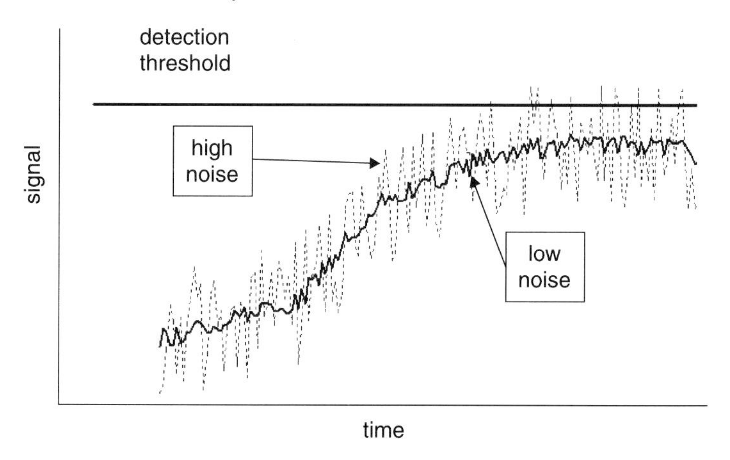

Fig. 3.4. The most important good noise: stochastic resonance. Stochastic resonance is the phenomenon in which a weak signal occasionally exceeds the otherwise limiting detection threshold with the help of noise. In this highly schematic illustration, an arbitrary example of this phenomenon is shown

Jaramillo, 1998). Without stochastic resonance, we would not hear well and most probably would not smell or see well either. Fish find more food (Russell et al., 1999), bones grow faster (Tanaka et al., 2003), and even memory retrieval is better (Usher and Feingold, 2000) in the presence of the appropriate noise. We may conclude that noise can be good, since noise is needed for all the pleasures of life.

If you want to learn, listen to Mozart, not Schoenberg. Memory retrieval has been shown to be higher in the presence of noise (Usher and Feingold, 2000). Various types of music from medieval songs through Mozart and the Beatles all have pink noise structure (Hsu and Hsu, 1991; Voss and Clarke, 1975). The combination of these two observations may explain why many people can learn better if there is music in the background. But does music always help us to learn? Pink noise is efficient at inducing stochastic resonance (Soma et al., 2003) which may help learning better than other noise structures. Schoenberg's music, and some other types of modern music do not have a pink noise structure (Hsu and Hsu, 1991; Voss and Clarke, 1975). Moreover, as the difficulty of a task increases, noise does not help, but disturbs (Usher and Feingold, 2000). This reminds me, to my great surprise, how my favorite Mozart pieces, which had helped me through all my chemistry exams, suddenly became a distracting disaster when I started to learn about Hilbert spaces in mathematics.

Optimal noise not only helps stochastic resonance, but also develops diversity in various networks. If identical networks have stochastic resonance, the time and probability when they reach the threshold and start to behave differently will not be the same. Moreover, noise amplifies selection in finite populations, increases fitness and may increase the chances of robust systems for evolution (Krakauer and Sasaki, 2002).

More reasons for needing noise. The housekeeping heat of steady-state thermodynamics. Oono and Paniconi (1998) suggested a very helpful thermodynamic description for the steady states of non-equilibrium systems. A testable prediction of their theory was suggested (Hatano and Sasa, 2001), and verified later (Trepagnier et al., 2004). A special feature of steady-state thermodynamics is housekeeping heat. This housekeeping heat is defined as the energy preventing non-equilibrium steady states from shifting to equilibrium. Self-organizing networks suffer various types of random damage. Therefore, if the network remained static, it would soon become dysfunctional. Some networks have developed highly specific screening systems which recognize and repair random damage. On the one hand, this process requires energy, which arrives in the form of perturbations or noise. On the other hand, noise-triggered network restructuring will repeat a few steps of the original self-organization and therefore constitutes a much cheaper way of providing a continuous repair function, with the additional advantage that it is always adaptive with respect to the actual environment of the network. The need for permanent noise for the continuous restructuring of networks resembles the housekeeping heat in the steady-state thermodynamics of Oono and Paniconi (1998).

In conclusion, optimal noise is an important condition for network stability, fitness and survival. To achieve optimal noise, a well-functioning, stable network should develop noise control. Networks have figured out many tricks for this. Special arrangements of a few network elements, called network motifs (Milo et al., 2002), may efficiently decrease noise. An excellent demonstration of this phenomenon was given by Becskei and Serrano in 2000, when they showed that, if they put negative feedback into a transcriptional system, the transcriptional noise would decrease. Network modules also protect against noise, since they restrict noise propagation. This happens via the weak links which connect modules and often break if an unusually large perturbation arrives. In the next three sections, I will show what happens in a network when noise arrives.

3.2 Life as a Relaxation Phenomenon: Dissipate Locally, Connect Globally

When the original status of a network suffers a perturbation, the network usually dissipates the disturbing effect, which means that the change is distributed over various elements of the network and relaxation occurs: the network returns to equilibrium conditions. To get distributed, the perturbation has to propagate through the links of the network. Here we have two basic scenarios. The first is when the perturbation can travel undisturbed, and the other, when the perturbation gets stuck at a given node. What are the chances of a perturbation getting around undisturbed? To find the answer, we must first describe the playground for this game. What are the possible paths that the perturbation may select for its round trip? I will discuss the various options, and whether this round trip goes smoothly or the perturbation is arrested at a given point. I will also explain the need for a dual action in the connectivity of networks. On the one hand, free travel of the perturbation has to be confined to a segment of the network causing a local dissipation. On the other hand, the reason for the existence of networks is a global communication spanning the entire network.[6] Finally, I will describe what happens if the perturbation piles up at the origin or any other point of the network. In fact, the following two sections will also elaborate on this latter problem, showing how perturbation-induced damage gets magnified as conditions worsen.

3.2.1 Confined Relaxation with Global Connection

Bearing in mind the above arguments which show that perturbations are necessary for the survival of networks, we shall examine how networks can survive the excessive damage they may cause from time to time.

Scene One

To examine how perturbations might get dissipated, we must first explore all possibilities as to where they may go. One of the most important properties of the network is the presence of a giant component. If a network has a giant component, than most of its elements are connected with each other. This property makes the network a network.

[6] I am grateful for the instructive title of the paper by Tewari and Toner (2005), which helped me to generalize this concept.

When the giant component is dissolved, the network becomes a set of isolated subgraphs which are not connected to each other and therefore cannot communicate. If the network was a living entity like a cell, animal, plant or human being, and if its giant component disappeared, the network would die.[7] *"There is something I do not understand here. Do you get rid of your favorite networks like this? They can die, period. When do they die? How can we protect them? I am a network! I want to know how I can survive!"* Wow, Spite, you got quite emotional! Do not worry. The next two sections will give you many examples of how to save your network.

Let me deal here with the other end. How is the giant component born? The answer is: abruptly. As the number of links increases in a network, we eventually reach the percolation threshold and the giant component appears suddenly.[8] Once there is a giant component, perturbations may start their trip. And now comes the really important question: How far can they go?

Scene Two

For the answer to the above question, we will examine the propagation of information in the network: rumors, infections, Pity's love letters during your history class, Spite (*"Wow, this is creepy. How does he know about this? Can he read my thoughts?"*), etc. The perturbation has to be dissipated fast. We therefore require efficient connectivity in the segment of the network that was hit by the perturbation. In other words, we need small-worldness. Lattice networks will have a hard time if a perturbation arrives. As an example, diamond is very hard. However, if it gets overwhelmed by perturbations, it never bends. It breaks. The diamond lattice has difficulty dispersing the trouble.

Scale-freeness also comes in again. In scale-free networks the perturbation can propagate easily, even when the probability of transmission is extremely small (May and Lloyd, 2001). However, in finite scale-free networks, there is a threshold (Dorogovtsev and Mendes, 2002; Park et al., 2005b), which means that below a critical connection (at low transmission probabilities), the perturbation may only have a restricted trip and may even get stuck somewhere in the network. The

[7] Complex network functions – all the characteristic features of a living organism – cannot be performed in the absence of widespread network communication. Life requires the integrity of the networks which form the living organism.

[8] The sudden appearance of the giant component and the emergent properties of the network may have formed the empirical background of the Taoist and later Hegelian dialectic concept of "quantitative changes becoming a qualitative change" (Hegel, 1989).

small worlds of scale-free networks seem to be small in a dual sense. They give long-range connections and they concentrate most of the important connections in a local environment (Lai et al., 2005).

"Your scale-free networks seem to lose out this time, Peter!" Not yet, Spite. We have not yet taken into account connection weights. The chances of the perturbation getting jammed in double scale-free networks, where both the degree distribution and the weight distribution exhibit scale-free behavior, are much smaller than in random networks (Toroczkai and Bassler, 2004). In other words, natural, weighted scale-free networks seem to be quite good at combining fast relaxation and restricted travel of the perturbation. In other words, scale-free networks are rather stable. This explains the earlier findings and assumptions that scale-free topology is an important element in the stabilization of many networks (Fox and Hill, 2001; Barabasi, 2003).

The positive side of traffic jams: spam protection and local dissipation. The relative difficulty in searching scale free networks increases even further if the network has a modular or hierarchical structure (Rosvall et al., 2005). Disassortative networks, where hubs are only seldom linked to each other, also restrict travel (Brede and Sinha, 2005). These difficulties confine damage due to perturbations and may also help networks to protect local areas from non-related communication (Rosvall et al., 2005). A carefully balanced combination of efficient local dissipation and global communication might be a key point in network design and survival.

What is the difference between the news and perturbations? *"Peter, a question has been repeatedly popping up in my mind for some time now. You said that we need a careful balance between local dissipation and global communication in the network. How does the network know which change is a perturbation that has to be confined and which is a piece of information that has to be transmitted?"* Spite, you have hit upon a good point. I am afraid the exact answer is not really known. I suspect the answer is similar to the one I gave as the difference between a signal and noise. The network considers as information only those perturbations which arrive often enough for it to develop a learned response, or which are important enough for the survival of the network. These perturbations get the network highways to go right round. All other perturbations are brought right round in local segments of a well-designed network, until they level off.

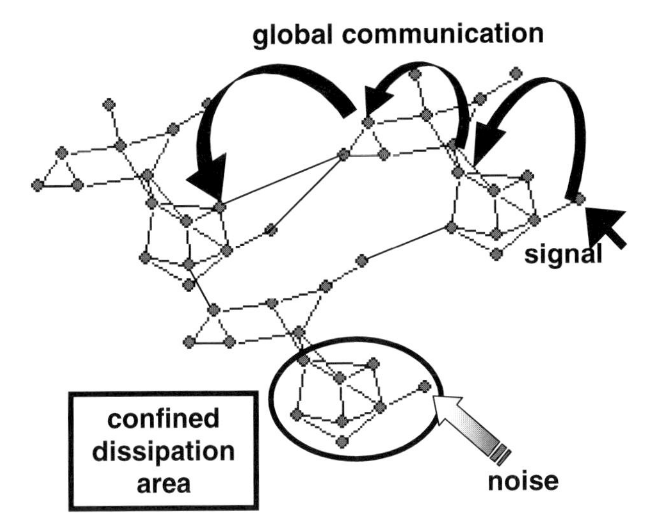

Fig. 3.5. Local dissipation and global communication. Networks should combine the benefits of local dissipation and global communication. Confined dissipation keeps the unspecific perturbation, called noise, in a restricted segment of the network. Global communication helps a few specific perturbations, called signals, for which the network has developed an adaptive response. These signals may reach distant elements of the network

Carefully confined relaxation may be an important element in the way the emerging complexity of self-organizing networks 'develops' stability (see Fig. 3.5). I will discuss an example of this in Chap. 6 in the context of complex gene networks, where a channeling behavior occurs without any extra additional mechanisms (de Visser et al., 2003). Self-organization often leads to the development of scale-freeness, modularity and network hierarchy, causing an 'automatic' stabilization of the network.

3.2.2 Self-Organised Criticality

We have seen that scale-freeness helps the local dissipation of perturbations. But what happens if a perturbation cannot be dissipated and gets stuck? Tension will develop. A very spectacular example of this scenario occurs when the perturbation is continuously repeated and the tension keeps on increasing. The development of tension does not lead to major problems for quite a while, since the individual perturbations arrive separately and get stuck at different points of the network. However, as more and more perturbations arrive, local tensions accumulate and may develop to a point when a propagating relaxation suddenly occurs like a kind of avalanche.

This behavior was called self-organized criticality by Per Bak (Bak et al., 1987; Bak and Paczuski, 1995; Bak, 1996). Unfortunately, self-organized criticality remained a rather loosely defined concept, describing a phenomenon in which, in a network with restricted relaxation, a gradual increase in tension is followed by sudden avalanches.[9] However, self-organized criticality is a very general phenomenon, and it is also very helpful for understanding network behavior. I will describe its meaning and give several examples of similar behavior in the following. At the critical event, a sudden relaxation develops, the effect of the perturbation starts to propagate, and finally a larger segment of the network will communicate. A netquake occurs. The extent of netquakes has a scale-free distribution in the given network, both in spatial extent and duration (Bak et al., 1987; Bak and Paczuski, 1995; Bak, 1996).[10] The frequency of occurrence also has a scale-free distribution. This shows that there is a correlation between netquakes, i.e., the netquake frequency depends on the whole history of the system (Lippiello et al., 2005). This is in fact true of most natural, self-organizing, critical events, and reflects the general behavior of pink-noise structures as described above (details will be given later).

Self organized criticality may lead to the development of scale-free networks. Self organized criticality reorganizes the network structure as the avalanches pass away. Elements will be disconnected, then reconnected again. Fronczak et al. (2005) showed that the network reorganization may find the equilibrium in which both the degrees and the avalanche properties exhibit a scale-free distribution.

[9] Originally, Per Bak and coworkers (1996) defined self-organized criticality for a rather well-characterized set of events involving sand piles, or rice piles. In this chapter, I will considerably generalize this concept to demonstrate its applicability in our everyday life. This may be misleading. I would like to ask the patience of those who prefer exact definitions and better defined concepts.

[10] Netquakes may be regarded as events in which the barriers confining perturbations to certain network areas start to break, and as perturbations start to communicate, more and more barriers cease to exist, or are overcome. Perturbations fluctuate continuously and get dissipated to different extents at a given instant of time. Consequently, a netquake is a rather stochastic process which soon stops in most cases. However, some exceptional netquakes propagate violently. The successive completion of distinct events during the occurrence of a netquake may explain the scale-free behavior of netquakes (see Sect. 2.2).

Netquakes are violations of the fluctuation–dissipation theorem. The fluctuation–dissipation theorem is a famous statement of statistical physics related to the phenomenon that a fluctuation in a system is dissipated as it returns towards equilibrium. A well-known form of this statement is the Stokes–Einstein relation between diffusion and viscosity, stating that $D = T/c\eta$, where D is the diffusion constant, T is the temperature measured in kelvins, c is a constant, and η is the viscosity. Here the fluctuation is Brownian motion, dissipated by resistance due to the viscosity of the surrounding material. Heterogeneous and/or slowly relaxing, glass-like phases can avoid equilibrium in the long term, thus violating the fluctuation–dissipation theorem. Here again, relaxation is confined and happens by directing the perturbation to the fastest relaxing segment of the network (Grigera and Israeloff, 1999). The extent of the violation of the fluctuation–dissipation theorem follows a scale-free distribution (Bonn and Kegel, 2003). This is in agreement with the scale-free behavior of netquakes, which probably occur in most cases here. In fact, the deviation from the fluctuation–dissipation theorem can be used as a measure of metastability (Bonn and Kegel, 2003), and maybe also as a measure of system complexity (the definition of complexity will be detailed in Sect. 4.4).

Netquakes can have a wide variety of forms. Earthquakes, landslides, forest fires, fractures, volcanic eruptions, avalanches, protein quakes, magnetization propagation (the Barkhauser effect), quasar emissions, solar flares, dripping faucets, rain, and many more examples mentioned in the last section, like the crackling noise we hear when we crumple a piece of paper, belong to the class of obstructed relaxation events and all conform to the rules of self-organized criticality (Alessandro et al., 1990; Bak, 1996; Bazant, 2004; Cote and Meisel, 1991; Gilden et al., 1995; Gisiger, 2001; Halley, 1996; Lu and Hamilton, 1991; Malamud et al., 1998; Milotti, 2002; Penna et al., 1995; Sethna et al., 2001; Turcotte, 1999).

Panic quake. One of the recent developments of self-organized phenomena came from Hungary (Helbing et al., 2000) and the Philippines (Saloma et al., 2003). Humans turn into a flock of sheep if a real danger arrives. This herding behavior is known scientifically as an allelomimetic tendency. Our most important decisions are made by our emotions (Damasio, 1994; Rolls, 1999), and this old thalamic system most probably takes over in panic reactions. We trample each other, form arches in front of exits, effectively block the only escape route, do not use alternative escape routes, etc. (Helbing et al., 2000). Saloma et al. (2003) studied mice, not humans. On the basis of the above notes, this would not seem to make a significant difference! There was a difference though. Unlike some of us, mice do not like to swim. The finding was that they left the pool in groups with a scale-free distribution. These mice seem to be as socially handicapped as we are. They probably watched the seniors. Would they dare to jump out? Nothing happened for quite a while, but tension built up. And then, suddenly one mouse equivalent of spiderman jumped and a whole flock followed immediately, causing an instant relaxation, or panic quake.

The nature of thunder and lightning. There are many more phenomena which behave rather similarly but have not yet been classified as networks reaching self-organized criticality. One of these is thunder and lightning. Here we may also discover all the items in the inventory: the series of events helping static to grow, thereby developing tension in the form of a voltage, and the sudden relaxation at the end (Nebuchadnezzar might have a story like this to tell!). It would be nice to establish the scale-free distribution of lightning intensities and the loudness of thunder. In fact, the $1/f$ bursts[11] of very low-frequency electromagnetic radiation coming from lightning discharges (Magnasco, 2000) strongly suggest that this is the case.

Fortunately, volcanic eruptions, landslides, and panic are not common everyday experiences. Protein quakes are too small and solar flares are too distant to observe. (So astronomers say. For the solar flare, you have only two opportunities with a telescope and bare eyes: one for your right eye and another for the remaining left!) However, there is one very important example of self-organized criticality at each moment of our life: lung quakes. Whenever we take a breath, the individual airways open in an avalanche-like fashion, producing the same scale-free statistics as all the other critical phenomena listed above (Barabasi et al., 1996; Suki et al., 1994).[12] A few more common examples of netquakes will be listed here.

Tick quakes. Self-organized criticality may be much more common than we think. Peterson and Leckman (1998) showed that ticks,

[11] This is the same as $1/t$ noise, an example of pink noise (see Sect. 2.1).

[12] Actually, the 'reverse phenomenon', coughing, may also be a self-organized critical event and asthmatic bronchoconstriction has recently been described as a self-organized event (Venegas et al., 2005).

which are rapid, brief, unintentional skeletal or vocal muscle movements in some people, like those with a Tourette syndrome, follow scale-free statistics. Ticks may in fact be muscle quakes, exhibiting bursts of the tension which has gradually built up in these patients and could not be gradually dissipated as it would in the rest of us.[13]

Gossip quakes. When a gossiper hears a great story, tension develops: "I should tell someone this!" The more stories arrive, the more likely a gossip quake is to occur. Eventually, the gossiper cannot resist any longer and picks up the phone. It may be worth asking whether time intervals between the transmission of two consecutive pieces of gossip obey scale-free statistics.

In all these events, both the probability and the magnitude of the event follow pink noise (Milotti, 2002; Sethna et al., 2001). What does this mean? If the event has $1/f$ noise statistics, there is a non-zero probability for any event which is an order of magnitude larger, but this probability is an order of magnitude smaller. We have a clear sense for this in the case of rain or earthquakes. Fortunately it is only very seldom that we have a devastating earthquake, or rainfall so heavy that it smashes everything or washes everything away. However, this is a general phenomenon. So watch out next time you start to crumple your candy paper at the movies. If you are truly unlucky, it may make a really big noise. *"I cannot believe that I will ever go deaf after crumpling a candy paper at the movies."* Yes, Spite, you are right. Natural scale-free events usually have an exponential cutoff. So there is no need to worry about going deaf due to a sudden noise explosion from your candy paper.

Our life is full of netquakes, from birth to death. Let me list a few examples here:

Ogling quakes. A girl and a boy travel on the bus. After the first glance, they both realize that the other is the most beautiful, most charming person they ever saw. Good manners, however, require them to stop gazing at each other after a second or so. "I need to see her again!" As the seconds pass, a significant tension develops. All of a sudden, good manners are forgotten, and they start ogling again. What is the distribution

[13] If you are a psychologist and after this sentence you want to throw the whole book into the fire, I ask you to consider the fact that this response might reveal a tension–relaxation problem.

of the length of gaze-abstaining periods? What is the distribution of gazing intervals? It is probably scale-free.

Woo quakes. Suppose the girl and the boy finally met. Now it is said that love makes one beautiful. If you have ever watched an amorous girl or boy, you will certainly agree that the statement is true. But why do we look more beautiful when we are in love? We do not look different. We start to behave differently. We become more playful. What does this mean? Among the many changes, we may observe sudden, unexpected movements and a kind of indolence. What is the driving force behind these movements? If everything goes well, a kind of continuously growing tension develops, pushing us to do things that are not really the right thing to do in the middle of the street, on the subway, etc. What is actually happening? Sooner or later a pseudo-action will follow. The tension relaxes by an unexpected small jump, by a sudden kiss, or a smile to a complete stranger. The relaxation is not complete, however. A large amount of this happy tension remains, and starts to grow again. What do we call these relaxations? They are woo quakes. What is the most probable distribution of the periodicity and extent of unexpected, playful actions? It is scale-free.

Sex quakes. Let me continue the above story. Imagine that the same couple finally reach a comfortable apartment, or a less comfortable glade in the forest, a boat, a space station, a live show, wherever their habits and possibilities may have brought them. Continuous – or perhaps it is better to say rhythmical – input of energy, gradual growth of tension and then, suddenly: BANG! The relaxation occurs like an avalanche, like the eruption of a volcano. Is this familiar to you? Evolution seems to figure out a self-imposed method for rehearsing scale-freeness in a joyful manner. Sometimes it is good, sometimes it is better. Can it be even better? If this is really scale-free, our happy couple always has a chance for an order of magnitude larger effect, although the chances of this are an order of magnitude lower. The scale here is probably rather limited. Are you sure? Our couple is like the inhabitants of Los Angeles. They can always hope that the next will be the Really Big One. I must admit, my example is perhaps not the best this time, since the emotional background of the two expectations seems to differ. *"Peter, let me remind you of what you said just a few lines above: natural scale-free events usually have an exponential cutoff. It might be disappointing to you but the same principle has to be applied here, I am afraid."*

Baby quakes. It is now nine months later. I see a cute baby with a comforter. Babies may use their comforters to relax a multitude of tensions which reach them continuously from the new and alien world. In the absence of comforter-induced relaxation, crying (a baby quake) develops. Both suckling of the comforter and crying may follow a scale-free distribution, although obviously in periods when the baby is awake.

Growth quakes. Let us take one more step. Our baby starts to grow. The young cells keep dividing. However, divisions themselves are multiply regulated and the growth of different tissues is not synchronized. A tension develops and growth becomes saltatoric (Lampl et al., 1992). Uneven growth brings us beyond self-organized criticality, since here tension development is not steady but follows a multidimensional pattern. Instead of exhibiting a simple scale-freeness, growth quakes may resemble the multifractal properties of heartbeats which are described in Sect. 7.2. Indeed, later studies (Thalange et al., 1996) showed that the rules for saltatoric growth can be very complex. Uneven growth is used by plants to change their shape or catapult their seeds, and has become an important category in evolutionary economy (Feenstra, 1996). I will spare you from a quake version of 'Buddenbrook House' to describe the life of the happy couple, but I hope I have convinced you that, until the very end (crying quakes of the same baby, now a little older, at their funeral), their life will be full of similar events.

In the self-organized critical state, weak links break first (behaving like the weakest link of the chain). In the netquake, the rearrangement of weak links helps to restabilize the system. In other words (Sethna et al., 2001): "Not all systems crackle. Some respond to external forces with many similar-sized, small events (for example popcorn popping as it is heated). Others give way in one single event (for example, chalk snapping as it is pressed). In broad terms, crackling noise is in between these limits: when the connections between parts of the system are stronger than in popcorn, but weaker than in the grains making up chalk, the yielding events can span many size scales. Crackling forms the transition between snapping and popping." Crackling needs weak links.

Culture quakes. When I wrote the first version of this book, we had the 40th anniversary of the Beatles avalanche in the USA in 1964. Why did they reach such a large segment of the society in such an incredibly short time? Why do cultural and technological innovations resemble self-organized criticality? Innovations should reach a percolation threshold (Ryan and Gross, 1943), after which they suddenly became a general custom of the majority of the social network. The acceptance of innovations is a cooperative action, which inherently contains the possibility of avalanches (Watts, 2002) and has indeed been shown to behave as a self-organized critical phenomenon for the change in pottery styles and for the propagation of the idea of self-organized criticality itself (Bentley and Maschner, 2000; 2001). Weak links may also contribute to the occasional burst of innovations, novel ideas or cultural changes to the whole society. These innovations often come from a module of the society[14] which is connected to the rest by weak links. Once the innovation has passed this bottleneck, as the Beatles crossed the ocean, it gets a free ticket to ride around.

Schumpeterian innovation clusters. The phenomenon of innovations breaking away from a state of relative isolation – combined with the notions of cooperation and the percolation threshold – may also help to explain Schumpeterian clustering of innovations (Schumpeter, 1947), i.e., a surprising onset of serial innovations behaving like bursts, and coming in clusters. As innovations spread in society, they will sooner or later reach someone who may add a new element to the existing design. The number of persons reached by the innovation probably follows scale-free statistics. Moreover, the innovative process itself consists of individual steps, where the next step depends on the one before. These processes also display a scale-free pattern.

To sum up, if we seek network stability (and so we should, otherwise we die), we need a highly efficient but local relaxation. To achieve this, the perturbation has to reach the elements of the given segment of the network rapidly enough. In other words, we need small-worldness for efficient local dissipation. We also need small-worldness for the global communication of signals. On the other hand, we need scale-freeness and nestedness (in the form of network modularity) to confine the perturbations. Efficient networks combine global connections with local relaxation.

If many perturbations arrive and get arrested one by one, avalanches will occur sooner or later. For this latter event, we need a constant inflow of perturbations, the subsequent development of a tension, and then its sudden relaxation through an avalanche. There is another

[14] I am grateful to Viktor Gaál for this idea.

Fig. 3.6. Crackling forms the transition between snapping and popping

important message here, if we put the last sentence in a slightly different form: for self-organized criticality, we need the constant inflow of energy, the subsequent development of tension, and then its sudden relaxation by an avalanche. For this exposes a surprisingly general meaning. Without a constant inflow of energy, there is no life. The arrival of energy drives the system forward, and by making the system thermodynamically open, allows a decrease in its entropy.[15] However, each energy packet is a new perturbation, and they must all be dissipated. Therefore, relaxation is necessary for life. And we had better ensure that it is fast and confined. Life behaves as a carefully confined relaxation phenomenon in globally connected networks.

[15]The free energy has to decrease for any change in the self-organizing networks. Since the free energy is defined as $G = H - TS$, where G is the Gibbs free energy, H is the energy, T is the temperature measured in kelvins, and S is the entropy, we need an inflow of energy (negative ΔH) to compensate the decrease in entropy (negative ΔS) and keep the free energy change negative.

3.3 Network Failures

Spite! You have become suspiciously silent. It is time to snap out of this dream-world relaxation. Wake up! Here is a question: What happens, if perturbations are not only temporarily, but hopelessly stuck in the network? I see your thoughts are still far away. I will give you the answer: if relaxation becomes impossible, than one or more elements of the network will receive a perturbation big enough to cause the disassembly of the bottom network below the unlucky element where the perturbation got stuck.

Let us consider an example. On 10 August 1996, the failure of a power line in Oregon led to a black-out in several states of the USA and Canada. What happened? Due to a combination of extremely high temperatures and a full power load at about 3:40 pm, the 500 kV line connecting the Keeler and Alston substations in Oregon sagged so much that it touched a tree and was immediately shorted. The resulting frequency oscillations knocked out 13 hydroelectric units at the McNary Dam. The disassembly of these bottom networks cut the main inter-tie, which runs between Oregon and Southern California, causing a chain of breakdown events as far as Nevada, New Mexico and Arizona. This triggered further outages throughout the whole western region of the USA leaving more than 4 million people without electricity in 11 states (O'Donnell, 2003). The national disaster on 10 August started due to relatively minor initial damage. However, this was enough to switch off the 13 hydroelectric units nearby and was immediately magnified. Such behavior is called cascading failure and is typical of network structures. Similar cascading failures have been observed in the economy, in earthquake aftershocks, and in other contexts (Bak, 1996; Moreno et al., 2002). One of the best examples of this is the domino effect. With only a slight touch, the long line of dominos all topple over in a fraction of a minute.

Power net quake. *"This cascading failure reminds me of the disturbing variety of quakes in the previous section. Don't you think you have forgotten to mention something?"* Well done, Spite! You learned your netquake stories very well! Indeed, the cascading power failures can be regarded as events of self-organized criticality in the power net (Carreras et al., 2004). The continuously increasing tension is the increasing load on the system. The continuous input of energy is rather obvious. Relaxation comes in the form of system failure. (Well, this relaxation does not seem to be valid for the controllers!)

Fig. 3.7. Permanent random damage can lead to a massive network malfunction

This is bad enough. If we cannot design systems wisely, to save our network from cascading damage, we may start to collect ice cubes on every hot day during the summer (or move to Alaska). But the situation is even worse. We do not even need a cascading failure to impair a network. Although scale-free systems are rather insensitive to random damage (which is why evolution did not throw them straight into the garbage can), they are much more sensitive if any of their highly connected nodes, the hubs, gets damaged (Albert et al., 2002). This is what pushed the situation from bad to worse and from worse to disastrous in August 1996, as the local damage disconnected the Oregon–Southern California inter-tie, causing an avalanche of further damage. Cascading failures reach hubs rather quickly, and sooner or later they are sure to reach a hub as they propagate in the network. As seen above, cascading failures may sequentially incapacitate the network up to a critical point where the giant component collapses and the network ceases to function (Moreno et al., 2002).

Error and attack tolerance of scale-free nets. If compared to a random graph, a scale-free network is surprisingly stable against random damage, but vulnerable against a planned attack (Albert et al., 2000; Bollobas, 2001). The following gives a few details concerning error and attack tolerance of various networks:

- **Exception One: When random damage is also bad.** As opposed to occasional random damage, permanent random damage can cause a massive malfunction of the network (Dorogovtsev and Mendes, 2001). In other words, aging can devastate our networks. Not abruptly, but efficiently. Some examples of this are discussed in Sect. 6.4.
- **Exception Two: When a planned attack is not that bad.** Assortativity makes the network resistant against planned attacks. In an assortative network, similarly connected elements associate with each other. Hubs

like hubs, smaller nodes seek nodes, and almost isolated elements will be associated with similarly isolated elements. Such an assortative network is quite typical in social nets. For the most assortative network, ten times as many hubs have to be removed to destroy the giant component than for the most disassortative one (Newman, 2003a). However, assortative networks are more unstable than disassortative ones. In assortative networks, perturbations can propagate further and a better synchronization leads to larger dynamical fluctuations (Brede and Sinha, 2005; di Bernardo et al., 2005).

- **Not all networks are scale-free.** As a final remark, I would like to warn you that it is only because of their wide occurrence and stability that we are talking so much about scale-free networks. There is a whole universe of unexplored network structures out there. To give but one possible example, a simultaneous optimization with respect to both random damage and intentional attack gives novel types of networks (Valente et al., 2004; Paul et al., 2004; Shargel et al., 2003).

Where does the perturbation get stuck? Do perturbations cause random damage? In principle, a perturbation may get stuck anywhere in the network. In practice, this is probably not true. I have not yet found an exact answer to this question. Perturbations may get stuck at narrow points of the network, where the connection is weakest.[16] Alternatively, perturbations may get arrested at hubs, where most traffic converges and causes a jam. In the latter case the perturbation behaves like a terrorist, attacking the most vulnerable points of a scale-free net. Do our networks direct the damage to their most vulnerable points, thereby making us hostages of terrorists? Fortunately not. The giant component of the network is well preserved even in the case of a planned attack. For example, the World Wide Web will remain connected and functional even after the removal of all of its nodes of degree higher than five (Albert and Barabasi, 2002).

The wisdom of our cells: low flux comes with high stability. Enzymes are parts of the metabolic network of our cells. Their stability is governed by the stability of their protein structure. If we introduce a few extra, stabilizing bonds, we will make the mutated protein more rigid. However, increased rigidity leads to diminished enzyme activity (Shoichet et al., 1995). Thus our stable protein will most probably be a 'weak point' of the metabolic net, since it will have a low flux and will offer

[16] These may be weak links between modules, which are often characterized by a high betweenness centrality (Gfeller et al., 2005).

better chances for the perturbation to stop. However, it is not bad if the perturbation stops here, since our 'weak point' is not in fact weak at all: with our stabilizing mutations, it has become more stable than the average protein in the original network. Thus the metabolic network of our cells may contain two types of protein: (1) high achievers, which have a high flux, ensure most of the tasks required for cell survival, but have a low stability and pass the perturbation to their colleagues, and (2) low achievers, which have a low flux, and consequently do not do much work, but have a high stability and absorb the perturbation, thereby saving the high achievers from major damage. If we look around at social networks, we may start to think that this division of labor is rather general in networks.

It is time to turn back to our cascading failures and confirm the initial statement: they are still most annoying. We build up networks at high cost, and yet they sometimes literally melt down in a matter of minutes. We have to make them safer. If the disconnected Oregon–Southern California inter-tie caused such big trouble, why do not we make another, parallel inter-tie? Actually, why do not we duplicate most of these sensitive lines?

Doubling the network seems to be a rather bad idea. According to a very interesting model worked out by Duncan Watts (2002), too many connections may make the extent of damage unpredictable. The Watts model is based on a scenario in which the status of each element of the network depends on the average status of its neighbors. This setup suits the description of any domino-type cascade, like the chain of events in August 1996, where neighbor influence is predominant. If we start to increase the number of connections in this network, we eventually reach a critical point where cascading failures become possible. Here a scale-free distribution of cascade size is observed. This means that most cascades will be small, whereas large, devastating cascades will occur only seldom. Paradoxically, if we then go further, cascades become larger but more seldom. This is due to the dilution effect. When a node has too many neighbors, the influence of each will be diminished. Interestingly, close to this second, critical point, a bimodal size distribution is observed, where large cascades have much greater chances of occurring than in the scale-free distribution regime. This implies a more extreme type of instability which is even harder to predict (Watts, 2002). Thousands of perturbations may arrive without any of them having any effect. And then the proverbial last son of the Perturbation Family gets in, smallest of all, and suddenly, in a single miraculous second, a gigantic cascade erupts. The instability of 'overconnected' networks is rather general and can be applied to food webs, trade networks and power nets (Fink, 1991; May, 1973; Siljak 1978).

Vulnerable points of networks. After the September 11 terrorist attack, a considerable research effort has been directed at identifying the most vulnerable points of our networks and figuring out methods for saving them. It turns out that the original ideas – that hubs are the most sensitive points of scale-free networks (Albert et al, 2000; Jeong et al., 2001) – were only partially true for real, weighted networks. Devastating global cascades are likely to occur if we remove a node with one of the highest loads in the network and if the network previously had a highly heterogeneous load distribution (Lai et al., 2005). Our cells offer an efficient and low-cost experimental field for studying network damage. Node removal here can be achieved by the deletion of a gene from the blueprint of the cell, the DNA, or by destroying the messenger RNA by a method called RNA interference. If the deletion was lethal, we have very likely struck out an important point of the network.[17] Comparison of the lists of lethal genes with their position in cellular networks has revealed that it is difficult to find simple correlations between a given network property and lethality, but a central position in cellular communication was a rather good predictor of the lethality of the deletion (Coulomb et al., 2005; Estrada, 2005; Schmith et al., 2005).

Our original plans for a giant network of parallel inter-ties are thus rejected. Should we give up? Should we live in uncertainty? No, there are ways to solve the problem of cascading failures.

Tricks to protect our networks against cascading failures. So far I have not mentioned the easiest trick: do not make more ties, make bigger ones! Increase the capacity of the whole network. As the capacity is increased, cascades will be diminished (Lee et al., 2005a). However, this requires enormous resources which are not often available. There are other solutions:

- **Redistribute the load.** If the traffic of the most central nodes is redistributed to other, non-central nodes, the network capacity can be increased by a factor of ten (Ghim et al., 2004; Yan et al., 2005).

[17] The interpretation of individual data is rather difficult here, since the deletion of an enzyme protein offering a unique reaction, which produces an essential molecule for cell survival, is certainly lethal, and yet this peculiar protein will never pop up in any network analysis as 'important'.

- **Rewire in the neighborhood of the damage.** If a node is destroyed, an emergency rewiring of its neighbors can save the network (Hayashi and Miyazaki, 2005). We may make these rewired substituting nets around the most vulnerable nodes as a precautionary measure.
- **Damage the network, if you want to save it!** This rather tricky method has been described by Motter (2004), who proved that the selective removal of network elements and links with either a small load or a large excess of overload also diminishes the size of the cascading damage.

I have left the really big invention until last on the list of rescue efforts. This is the modular structure of the networks. In fact, this is what was actually implemented after the power-failure disasters in the USA (O'Donnell, 2003). Cascading failures can be stopped at intermodular borders. *"Wow! Do cascading failures arrive at the bridge connecting the two sides of the border, and start to shout: 'Damn it! I left my passport at home again!' "* No, Spite, the real situation is more striking than that. When cascading failures arrive at the bridge connecting the two sides of the border, the bridge starts to shout: "Damn it! Cascading failures!", and then collapses. The central idea of the book, weak-linkness comes in again here. Intermodular contacts are often formed by weak links. These contacts may behave like a fuse and melt when the damage arrives. *"Peter, your brain seems to have got boiled in the hot August of 1996. You just made a big issue about the break of the Oregon–Southern California inter-tie and now, what do you want to sell us? That your favorites, those fabulous weak links are good because they break."* I know, this is the point when you throw the book into the trash can for the 67th time.[18] The key point is this: after breaking the inter-tie, the two electrical systems were still connected and the damage could still propagate from one to the other. If I can *really* disconnect the module – because it is linked to others *exclusively* by weak links, and not by strong inter-ties, where one may melt but the others still remain as viable connections – the module may become self-sustaining and still be able to function.

Weak links and microcracks. Intermodular weak links help a number of natural networks. As one example, the fracture process in various materials depends strongly on their heterogeneity. Heterogeneous materials produce microcracks which begin to merge to reach a critical density. Here the microphases (modules) are connected by weak links (weaker forces) allowing a

[18]This actually means that my beloved work has already been fished out of the trash 66 times: I thank the reader! If it is getting too dirty, you may just go and buy another one. I promise the publisher will love both of us.

selective relaxation of the tension by the disconnection of the modules. Perfect crystals have no weak links and no modules, so that they break unpredictably, causing a sudden and devastating decomposition of their structure and often producing a scale-free fragment size distribution (Bazant, 2004; Kun et al., 2005; Sornette, 2002).

In conclusion, if you witness a constant inflow of perturbations, prepare for the worst. A cascading failure may start, leading to devastating damage in your network. A scale-free degree distribution will help a lot to fend this danger off. However, perturbations do not cause random damage. They act like terrorists, often preferring hubs and critical connective elements between distant segments of the network. Therefore, modules become critical for stopping a damage avalanche. Weak links help to control cascading damage, and thus save the connectivity and the life of the network.

3.4 Topological Phase Transitions of Networks

We may feel better after the previous section. If we have any trouble with network construction, or a number of perturbations get stuck, we do not necessarily have to find a new network. Our old network can be rescued. But what happens if our network gets completely saturated with perturbations? Should we give up?

Why are more perturbations worse than a few? So far I have been unable to find an elegant proof of this instinctive truth. Networks may have a characteristic relaxation time. If the next perturbation arrives sooner than this, it cannot be dissipated and the new perturbations just pile up.

However many perturbations arrive, a good networker – and let me invite you to join the club – never gives up. There is still one more way to escape: a topological phase transition. Networks may undergo a series of interesting transformations called topological phase transitions. A topological phase transition occurs if the continuous increase in the number of perturbations provokes a singular change in the global topology of the network. The global topology is best monitored by the measure G/N, where G is the size of the largest connected component

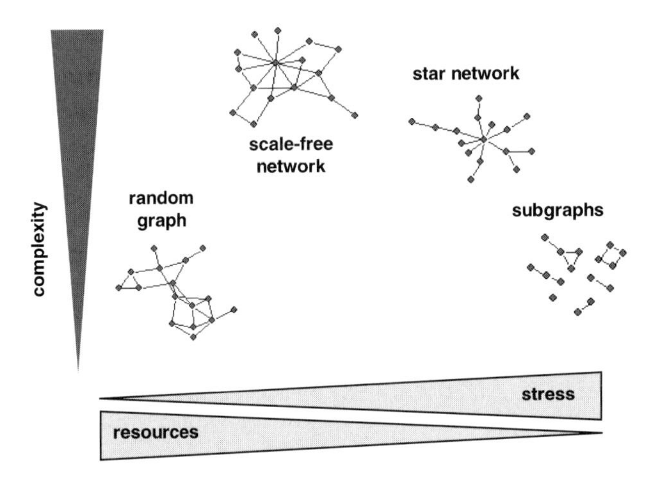

Fig. 3.8. Topological phase transitions of networks. The figure shows the topological phase transitions random graph → scale-free → star phase → disintegrated, fully connected subgraph (Derenyi et al., 2004; Palla et al., 2004) as resources become more and more limited or stress grows. Note that resources and stress have been substituted for the network temperature used in the original publications. The level of complexity, and also the random and scale-free graphs, are merely illustrative

of the network and N is the total number of its links (Derenyi et al., 2004; Palla et al., 2004).[19]

If we keep the network at a high temperature and perturbations arrive continuously, new connections can easily be formed. Under these conditions, the whole network will resemble an Erdős–Rényi-type random graph (Erdős and Rényi, 1959; 1960). In such networks all connections are formed at random, and all elements have the same probability of being connected. If we lower the temperature, a condensation occurs, where compactness is increased and higher degrees are preferred. The network will first develop a scale-free degree distribution. Finally, it will arrive at a star conformation where one or very few mega-hubs dominate the whole system (see Fig. 3.8).[20] The star network resembles dictatorships in social networks. If we decrease the temperature further, the star phase will condense even more, which is possible only

[19] Alternatively, the measure $k_{\max}/M$ can also be used, where $k_{\max}$ is the largest degree of the network and M is the number of edges in the network (Derenyi et al., 2004; Palla et al., 2004).

[20] A similar emergence of the scale-free network together with transitions between graphs with variable degrees, star phase and random structures has been described by Biely and Thurner (2005).

by disintegrating the original network to produce fully connected subgraphs. This means the disappearance of the giant component, and if the network was a living system, this topological phase transition would be called death.

The ultimate peace: a single, fully connected graph. If we decrease the temperature even further and the network approaches an absolute, undisturbed peace, the isolated, fully connected subgraphs will condense to a single, fully connected graph in the physical model (Derenyi et al., 2004). The meaning and significance of this condensation is currently unknown in biological or social networks. Most probably, the fully undisturbed phase is so hypothetical in these systems that we actually never experience the fully connected full graph phase. We can only speculate! If the subgraph phase corresponds to the death of biological networks, what is the phase which comes after it, when things get unrealistically peaceful?

Topological phase transition reflected by the scaling exponent. Let us return to the general formula for the degree distribution in scale-free networks: $P = cD^{-\alpha}$, where P is the probability of the given degree in the network, c is a constant, D is the degree, and α is the scaling exponent. Using the expression above, we may describe the topological phase transitions as a gradual change in the exponent α. The exponent starts from 1 (denoting random networks), grows until about 4 (denoting scale-free networks, see Table 2.1), and then grows further to even higher numbers showing the presence of fewer and fewer hubs with more and more connections. As the scaling exponent α becomes larger, the degree distribution will shift towards an exponential decrease, implying a rapidly decreasing number of highly connected elements and reaching the star phase as an extreme case.

We can transpose the driving force of the above topological phase transitions, the network temperature, to a different context. At high temperatures, our lucky network enjoys a large amount of energy and has a high number of connections. Even if the perturbations cause rather significant damage, there is a good chance that the giant component of the network will still be preserved. Moreover, the high energy contributes to a fast rebuilding of destroyed connections. On the other hand, if the network is in a low-temperature, low-energy environment, the average connectivity is lower. The chances grow that the perturbations will hit vital connections which do not have a backup version in the network. As an additional angle, if we have a large amount of energy around, perturbations arrive almost continuously. They are smooth and predictable. As the outside energy becomes lower, the distribution of perturbations becomes uneven and unpredictable. This is a phenomenon I will call stress in this book. In this context increasing stress provokes topological phase transitions (see Fig. 3.8).

Definitions of stress in physiology and physics. In physiology, the word 'stress' was coined by Hans Selye (1955, 1956), who used the term for a wide range of strong external stimuli, both physiological and psychological, which cause a general protective response, the stress response. In agreement with his definition, from the network standpoint, stress is any large, unexpected, and sudden perturbation of the cellular network, to which the network (1) does not have a prepared adaptive response or (2) does not have time to mobilize its adaptive response. In the latter case, the strong external stimuli overwhelm the elements of the biological networks and fail to provoke the learned, adaptive response. Stress in this book will be used differently from stress in the usual sense in physics, where it is a force that produces strain in a physical body.

Topological phase transitions follow the parsimony principle. Star networks involve a lower cost than either random graphs or scale-free nets, since star networks require less wiring than either of the other two configurations, but still provide a full connectivity of the elements. Bentley and Maschner (2000) showed a random → scale-free transition in the development of a citation network as resources became more sparse, implying that fewer journals wanted to publish the articles on those topics. Similarly, Stark and Vedres (2002) described random → scale-free → star topological phase transitions of business consortiums as the economy worsened. A decrease in connections as a source of stabilization has also been demonstrated with a simpler network model in early work by Gardner and Ashby (1970).

It is now time to go back to our starting point: an overwhelming number of perturbations. When a perturbation gets arrested in the network, the unlucky element at which this happens receives a lot of focused energy. In extreme cases, whatever old connections there were around this element stop functioning and the element is free to seek out and build up new connections. In principle, the underlying scenario for a topological phase transition is created. If the energy level is generally low, more connections around the unhappy element will be broken, than reformed. Connections are lost, condensation occurs, and we start to go from random to scale-free, from scale-free to star and from star to disconnected subgraphs.

Topological phase transitions of networks seem to be a fairly general phenomenon. Let us consider two examples here. Some of their elements may seem rather wild, but they nicely demonstrate the imaginative power of generalization along the lines of similar network properties. Obviously, many further experiments and studies are needed to validate either of these examples.

Topological phase transition 1: Cell death

- **Random graph phase.** In this cellular state, an abundance of outside resources provokes a shift in the metabolic network towards the random graph pattern. The cell has exponential growth, low noise and uniformity.
- **Scale-free phase.** If resources are reduced and the cell experiences a low level of stress, a scale-free metabolic network develops. We have higher noise, some proteins – as elements of the cellular network – become damaged by a few perturbations, and the repair system (provided by chaperones) is gradually overloaded, leading to several deviant responses. Consequently, cellular diversity starts to develop (for further details, see Sects. 6.2 and 6.3).
- **Star phase.** With higher stress levels, system resources grow critical. The cell has to concentrate its energy in the form of ATP consumption for a minimal set of vital functions, and the metabolic network will shift towards the star phase.
- **Disintegration to subgraphs.** If system resources go below the critical level or noise becomes too great, too many damaged proteins develop, the system begins to disintegrate and the cell dies from apoptosis or necrosis (Sőti et al., 2003).

Topological phase transition 2: Ethology

- **Random graph phase.** The random graph phase corresponds here to parallel cooperation between members of the animal group. Every animal does the same thing and there is no sharing out of jobs and tasks.
- **Scale-free phase.** As resources are reduced, a sudden topological phase transition occurs towards complementary cooperation, and task distribution develops between various members of the animal community (Le Comber et al., 2002; Theraulaz et al., 2002).
- **Star phase.** If system resources become close to critical or the level of stress increases (e.g., carnivores arrive in great numbers), a star-phase network develops with a dictator, an alpha male (Hemelrijk, 2002).

- **Disintegration to subgraphs.** Extreme danger disorganizes the larger network, and subgraphs – core families – try to escape and survive together.[21]

Topological phase transition of prebiotic networks. Shenhav et al. (2005) described the idea that prebiotic catalytic networks might have a random configuration and raised the question as to how this early molecular ensemble changed to the scale-free metabolic networks we observe today. A topological phase transition may give a clue here. As the number of prebiotic networks grew and system resources became sparse due to mutual competition, the random network may have been forced to change configuration to the scale-free degree distribution we observe today. The differences in the reports on the degree distribution of metabolic networks may also be partially derived from the different configuration of these networks under various levels of cellular stress.

Striking similarities, are they not? I will list some more examples of topological phase transitions in firms and in human society later. But let us stay with our perturbations for a while. I have shown above that they help topological phase transitions by local disintegration of the old connections around the unlucky, perturbation-hosting elements. I also raised the point that with low system resources, not all connections will be reformed. However, I did not ask which of the links would be lost: the strong links or the weak links?

The parsimony principle would require the strong links to drop, since they are more costly. However, if the energy is lower, we need condensation. This purpose is better served by strong links. As supporting evidence, in the case of high unemployment or any other type of increased everyday stress, weak links are broken and people tend to rely on strong links (Granovetter, 1983). In the case of low system resources, implying a large environmental stress, the link strength and degree distributions may condense in parallel. The network sheds all non-essential weak parts and a core of strong links remains. The scale-freeness of degree and link strength distributions is born and becomes lost in parallel.

3.5 Nestedness and Stability: Sync

Networks are rather resistant to attack by perturbations. As we have seen in the previous sections, relaxation, topology-based resistance

[21] I am grateful to Péter Száraz for these ideas.

against network damage, and as a last resort, reconfiguration of the network topology, all contribute to network survival, e.g., the preservation of the giant component, if trouble arrives. In terms of network properties, fast relaxation requires small-worldness so that perturbations can reach the elements of the local network region as fast as possible. Confinement of perturbations to this local network region and resistance against random damage require scale-freeness. Only one important network property has been left out so far: nestedness.

We have already noted several times that we run into trouble if the perturbation gets arrested somewhere in the net. It is time to ask why we then run into trouble. This is where we need nestedness to provide an answer. If the perturbation gets stuck at a given element, the bottom network that this element actually represents will be overwhelmed by the large amount of energy it receives, since these energy packets usually just go very quickly through this particular element.

Is a transient perturbation inefficient? Why is a bottom network not affected by a large energy packet, if the packet goes quickly though this bottom network?

There appears to be a delay in the response of bottom networks, leading to a form of 'laziness', or the idea of a time window for escape, that have hitherto remained uncharacterized. During this time they can put up with the presence of the perturbation. However, if the perturbation stays any longer, the bottom network will be in trouble. One quite widespread form this trouble can take is that the affected bottom networks decouple from a synchronous oscillation. *"What do you mean when you speak of the synchrony of oscillations?"*

Oscillation synchrony was discovered by Christian Huygens when he was sick in 1665. As he lay in bed, he watched his two newly constructed pendulum clocks on the wall. He had invented these clocks himself to win the fabulous prize offered by the Royal Society in London for a clock which would be precise enough to keep time on ships sailing long distances across the sea. *"With all due respect to all their noble deeds, these Brits are a bit crazy. Did they pay such a tremendous amount of money just to have their five o'clock tea exactly on time aboard their ships?"* No, Spite, this was not the case. If you know the Greenwich mean time (or any other time at a given point on Earth) exactly, you can work out your longitude from the time difference. The British sailors needed an exact time to know where they were, and to situate the treasures they found.

But let me return to Huygens. We left him sick in bed. Well, he is still sick, and actually gets much sicker. As he lay there, he observed that the two clocks he had on the wall gradually became synchronized and finally oscillated together. Now that is interesting! Forgetting about being sick, he jumped from bed and started to reposition the clocks on the wall. After many experiments (which probably prolonged his sickness by delaying the resynchronization of the oscillators inside his own body), he realized that it was the wall that was transmitting weak signals from one clock to the other to cause the miraculous synchronization. He was proud to report his findings to the Royal Society (Huygens, 1665). The honourable members discussed his new experiments at their next session of 8 March, and arrived at an unexpected conclusion: "Occasion was here by some of the members to doubt the exactness of the motion of these watches at sea, since so slight and almost insensible motion was able to cause an alteration in their going" (Minutes of the Royal Society meeting, 8 March, 1665). Huygens never got any money for his pendulum clocks. The million dollar prize had to wait a hundred years. Huygens got very disappointed and never dealt with the synchrony phenomenon again (Strogatz, 2003).

Fortunately, others did. Steven Strogatz gives a wonderful survey of the synchronization of various oscillators in his excellent book *Sync* (Strogatz, 2003).[22] Chemical oscillators like the 64 nickel electrodes of Kiss et al. (2002) can vary their potential in synchrony; temporarily synchronized neurons ensure successful memory formation (Fell et al., 2001); and our level of synchronization with circadian rhythms (Ogle, 1866) can be conveniently studied on sleepless nights after a transcontinental flight. Visual and acoustic interactions make fireflies flash (Buck, 1938), crickets chirp (Walker, 1969) and audiences clap (Néda et al., 2000) in synchrony. Women living together synchronize their menstrual cycle (McClintock, 1971). Hare and lynx populations of Canada are even better. Their population cycle can manifest a synchrony over millions of square kilometers (Blasius et al., 1999). As a less glorious, but equally large-scale example, the occurrence of syphilis was synchronized in the whole of the United States from New York City to Houston between 1960 and 1993 (Grassly et al., 2005). As a final example, unconscious synchronization of fine movements during the steps of the celebrating crowd caused a violent wobbling of the 690 ton London Millennium bridge on 10 June 2000, on the day of its

[22]Synchrony is also called entrainment, especially in music.

Fig. 3.9. Synchrony seems to be a joyful phenomenon. It tends to make us feel stable and safe

opening (Strogatz, 2003). In spite of the last few examples (and Huygens' despair), synchrony seems to be a joyful phenomenon. It tends to make us feel stable and safe (where 'us' refers to fireflies, crickets, hares, lynxes and even nickel electrodes – a fine brotherhood, indeed).

Music and learning again. Is the beneficial effect due to synchronicity? You may remember my remark in Sect. 3.1 that listening to the right kind of music – I am not referring to quality here, but simply requiring that it has a pink-noise-type, scale-free structure – may help your learning abilities. Here I would like to put forward the idea that the scale-free pulses of the external noise may help neuronal synchronization, which is responsible for memory formation (Fell et al., 2001). Next time you forget something and start to scratch your head, think about this. External noise may help your internal oscillations.

Synchronization is a network phenomenon, not only in the sense that it requires a network of oscillators to happen, but also in the sense that it has many properties characteristic of networks. Here are two examples from among many:

- Similarly to the percolation threshold of networks, synchronization also has a phase transition. As the difference between the frequencies of different oscillators is decreased below a certain threshold, all of them will suddenly become synchronized, achieving syntalansis (see Fig. 3.10) (Winfree, 1967). Synchronization has nestedness. Network elements at all levels may behave as synchronized oscilla-

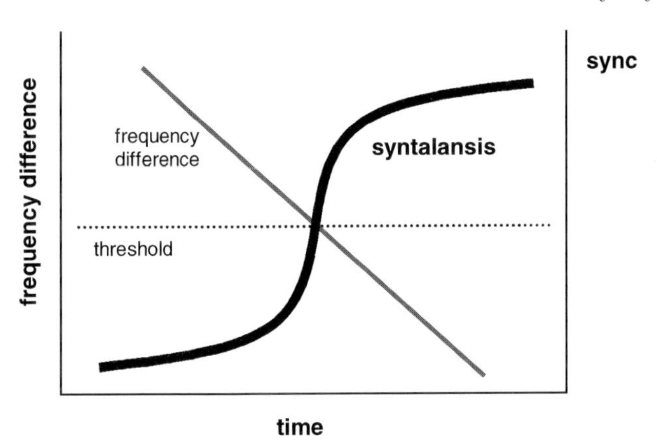

Fig. 3.10. Achieving syntalansis, the sync phase transition. The phase transition of synchronization is shown schematically here. As the difference between the frequencies of different oscillators decreases below a certain threshold, they are all suddenly synchronized, achieving syntalansis (Winfree, 1967). The *grey line* denotes the frequency difference, while the *bold curve* illustrates the relative extent of synchrony

tors. Cells are synchronized like the neurons in our brain to make recognition possible. Organs are synchronized to help lampreys and leeches to swim as well as to produce the peristaltic movement helping our digestion. Finally, organisms are synchronized with each other (Bressloff and Coombes, 1998; Strogatz, 2003). However, synchronization also has modularity, which means that the synchronized state may only extend to some of the networking oscillators (Winfree, 1967).

- Synchronization has self-organized criticality. The roughly 10 000 pacemaker cells in the sinoatrial node of the heart can be efficiently modeled by a network of oscillators, where each of the oscillators gradually increases its membrane potential and then, after reaching a threshold, becomes discharged. When the discharge occurs at any particular oscillator, all neighboring oscillators will become slightly more depolarized. Mirollo and Strogatz (1990) showed that these oscillators produce an avalanche as they subsequently become synchronized with one another. The final phenomenon is very similar to earthquakes or other avalanches discussed in Sect. 3.3. The extent of synchronization is also important, since disruption of local synchronization prevents efficient relaxation and may cause self-organized criticality and subsequent avalanches (Ponzi and Aizawa, 2000).

Synchronization depends on the network properties. Everything making the network better connected helps synchronization. The seminal paper by Watts and Strogatz (1998) showed that small-worldness helps synchronization. Actually, small-world topology is extraordinarily effective for this purpose (Barrahona and Pecora, 2002). Small-worldness reduces the divergence between extremes of the individual oscillators, and this helps to preserve the synchronized state (Guclu and Korniss, 2004).

Scale-freeness of the top network has a rather adverse effect on the extent of synchronization. Hubs connecting many nodes to each other have to be avoided, since these 'center' oscillators interacting with a large number of other oscillators tend to be overloaded by the traffic of communication passing through them (Nishikawa et al., 2003). When the hubs of a scale-free network of oscillators were replaced by triads, the level of synchronization was remarkably increased (Zhao et al., 2005). Assortative networks, where hubs are coupled to hubs, were found to synchronize even less well due to the mutually disturbing effects of the overloaded hubs (di Bernardo et al., 2005). As we have seen in previous sections, scale-free structure tends to confine the effects to smaller segments of the network. Here again, small-worldness and scale-freeness interact to keep a balance of optimal synchronization.

Jung revisited: A possible example of nested sync. The former remark on nestedness brings me to the various and sometimes rather vague interpretations of sync. In the famous essay by Carl Jung (1969), synchronicity was perceived as the opponent of "constant connection through effect" (causality) and meant an "inconstant connection through contingence, equivalence or meaning". Among the many examples Jung gave, I would only list here fulfilled dreams and prayers.[23] Thinking about synchronicity in the context of the present chapter, i.e., the synchronicity of oscillators and networks, many of the Jungian examples can actually be interpreted, if we suppose that strong synchronization in a given network may induce a synchronization of the elements of the nested network one level up or down (called here nested sync). Thus the fulfilled dream Jung mentioned on a disaster in a remote island may be perceived as synchronization propagation from the top to the bottom network. Here a synchronization event in the top

[23] I would like to note here that most of the other examples, e.g., Jung's extensive experiments on astrology, are much more difficult to accept in the present context of nested sync, and in the current state of scientific knowledge.

network (the simultaneous death of several people on the island) may induce the synchronization of the elements (neurons) of one of the bottom networks (the dreaming person) and cause the dream.[24] Conversely, a fulfilled prayer may be perceived as a synchronization propagation from the bottom to the top network. A high level of synchronization of the neurons in one of the bottom networks may cause a synchronization of the top network and the prayer becomes fulfilled.[25] What do we call the top network in this case? Well, it depends on your religion! I think, this is the point where I have to remind you that we are in a triple smiley box, which means that the content is fiction and not science. However, I have one further remark: an increased coherence of electroencephalograms (EEGs) has been repeatedly observed during meditation (Aftanas and Golocheikine, 2001; Orme-Johnson and Waynes, 1981). This is by no means a proof for nested sync (no one is able to measure the EEG-like phenomenon one or two networks higher!). However, it shows that intensive mental states can indeed lead to more coherent brain functioning. My last remark is about Mozart. Did the neurons in Mozart's head know that they were in sync because the Master had just conceived the Requiem? Moreover, did the Master know that the Requiem would be in sync with his own death? It is a bit difficult to grasp trans-network connections, especially from the bottom network. We have to be patient with the explanations given by Mozart's neurons. I hope I have convinced you that you also have to be patient with my explanations.

My first encounter with nested sync. *"Peter, if you start talking about your fulfilled dreams here, I'm leaving."* Do not worry, Spite, *this* dream of mine will never be fulfilled! I just want to offer an explanation as to why I might be so sensitive to the above, rather unusual ideas. When I was around four, I had a terrible ear inflammation. Suffering unbearable pain, an idea suddenly came into my mind: the fact that I had to suffer had a purpose. Something much bigger was using me to have good ideas, and it was the unusual intensity that was giving me the pain. I agree that this may be even less well grounded than any of the ideas in the comment above, but it worked! When I got to this point, the pain had stopped.

So far I have outlined some of the possibilities for the way synchrony develops. We have seen that network properties like small-worldness

[24] If the fulfilled dream is about a future event, well, let's keep the 'explanation' for my next book!

[25] Actually, it would be very interesting to measure the difference in the synchronization intensity of the affected neurons during an ordinary and an emotionally involved, intensive prayer.

and scale-freeness assist synchronization. We even had an esoteric example of nestedness in sync. However, in all these examples, sync was either there or it was not. The question arises as to whether there are levels of synchronization? Well in fact there are. We have basically three levels. If synchronization is weak, the frequencies will be uniform. At the next level, the phases of the waves will also be synchronized. At the tightest coupling level, the amplitudes of oscillations will also be the same. How are the various levels of synchronization achieved? To a first approximation, we should remember Huygens and his wall. Indeed, there must be an interaction between the individual oscillators to get them synchronized. How strong should this interaction be? *"I guess we will see your weak links again."* Yes. Fasten your seat belts! The weak links are on their way.

Strong links lead to amplitude synchrony, while weak links induce only a weaker, phase synchronization (Blasius et al., 1999). *"Well Peter, your weak links seem to lose out this time."* There is no need for that glorious smile, Spite. Weak synchronization intensity does not mean weak benefits. Several famous models of synchronization, like the Josephson effect in superconductors, the Winfree model, or the Kuramoto model use weak links to achieve synchrony (Ariaratnam and Strogatz, 2001; Feynman et al., 1965; Kuramoto, 1984; Strogatz, 2003; Winfree, 1967). Oscillators are not usually identical and their coupling is therefore weak. As an example of this, individual hamster clock cells are very diverse, and their coupling is weak. Hamsters still have a rather good circadian rhythm (Liu et al., 1997). Moreover, stochastic resonance[26] of coupled oscillators is best achieved if oscillators are coupled by weak links (Gao et al., 2001; Lindner et al., 1995; 1996).

The Winfree and Kuramoto models: Examples of weak-link-induced sync. The Winfree model describes the synchronization of N coupled phase oscillators with the formula

$$\theta_i = \omega_i + \frac{\kappa}{N} \sum_{j=1}^{N} P(\theta_j) R(\theta_i) \ ,$$

where $\theta_i(t)$ is the phase of the ith oscillator, κ is the coupling strength, and the frequencies ω_i are drawn from a symmetric unimodal density $g(\omega)$. The model assumes that the mean of $g(\omega)$ equals 1. $P(\theta_j)$ is the influence function of the jth oscillator, while $R(\theta_i)$ is the sensitivity function of the

[26] In stochastic resonance, noise helps in the detection of sub-threshold signals. For a proper description of the phenomenon, please turn back to Sect. 3.1.

i th oscillator with respect to the average influence of all oscillators (Winfree, 1967). The Kuramoto (1984) model is a refined model of the Winfree model. Here the same N coupled oscillators are described by the formula

$$\theta_i = \omega_i + \sum_{j=1}^{N} \kappa_{ij} \sin(\theta_j - \theta_i) \ ,$$

where the symbols denote the same as above. Both models display synchrony at relatively low coupling strengths κ.

Proper weights improve the synchronization of scale-free networks. As mentioned earlier, networks with scale-free degree distribution generally disturb synchronization. However, if the scale-free degree distribution is accompanied by a scale-free weight distribution, where the weights are distributed in such a way that all the nodes receive the same input signal, the synchronization of scale-free networks is greatly improved. In fact, this weight distribution corresponds to the minimum cost, which is related to the total strength of all directed links (Motter et al., 2005).

In summary, we may conclude that there are many examples in which weak links couple individual oscillators and help them to develop synchrony. Is there any special benefit from synchronization in general, and weak-link-induced synchronization, in particular? The early work by Enright (1980) already put forward the idea that coupled oscillators are more stable. Later it turned out that an optimal level of sync is the best to achieve the highest stability (Yao et al., 2000). If we have complete sync by strong links, we have lower stability than with partial sync by weak links.

Analogy between the optimal level of sync and the notion of local dissipation and global connection. The main message in Sect. 2.2 was that properly designed networks should keep a balance between rather free local traffic to ensure the proper dissipation of perturbations and global connectedness to help system-wide responses of the whole integrated network. This idea is strongly analogous to the optimal level of synchrony we observe here. Complete synchrony would help the propagation of perturbations in the whole network, leading to a continuous risk of cascading failures. Zero synchrony would isolate various segments of the network from each other and prevent system-wide responses.

How is the 'extra stability' of partial sync achieved? For the answer, let us turn back to the starting point, i.e., to the perturbations. Strogatz et al. (1992) showed that, in cases where the frequency distribution of the oscillators is restricted to a frequency interval, the dissolution of synchrony after a perturbation is slower than exponential. In other words, partial sync is better for sticking oscillators together. If the perturbation arrives disguised as a traveling wave, only those oscillators stay coupled which were coupled weakly (Bressloff and Coombes, 1998). As a third example of the same phenomenon, phase synchrony, the specialty of weak-link-induced synchronization, seems to be particularly resistant against the attacks of various perturbations (Blasius et al., 1999).

Weakly coupled oscillators seem to relax faster. Obviously, if the coupling is too weak, it becomes ineffective in the sense that extremely weak coupling between oscillators gives no synchrony at all. However, weak coupling seems to be quite profitable. If you want to get a stable joint oscillation, a Winfree syntalansis (1967), then seek weak links. The following advice emerges: stay away from authoritarian groups requiring full synchrony. In Sect. 10.3, I describe in detail the study by Kunovich and Hodson (1999), who showed that after massive stress (the civil war in Croatia), psychic recovery was better helped by informal organizations than formal ones. The explanation is rather straightforward: informal organizations allowed better sync and faster relaxation of the prolonged tension.

Why do we like sync? Synchronization certainly gives us pleasure. To be honest, I have no idea whether the nickel electrodes, our neurons, the fireflies or the crickets were happy producing synchrony. Moreover, it is difficult to measure whether groups of women feel more secure emotionally when their menstrual cycles are synchronized. However, when audiences clap in unison (Néda et al., 2000), people make waves in stadiums (Farkas et al., 2002), a crowd sings the national anthem, people laugh, or I keep my knees hitting in the same rhythm as my dog runs when we are out together in the park, there is no doubt that all these things give great pleasure.

Synchronized laughter quakes. Laughter is a rather joyful act. We spend countless hours of our life in search of laughter and involved in attempts to make others laugh (Dunbar, 2005). We are not alone. Even rats laugh – though in the ultrasound region (Panksepp and Burgdorf, 2000). I would like to put forward the idea that laughter is in fact a synchronized self-organized criticality phenomenon. Before we start to laugh, there

Fig. 3.11. Tensions that any of us are unable to cope with alone are transferred to others by the sync we all enjoy

is a gradual input of information causing tension, which finally bursts into laughter. As an example of the gradual input, think about the 'annoyance' of repeated unlikely events, or the cognitional tension of an outright absurdity. Laughter is certainly a relaxation phenomenon. It would be nice to see whether the intensity and duration of laughter display scale-free behavior. However, we seldom laugh alone. Laughter is provoked by others' laughter in a rather contagious fashion. Laughter is a very good example of our inherent love of sync.

Why do we like sync? Sync may give us an additional level of stability by helping relaxation. Fast relaxation stabilizes systems and sync helps relaxation. Optimal sync helps local relaxation. Relaxation has been conditioned to cause joy so that we learn to seek it.[27]

Let me ask you to take a deep breath, to drink a glass of crystal clear water, to relax, and most importantly, to think. Is this not beautiful? For our life we need a continuous inflow of energy. However, this generates ever more tensions, which may cause our death. Relaxation is indeed a question of life or death for us. This is a formidable task, but we are not alone! We are in sync. This gives a safety net to all of us. The tension that any of us is unable to cope with alone is transferred to others by the sync we all enjoy. The most beautiful moments and feelings of our life: joy, laughter and happiness are all tied to relaxation and sync. This is our reward, if we successfully learn about one of the most general laws nature has yet made for collective survival.

[27]Relaxation comes only after the sync has been set and an unexpected perturbation is experienced. Therefore it is better to be conditioned by another feeling, joy to seek the sync, which will ease future relaxations. I will return to the connections between joy, relaxation and stability in the concluding chapter of the book.

"Peter! Don't you think it's time to put your ideas into practice and stop writing NOW. Then you can start that well-deserved relaxation?" Spite, it is truly kind of you to be so considerate about my health, but now you have spoilt a great moment for me. Here I am, sitting with a glass of crystal clear water in front of me, with my soul in heaven, and you drag me back down to Earth. But don't be sad, Spite. You have only done what you are supposed to. Returning to your question, believe me, there are so many exciting thoughts in my mind that if I stopped writing NOW, I would end up in the madhouse and never achieve any form of relaxation. Writing is my relaxation right now, and I thank the publisher and the reader for giving me this opportunity.

The emergent property of human sync. Superconductivity, heart beat, peristaltic movements and our thoughts are all emergent properties of sync at the level of the top network. We can recognize the emergent property if network members are particles or cells. What is the emergent property when we humans start to make sync? Do the millions of synchronized exclamations after a goal in the final of the world soccer championship mean that Gaia has remembered a great joke she heard in the Cambrian epoch?

In the last few sections we have learned several recommendations about how to save our networks. We have seen how to discriminate between good and bad noise. We now recognize the degree of danger if a perturbation arrives alone or with a number of other perturbations. We have learned how to confine relaxation to a segment of the network. We have studied avalanches, network failures, topological phase transitions and finally sync. How should we put all this into practice? How can we construct a network which will resist against all perturbations? Stay tuned! The next section will try to give a clue.

3.6 How Can We Stabilize Networks? Engineers or Tinkerers

Network design and network stabilization are engineering tasks. Stability can be achieved in connected systems by negative feedback, for example, a typical element of engineered systems. Indeed, highly optimized engineered systems display a considerable level of stability. These systems have many highly optimized feedback regulations called integral feedback control or more generally, highly optimized tolerance

(HOT). Modern machines demonstrate a very high level of organization. In a Boeing 777, a hundred and fifty thousand different subsystem modules can be found, piloted by close on a thousand computers. The final testing of such an airplane generates data almost equivalent to the human genome every minute (Carlson and Doyle, 2002; Csete and Doyle, 2002).

However, our sophisticated machines were not the first complex systems on Earth to show this level of stability. We ourselves are also pretty good examples of complex and stable networks, and date back somewhat further in evolution than the Boeing 777. Francois Jacob (1977) pictured evolution as a tinkerer, "who does not know exactly what he is going to produce, but uses whatever he finds around" and "gives his materials unexpected functions to produce a new object". Indeed, evolution does not optimize the system in advance making a blueprint, but assembles interactions until they become good for the task (Maynard-Smith and Szathmary, 1995). Steven Rose put the same idea in his *Lifelines* (1997): "We carry the burdens of the past with us."

What are the common features of the engineered and evolutionarily tinkered systems? Just to name a few of the most important ones: modularity, robustness, and on the other side of the coin, failure avalanches (Carlson and Doyle, 2002; Csete and Doyle, 2002). However, there are major differences between the results of engineering and tinkering:

- **Evolution must make all intermediates viable.** As opposed to an evolutionary system, an engineered system has been optimized for the purpose at hand. In engineering, there is no need to optimize all predecessors and there is not such a strict requirement for continuity amongst these predecessors. Finally, the engineered system is not forced to change by introducing just a few small changes at each point in its development. Though the concept of punctuated equilibrium (Gould and Eldredge, 1993) introduced discontinuity into the evolutionary process, and later a few molecular mechanisms (Rutherford and Lindquist, 1998; see Sects. 6.2 and 6.3 for details) were also uncovered to explain jumps in evolution, the required level of continuity still makes a difference between engineered and evolutionary systems.
- **An engineered system is complicated, while an evolutionary system is complex (Ottino, 2004).** In the case of complicated systems, the pieces can be disassembled and reassembled again, and the function of the whole can be guessed quite well from the functions of the parts. In the case of complex systems (for a

discussion of complexity, see Sect. 4.3), the function of the whole is an emergent property of the parts, and in most cases we cannot make a straightforward guess about the function of the top network if we only know the function of the bottom networks (modules) in a piecewise manner.

- **As opposed to engineered systems, evolutionary networks are integrated and their parts cannot be optimized separately.** In their famous essay against the Panglossian Paradigm, which considers each part of a complex organism as the result of evolutionary optimization, Gould and Lewontin (1979) wrote: "Organisms are integrated entities, not collections of discrete objects." Although engineered systems are also integrated, the function of individual parts in these systems is better described and this function is usually close to optimal by itself.
- **Evolutionary networks have greater designability.** Evolutionary networks have a higher capacity for combinatorically different setups of their components than engineered networks (Changizi et al., 2002). This property is also called designability (Tiana et al, 2004).
- **The evolutionary design is stable under many conditions.** As a result of the design process, engineering gives stability only with very finely tuned parameters, while evolutionarily tinkered networks are stable under a much wider range of initial conditions (Aldana and Cluzel, 2002).
- **Link strength difference between engineered and evolutionary systems.** As shown in Sect. 2.4, evolutionary systems develop a continuous range of interaction strengths between their parts. In engineered systems, reliability is a crucial factor. Two parts either interact, or they don't. Probabilistic, vague, 'almost' interactions do not reflect a skillful design. Though interaction strength can be defined by the duration of interaction, and in this way engineered systems also contain weak links, link strength diversity is rather limited in engineered systems as compared to evolutionary networks.
- **The evolving system grows.** Finally, an engineered system does not necessarily grow, whereas the evolutionary system grows by definition.

What happens if a self-organized system cannot grow? Is growth arrest a source of aging and death for networks? Is growth arrest a serious form of stress which leads to a series of topological phase transitions of the network (see Sect.3.4), resulting finally in the disintegration of the net and death? Are we sentenced to grow in running away from our own death?[28]

Evolution can go backwards. Originally the above list had an additional point: it was thought that, in contrast to engineering design, evolution cannot restore information that has already been deleted. In a way this statement was not true even at that time, since the deleted genetic information might have been stored in another species, in such a way that it could be regained. However, the recent paper by Lolle et al. (2005) has described an even more elegant way to save the blueprint of a good old design, until the new one proves that it is indeed better. The blueprint is actually pale-blue here, since the old and discarded genetic information is saved, not in the form of DNA, but in the form of RNA. The reversion of the particular gene in the plant *Arabidopsis thaliana* to the older and discarded version was as high as 10% in some cases. The process was governed by an RNA segment which was retro-transcribed to the DNA. The mechanism may actually be quite general. However, many more experiments are needed to assess the importance of this non-Mendelian inheritance.

Although I have shown that engineered and evolutionary developed systems differ in many respects, the contradiction between the engineer and tinkerer types of development is only apparent. In Sect. 9.5, I show the convergence of the two developmental schemes in highly sophisticated, modern designs.

[28] I am grateful to Bálint Pató for these questions, which are related to the necessity of housekeeping heat (Oono and Paniconi, 1998) mentioned in Sect. 3.1.

4 Weak Links as Stabilizers of Complex Systems

4.1 An Emerging Synthesis: Weak Links Stabilize Complex Systems

In the previous chapters we learned how weak links stabilize complex systems. The classical study of Granovetter (1973) demonstrated that weak links help the cohesion of society (see Chap. 1). Weak links are necessary for small-worldness, emerging in parallel with topological scale-freeness, and making a key contribution to the formation of nestedness (see Chap. 2). Weak links buffer noise, help relaxation, form barriers against cascading failures, and stabilize the coupled oscillators of bottom networks (see Chap. 3). Table 4.1 summarizes the effects of weak links on network behavior.

Summarizing the previous chapters, Table 4.1 gives an initial synthesis and several hints to suggest that weak links may play a prominent role in defining network stability. The generality of network topology (see Chap. 2) and the imaginative power of the applications of this generalization prompt me to take up the challenge and form the main hypothesis of this book:

Weak links stabilize all complex systems (Csermely, 2004; 2005).

I cannot (yet) formally prove the general validity of weak-link-induced network stabilization but hope to show its generality and power to explain and regulate the world around us. However, a statement cannot be judged without definitions. When do I call a link weak? When do I call a system stable? Why do I speak of complex systems rather than networks and what is the definition of complexity? The rest of this chapter gives starting definitions for all these notions and discusses some of the weaknesses and strengths of the basic statement. Before attempting any definitions, a few comments are probably in order.

Table 4.1. Network behavior in the presence and absence of weak links

Network has many weak links	Network has few weak links
Long-range contacts give small-worldness, modules are well-connected	Average distance between elements is large, modules are sparsely connected
Behavior of bottom networks is optimally synchronized, giving small fluctuations	Bottom networks are either tightly coupled with large fluctuations or behave independently of each other
Communication is good in the network, relaxation goes smoothly	Communication is restricted in the network, relaxation is disturbed, relaxation avalanches may occur
Noise is easily dissipated or absorbed in the network	Network is noisy and noise stays in segments of the network
Network is integrated and behaves as a whole	Network is segregated and behaves as an assembly of its constituent modules and bottom networks
Changes are dissipated and occasional errors are isolated, so that the network is stable	Changes and noise persist, the network is error-prone and unstable

Misfortunes of the Statement: 1, Weak Links

Weak links are both elusive and overwhelming. Science has grown used to examining strong links. Strong links are always there. Strong links are reproducible. Strong links are few in number and hence comprehensible. Strong links are already known. Strong links are scientific. Strong links are exciting. In short, strong links are like friends to us. In contrast, weak links are transient. Weak links are undetectable. Weak links are overwhelmingly numerous. Weak links are unknown. Weak links are unscientific. Weak links are hopeless. In short, weak links are like foes to us.

Misfortunes of the Statement: 2, Stabilize

Stability can be defined at various levels (see Chap. 5) and may also behave as a rather elusive concept. To measure stability, we have to follow and quantify system dynamics. Non-equilibrium simulations are still too complex to be fully resolved in many cases. Moreover, stability is not easily measured. Fortunately, there is another way to approach it.

The level of system noise may give us valuable information on system stability. However, for the traditional mind, noise is a nuisance which is better avoided than measured. For the traditional way of thinking, we can offer yet another choice. As I showed in Sect. 3.1, increased noise is linked to increased diversity. However, monitoring diversity is not a usual feature of experimental documentation either. Odd findings are a matter of shame that should be hidden. Ignoring the advice of Bacon (1620): "Whoever knows her [Nature's] deviations will more accurately describe her ways", science deals with the average and ignores the exceptional. Exceptions are not mentioned in titles, abstracts or key words. I had to search for months for papers noting peculiarities. This book is not only about networks and weak links. It is also about the scientific method. We have to change our attitude. In writing this book, I would like to stress that irregularity is not an annoyance, but gives the true flavor of the world's richness. It gives us strength and stability to regulate and preserve all the complex systems inside and around us.

Misfortunes of the Statement: 3, Complex

Complexity is related to integrated behavior (see Sect. 4.4). To observe the stabilizing effects of network elements, one has to think about the whole network and its function. In some cases, the situation gets even worse. We cannot stay at the bottom network level, but have to go one level higher, to the level of the top network, to understand the emerging network function which is stabilized by the weak links. As an example, words neither stabilize themselves nor the sentences they form. Words may only stabilize the *meaning* of the whole textual network they belong to. With our deductive socialization in science,[1] it is not always easy to find the emerging, synthetic thought pointing towards the beneficiary of the stabilization.

Misfortunes of the Statement: 4, All

Since I am unable to prove the generality of the statement for the moment, it is rather wishful thinking to use the word 'all' here. Feel

[1] Deduction became the main scientific method after the end of the 19th century, when scientists got enough experimental tools to analyze the smaller details and mechanisms of network properties. Although deduction is an incredibly useful tool to understand how nature works, for the analysis of large data sets and the understanding of nested networks, we need a better training in the reverse of deductive thinking, i.e., induction.

Fig. 4.1. Irregularity is not an annoyance, but gives the true flavor of the world's richness. It gives us strength and stability to regulate and preserve all the complex systems inside and around us

free to substitute 'many' in the place of 'all', if you prefer precise statements.

Misfortunes of the Statement: 5, Period

You may have recognized that by now all parts of the original statement have been questioned. Only one is missing, namely the period at the end. To complete the job, I am not quite sure that a period is the most appropriate sign at the end of the sentence. Let me suggest that you imagine a light grey, vanishing period, at least for the time being. I hope that by Chap. 12, when I return to this sentence again, we will agree that the period should be replaced by an exclamation mark.

"Dear Peter, if your favorite statement is even weaker than the links it is about, why don't you leave this book and go for a swim?" Well, to begin my answer, I *will* go for a swim in a minute. But before doing so, I will end this section by listing some of the strengths of the same statement.

Strengths of the Statement: 1, Weak Links

Being weak is a relative category. When there is any difference between network elements, weak links emerge. In real networks, there are always differences between network elements. Consequently, we always have weak links. Moreover, real networks are not static. Links form,

Fig. 4.2. If a system becomes too stable, it cannot change, cannot develop

then vanish. Weak links may also mean links with short duration or low probability. This gives another chance for weak links to emerge. Indirect, higher-order interactions are often perceived as weak links, especially in ecology. Lastly, both intermodular and long-range links are usually weak. These links are very important in network stabilization. Weak links emerge at critical points of the network. Weak links are not unique, they are general. Weak links are not just the leftovers when strong links have been taken into account. Weak links are important.

Strengths of the Statement: 2, Stabilize

All those networks we know about are more or less stable. Highly unstable networks cannot be studied. The time resolution of our methods defines the level of instability we are able to study and describe. Moreover, by definition, a network contains links between multiple elements. The simultaneous presence of multiple links also implies at least a minimal level of stability. Network stability is not just born from nowhere. Weak links do not need to establish it, but they do help to increase existing stability.

Strong links are good too! After one of my lectures on weak links to high school students, a young student researcher approached me and asked: "Peter! You have have been talking all the time about the benefits of weak links. Does this mean that there is something wrong with strong links?" No! Not at all. A lot of weak links lead to over-stabilization. The system becomes too stable, too lazy, and cannot change or develop (for a more detailed description see Chap. 12). As with noise or synchrony, we need an optimal level of weak links. Moreover, if we have *only* weak links, we lose them. Weak can only be weak in the presence of strong. When I finished my answer, the young man sighed: "I am relieved. I thought perhaps you were suggesting that I should leave my best friend to get stability." Do not worry. Best friends can stay. Indeed, best friends *must* stay. Without best friends, simple acquaintances are useless (at least in the sense of network stability). However, acquaintances are *also* part of our stability. We should cherish both. I will return to this in Sect. 10.2 describing the need for *both* conservative and liberal thinking to build a stable society, as well as the advantage women obtain with regard to stability by building up more weak links in their contacts than men.

Are strong links stabilizing? To complete the rehabilitation of strong links, let us note that strong links are not only there to allow weak links to be weak. Strong links are the essence of all that makes a network a network. (I know, it is time to start my next book on strong links!) Strong links define the network. If I knock out a strong link, the network will generally behave differently. Strong links do contribute to network stability. The peculiarity of weak links is that, by removing any of them, the network does not necessarily change its main parameters (in contrast to the removal of strong links). However, there is still a change: after the removal of weak links (the more the better), the network will become unstable. This notion will be the core of the starting definition of weak links in the next section.

4.2 Weak Links: A Starting Definition

In Sect. 2.4, I showed that widely different natural networks develop a scale-free distribution not only in topology, space and time but also in the distribution of link strength (Almaas et al., 2004; Barrat et al., 2004a; Caldarelli et al., 2004; Garlaschelli et al., 2003; Ghim et al., 2004; Goh et al., 2001; Leland et al., 1994). If there is a continuous growth of link strength from vanishingly weak to extremely strong, it is rather difficult to define a discriminating value below which a link can be said to be weak. The examples in the following sections which illustrate weak-link-induced stabilization are not very helpful either. Most of these exciting examples are not detailed enough to use for a definition of the threshold link strength for network stabilization. I will nevertheless make an attempt and say, in analogy with the Pareto law, that all interactions falling below the strongest 20% will from now on be defined as weak. But my feeling is that this threshold is context-dependent, and cannot be generally defined.

Aware of the difficulties in marking a threshold value for weak links, I resort to a functional definition. I therefore begin with the definition due to Berlow (1999):

> **Definition of Weak Links.** A link is defined as weak when its addition or removal does not change the mean value of a target measure in a statistically discernible way.

I am aware that, like all functional definitions, this one is also highly context-dependent. For this definition, we have to set a target measure, we have to be able to add or remove the link, we have to be able to repeat the determination of the measure several times and, the most difficult condition, we have to maintain all conditions of the network intact (apart from the addition or removal of the link) between these measurements.

Ladies and Gentlemen! May I start my round with my empty hat for your generous contributions? The show is over: you have seen the sad life of the experimentalist, when asked for definitions.

Are all links weak? Remarks on a suspicious definition. *"Peter, if you delete a 'strong' link in a smaller network, like some of the ecological networks, you will obviously have a big change in the network parameters. What will be the behavior of large and highly redundant networks like those in cells and societies? Eric Berlow was certainly right to state his definition for small eco-webs, but your generalization is wrong. Large and redundant networks will not have a single strong link using this definition."* I think I am safe here. Even large networks contain some links which are essential. 20% of yeast genes seem to be essential for viability. An additional 40% of yeast genes may become essential in various conditions (Papp et al., 2004). The number of unconditionally strong links is small (20%), but if you remember from Sect. 2.4, link strength has a scale-free distribution, and this means precisely that there are a small number of strong links. Actually, even the percentage is familiar. Indeed, the 80–20 rule of Pareto (1897), the archetype of scale-free distributions, also drew the line for strong contributions at 20%. The Berlow (1999) definition seems to behave quite well in general terms.

Indirect effects as weak links. Weak links need not always be direct. As an example of an indirect effect, the effect of a neighbor's neighbor can also be calculated. If all interactions are of equal strength, these second-neighbor effects will obviously be weaker than those due to direct neighbors. Indirect effects are often considered in ecology, where the effect of each participant of an ecosystem can be important to a given species (McCann, 2000).

Weak links are cheap. Both the formation and maintenance of weak links come much more cheaply than they do for strong links. Weak links are formed easily and do not constitute a great loss when they are thrown away.[2]

Weak links are undirected. Strong links are formed between stable network elements. If an interaction has been refined to the point where the participating elements are reproducibly and often engaged in it, than this strong interaction has a greater chance of being directed than a weaker interaction. Strong links are predictable, whereas weak links are transient, meaning here that the direction of the interaction may just reverse from time to time.[3]

Weak links are remnants of our past. Were 'the first weak links' developed after strong links to maintain the stability of the networks determined by the strong links? This sounds rather unlikely, if not impossible. A strong link, a high affinity binding, requires a developed and mutually adjusted structure of the two partners. At the very beginning of life, these conditions were not present on Earth. Life started from weak links.[4] Strong links were only added later. This also means that originally we had no networks on Earth, in the current sense of the term. Networks were not solid, stable assemblies, but were continuously formed and reformed. The proposed widespread lateral gene transfer (Rivera and Lake, 2004; Woese, 1998), implying that at the very beginning no living creature on Earth possessed stable genetic information and that all life could be regarded as one single organism from the genetic point of view, is one of the many suggestions pointing towards this view. In other words, early networks had links of

[2] I am grateful to István Molnár for this suggestion.
[3] I am grateful to Attila Steták for this suggestion.
[4] I am grateful to György Buzsáki for this suggestion.

equal (weak) strength. They were much closer to a random network than our networks today. During the preparation of this book, this hypothesis was also formulated by Shenhav et al. (2005), as already mentioned in the context of network phase transitions in Sect. 3.4.[5]

4.3 Stability: A Starting Definition

Stability of networks can be assessed minimally at two levels. These two levels of stability are defined here to help discriminate between them in later chapters of the book:

> **Definition of Network Stability (or Parameter Stability).** A network is stable if it shows a tendency to return to its original parameter values after a perturbation.[6]

This definition resembles the Le Chatelier principle[7] with the important difference that complex systems are almost never in a traditional equilibrium. It would thus be better to talk about robust behavior gravitating towards certain parameter sets, or attractors of the network, after perturbations.

> **Definition of Netsistance (or Network Persistence).** A network has netsistance if it can preserve its giant component and percolation by keeping most of its elements connected to each other.[8]

At this level of stability, the network may leave the original attractors and shift to new ones, or even undergo a topological phase transition,

[5]Random networks are stable when resources are plentiful. At the very beginning, there were quite a few self-organized systems on Earth, so they probably enjoyed a relative abundance of resources in their environment.

[6]The network parameters in this definition are often emergent properties of the network, which do not exist without the formation of the network and cannot be measured by knowing only the constituents of the network.

[7]The Le Chatelier–Brown principle describes the behavior of systems after their equilibrium has been perturbed. When a system in equilibrium suffers an effect which changes its conditions, the system will adjust itself to minimize this change.

[8]Some readers may feel uncomfortable about seeing the clause 'most of its elements' in this definition. The undefined ratio of connected elements might make the definition of the giant component and netsistance useless. Do we need to connect 5% or 95% of the elements to get the giant component? The necessity of percolation for the existence of the giant component saves the definition. Percolation develops as a phase transition in most networks. With a little approximation

e.g., change its scale-free degree distribution to a star, or random degree distribution as the outside conditions become harder or easier, respectively (see Sect. 3.4). At netsistance, the stability criterion is to keep the network functional, i.e., to preserve its giant component, percolation and emergent properties. In the case of cells or organisms, this means that the network stays alive. In the case of supra-individual networks, netsistance is often called resilience, to discriminate it from the 'simpler' chemical type of stability (Holling, 1973).

Relaxation as a measure of stability. Stability is preserved by efficient relaxation. If relaxation is fast, e.g., exponential, the network most probably has quite good stability.

Noise as a measure of stability. From the network point of view, noise can be regarded as a perturbation. If stability is high, the observed fast relaxation helps noise dissipation or noise damping. A high average level of noise in the network is a good indicator of low stability.

Diversity as a sign of instability at the bottom network level. Low relaxation usually accompanies parameter instability. If relaxation is low, perturbing energy may stay at a network segment, helping the whole network to reach a novel local energy minimum which could not have been reached before due to the prohibitively high activation energy. Once the network has jumped to the new local energy minimum, it stays there and becomes different from the original network. Now let us imagine

---

we may say: percolation is either there or it is not. Whatever segment of the full network exhibits percolation will be what constitutes 'most of its elements' in our definition. *"This definition is still not good enough for me. What if 95% of the network is left as a lonely element and the residual 5% forms a VIP club and starts to percolate?"* In principle, you are right, Spite. This may happen and may be rather disturbing. But in practice, if 95% remains as a lonely element, why do you want to include it in the definition of the network? *"Okay, you win. But I have another objection: what if you have 5 fully connected happy elements, which percolate, and the residual 5 000 elements have not joined the network yet?"* In principle, you are right again, but your 5-clique is too small to produce a phase transition. Moreover, when percolation is born, the network usually starts to show most of its emergent properties. As a possible example, if your 5 fully connected happy elements can reproduce the network, show a large number of adaptive responses and save this information from the dissipative changes, then I will be happy to call them a network without a phase transition.

several clones of the original network. Since both the site and magnitude of perturbations differ from each other in the various network copies, changes of these networks will also differ. Therefore, the different copies will develop and explore entirely different new local energy minima which did not exist previously or had prohibitively high activation energies. After a while many of the clones will end up in different states and will all differ from the original network and from each other. Parameter instability develops diversity. Diversity reveals the instability of the diverse (bottom) networks. However, the same diversity will stabilize the network of these networks, the top network, as I will show later.

Nestedness: stability from top to bottom. In Sect. 2.3, I showed that elements of the top network are themselves networks, i.e., networks show nestedness. If the bottom networks constituting the elements of the top network are not stable, they cannot make strong links, and therefore many of them cannot be used for building the top network. The top network has to figure out mechanisms to (a) stabilize, (b) segregate, or (c) disassemble unstable bottom networks. Chaperones perform precisely this task in the cell, as will be described in Sect. 6.2. As another example, in societies (a) stabilization is provided by psychologists, physicians, teachers, laws, rules, norms, gossips or closed communities; (b) segregation is achieved by prisons, madhouses, hospitals, quarantines, and last but not least, research institutes. Fortunately, the methods of purposeful disintegration (c) have been mostly outlawed in civilized societies by the 21st century, with the exception of some third world countries and a few states of the USA.

Nestedness: stability from bottom to top. Networks try to stabilize not only their bottom networks, but also their environment.[9] Stabilization of any environmental parameter (think about your house, heating and air conditioning systems) gives an advantage for survival. Symbiotic relationships (Margulis, 1998), the formation of top networks, species diversification, and the whole process of self-organization are also signs of environment stabilization efforts.

When does the top network kill its bottom networks? When are the top network and the bottom networks combined to such an

[9] I am grateful to Péter Száraz for this suggestion.

extent that the disintegration of the top network automatically leads to the disintegration of its bottom networks?[10]

Network stability as a source of the scientific method. As already mentioned in Sect. 4.1, without network stability, science would not have developed as a form of human cognition. Without network stability we would have had no reproducible experimental results and no chance of generalizing any of our constantly changing observations.

4.4 Complex Systems

Complexity has been the focal point of extremely powerful thought and concepts. However, complexity is often poorly defined (Tononi et al., 1998): "While we think that we recognize complexity when we see it, complexity is an attribute that is often employed generically without any attempt at conceptual clarity or, even less, quantification." Trying to avoid these problems, I will give a very brief overview of the most important concepts concerned with the numerical definition of complexity, as well as the most important properties of complexity.

Definition of Complexity as a Measure. I begin the references to numerical definitions of complexity with the definition due to Kolmogorov (1965), which says that the *algorithmic information complexity* of a string of characters computing x equals the length of the shortest program that computes x and then stops. The erroneously high complexity value attributed to random processes by this definition led Murray Gell-Mann (1994; 1995) to formulate the concept of *effective complexity*, which is the length of a highly compressed description of the regularities of the entity. A similar definition concerns *statistical complexity*, which is related to the amount of information required to produce optimal forecasts of the system (Crutchfield, 1994).

For a system with random behavior, the regularity is zero and the characterization of future states may also require only one parameter.

[10] This is certainly true for a living organism, like ourselves: our cells die with us. However, individual power plants will not necessarily die if the power lines are cut.

Random systems are simple in this sense. Similarly, periodic (lattice-type) systems are also simple. If I know the frequency, amplitude and phase of a periodic signal, I can both describe its regularities and predict any of its future states. Complex systems lie in-between these two extremes, mixing random behavior and periodic structure (Gell-Mann, 1994; Tononi et al., 1998). Complexity can arise in the topology, link strength, dynamics and numerous other features of the network. In this book, complexity is not used in the strict, numerical sense, since the stabilizing role of weak links has never been systematically tested as a function of numerical complexity. This exciting task awaits future work.

Do weak links increase complexity? The use of complexity in a strict numerical sense as a system property to study weak-link-induced stabilization may lead to an even more complicated task, since the system complexity itself may be changed by changing weak links. As summarized in Table 4.1, weak links make networks integrated. If the system is an irregular, but not random network (and not a regular lattice, for example, where integration of two parts will not increase the complexity at all) the integrity and complexity of the network are strongly related. Deletion of weak links will most probably induce a decrease in system complexity. A formal proof for this assumption represents another nice challenge.

Definition of Complexity as a Property. If we do not require a numerical definition of complexity, what can be said about the 'signature properties' of complexity? Gerald Edelman gave the following brief definition (Wilkins, 2004): "A complex system is a system which has heterogeneous smaller parts, each carrying out some specialized function, not necessarily exclusively, which then interact in such a way as to give integrated responses." 'Complex' is not synonymous with 'complicated'. In contrast with complicated systems, the function of the whole in a complex system cannot always be guessed from the function of the parts, and the reassembly of the parts does not always give back this function (Ottino, 2004).

Weak links stabilize complex systems. Weak-link-induced stabilization refers to complex systems. I purposefully did not write: weak links stabilize networks. Weak-link-induced stabilization is not a network property in the sense that it would be true of all networks. Networks with very low complexity do not have weak links. The parallel presence of weak and strong links excludes both fully random and fully regular networks. Weak links can only help the stability of complex systems, since the presence of weak links already attributes an element of complexity to the system.

How does weak link-induced stabilization depend on system complexity? Is there a complexity threshold below which weak links do not stabilize appreciably? Does weak-link-induced stabilization level off as complexity exceeds another threshold?

4.5 Weak Links and System Degeneracy

Degeneracy is a property of most complex systems, in which a system function is performed by two different system components. "Degeneracy is not a property simply selected by evolution, but rather is a prerequisite and an inescapable product of the process of natural selection itself" (Edelman and Gally, 2001). In this section, I show that degeneracy is linked to the emergence of scale-free networks, co-occurring with the emergence of weak links and helping to stabilize complex systems. Occurrence of degeneracy will provide a tool for demonstrating the generality of the stabilization induced by weak links at all levels of development.

Degeneracy gives rise to weak links. Degeneracy is caused by network elements, motifs, or modules performing a similar function. Elements, motifs or modules displaying a similar function will all be linked to the same set of other network elements, but will certainly have a different affinity towards them. Equal strengths of these interactions will be the exception rather than the norm. With each pair of degenerate elements, motifs or modules, new weak links are born. Degeneracy also gives rise to modular structure. Intermodular contacts are yet another source of weak links.

Degeneracy leads to the stabilization of the network. As a typical starting point in the development of degeneracy, gene duplication has been shown to induce developmental stability (Wilkins, 1997). Degeneracy is also a stabilizer of complex systems (Edelman and Gally,

Fig. 4.3. With each pair of degenerate elements, motifs or modules, new weak links are born

2001; Sole et al., 2003a). In fact, degeneracy stabilizes genetic networks more than simple gene duplication (Wagner, 2000; Kitami and Nadeau, 2002). Table 4.2 lists some examples of degeneracy-linked weak links and the functions which may become stabilized by both.

I am happy to inform you that half of our job is completed. I have now summarized many important properties of networks, listing the relevant effects of weak links. The central hypothesis of the book has been formulated and an initial definition has been provided for all its elements. Lastly, using system degeneracy, I have introduced the astonishing variety of networks which may be stabilized by weak links. In the second part of the book, let me invite you on a great journey through Netland. We shall use weak links as a thread (do not worry, I hope to convince you that this thread is even stronger than its proverbial Cretaean ancestor helping Theseus out of the Labyrinth) to visit a number of networks at different levels of complexity. Weak links will give me a good excuse for introducing these networks and exposing their wonderful unity.

Table 4.2. Degeneracy in various networks as a source of weak links and network stability. Partner A denotes the degenerate component which has multiple forms performing a highly similar function. Partner B denotes the component which is the common partner of the multiple degenerate components. Since both partners A and B are members of networks, their 'binding' is often not physical

Source of degeneracy[a]	Emergence of weak links between:		Stabilized function[b]
	Partner A	Partner B	
Multiple nucleotide triplets coding the same amino acid	t-RNA	Ribosome	Translation
Multiple transcription factors inducing the same gene	Transcription factors	Promoter region	Gene transcription
Multiple protein sequences with the same folds	Similar proteins	Binding partner	Cellular responses
Multiple iso-enzymes catalyzing the same reaction	Iso-enzymes	Metabolic pathway	Metabolic networks
Multiple structural proteins with similar binding properties	Structural proteins	Cyto-architecture	Cellular structure
Proteins with multiple subcellular localization	Organelle-specific docking sites	Protein stabilization	Subcellular organelle
Non-identical subcellular organelles	Organelles	Organelle function	Cellular responses
Multiple mechanisms of synaptic plasticity	Protein complexes	Plasticity	Memory
Non-identical cells in tissues	Cells	Cellular function	Tissue function
Parallel signaling pathways	Pathways	Response	Signaling networks
Multiple immune cells against the same antigen	Immune cells	Effect	Immune response
Parallel neural networks performing the same function	Neurons	Function	Neural response
Parallel non-identical muscle fibers with similar functions	Motor units	Contraction	Body movement
Multiple bone trabecules stabilizing against the same pressure	Bone trabecules	Pressure	Bone stability
Multiple stimuli provoking a similar sensory response	Stimuli	Sensory output	Sensation
Multiple behavioral elements with the same final effect	Behavioral elements	Environment, stimulus	Full behavioral response
Multiple words with similar meaning, ambiguity	Words	Meaning	Message

[a] Most examples are from the seminal paper by Edelman and Gally (2001). Bone asymmetry was included after Fox and Keaveny (2001) and da Fontoura Costa and Palhares Viana (2005).
[b] Most of these examples are hypothetical.

5 Atoms, Molecules and Macromolecules

If you are ready, our journey will now begin. After entering Netland, we will first visit 'simple networks'.[1] In the next section, I discuss proteins, essential building blocks of our cellular networks. They carry out almost all tasks in the cell and provide most of the cell's shape, too. We shall begin with the enormous difficulties our proteins have to overcome to reach their final structure. In the next step, we examine their complex stability patterns using the idea of energy landscapes. Finally, we discover how weak links help them in all these processes.

5.1 Protein Folding Problems

Most proteins in our cells are active only in their native structure, which represents the minimal energy state of all possible shapes. These shapes are called conformations.[2] A random search for the absolute energy minimum (the native conformation) or in other words, protein folding, would require an astronomical number of approximating steps if all conformational states were probed. This has led to the famous paradox due to Cyrus Levinthal, stating that the Universe has

[1] The expression 'simple networks' reminds me to stress that all self-organized, natural networks are rather complex. In particular, the level of complexity may significantly increase with the number of bottom network layers below the level of the examined top network. From this standpoint, networks regarded as simple by common sense are indeed simpler. However, these 'simple' networks may have a much more complex structure than their 'complicated' counterparts.

[2] In a few cases, as in prion proteins, the shape with minimal energy cannot be reached by the protein, since an energy barrier prevents the protein from adopting this favorable conformation. However, in the specific case of the prion protein, we are very lucky to have this energy barrier. Without it, the prion proteins in our brain, which are protecting both me and the reader from brain damage at this very moment, would fold to their minimal energy state and aggregate. This extensive self-assembly would cause the death of our neurons.

not lasted long enough for the folding of a simple protein (Levinthal, 1968).[3]

The Levinthal paradox. To understand the details of the Levinthal paradox, we have to imagine a polypeptide chain which contains one hundred amino acids, bound one after the other by peptide bonds. Due to rotation around single bonds, each amino acid residue has many distinct conformations, depending on its main chain and side chain dihedral angles. Geometrical constraints restrict the number of these conformations. The reason for this is that two atoms cannot approach each other beyond a certain limit, and large overlaps are not permitted. Although this does indeed restrict the number of possible conformations per amino acid, the number is still higher than unity. Taking the very moderate estimate of two, the whole polypeptide chain must have 2^{100} possible conformations, which is an extremely large number. If we let the protein fold and sampling is random, in the worst case the protein will try every conformation during its folding before finding the single solution among the 2^{100} possibilities. If we allow 1 picosecond (10^{-12} seconds) for each transition between two conformational states, the time required for the folding process would be slightly more than 2^{18} seconds, which is close to 40 billion years. This time is more than double the currently estimated age of our Universe. Levinthal's paradox clearly shows that the sampling of the conformational states cannot be random and has to be guided (Levinthal, 1968). The details of this guidance will be revealed in the rest of the chapter.

The Levinthal paradox already shows that proteins have to solve a formidable problem in order to find their unique native structure. To get a visual idea of the task, imagine several hundred, rather different pieces of nano-LEGO. These will correspond to the building blocks of the proteins, the amino acids. Among the 20 types of native amino acids, the smallest amino acid, glycine has only one proton and a pair of electrons as a side chain, while the largest, the tryptophane has two complex ring systems. If you were asked to assemble several hundred of

[3]The absolute energy minimum here is in fact the lowest free energy. The free energy is defined in thermodynamics as $G = H - TS$, where G is the Gibbs free energy, H is the enthalpy (which is the energy gain due to the formation of chemical bonds during the folding process), T is the temperature measured in kelvins, and S is the entropy, the measure of disorder in the system. Therefore, reaching the free energy minimum during the protein folding process requires a simultaneous optimization of the energy gain by the formation of various bonds, and the minimization of the entropy decrease due to the organization of the protein structure.

these vastly different molecules in such a way that there was no space left between any two of them, you would already start to scratch your head. Suppose now you are told that additional rules exist: some of the nano-LEGO pieces are white (these are called hydrophilic amino acids, e.g., glutamic acid) and they have to stay on the surface all the time. Other nano-LEGO pieces are grey (these are called hydrophobic amino acids, e.g., leucine) and they all have be buried inside the molecule (see Fig. 5.1). Additionally, all amino acids are tied with a string in a specific order, and the string must not be cut since it represents the protein backbone, where the linking chemical bonds,[4] the peptide bonds, are very stabile and will not break. Finally, you are informed that you have less than one second to complete the job. I think by this time you would start to laugh (thereby wasting your precious second) and you would give up. Not the proteins! Most of them complete the assembly, which is called protein folding, in a fast and highly efficient way.

How do proteins fold? Protein folding is characterized by two main steps in vitro (see Fig. 5.1). In the first few milliseconds most of the secondary structure, i.e., individual α-helices and β-sheets, has already formed. In most cases folding starts with the formation of α-helices, since the participation of adjacent amino acids is required here. β-sheet formation establishes hydrogen bonds[5] between amino acids, which are far from each other in the primary sequence. For this reason, when β-sheets are formed, a greater entropy decrease occurs than during the formation of α-helices. At the end of this first step the hydrophobic segments are segregated by the surrounding water and form the hydrophobic core of an intermediate which is often called the molten globule. This process is known as hydrophobic collapse since the volume of molten globules becomes almost as small as that of the final,

[4]Chemical bonds are made of one or more pairs of electrons which interact simultaneously with the two atoms bound together. Electrons from the two interacting atoms will leave their original atomic orbitals and form a bond if their energy is lower in the molecular orbital.

[5]Hydrogen bonds, or hydrogen bridges are low energy chemical bonds, where the nucleus of a hydrogen atom, a proton, can be found simultaneously in the vicinity of two atoms. In biologically important molecules, these two atoms are usually nitrogen or oxygen, both having a free electron pair. The proton uses its wave properties to occupy the space beside both atoms, even if they are relatively far from each other so that a 'forbidden' region occurs between them. Besides electrons, the proton is the only building block in our molecules which still has this property to any appreciable extent. Quantum mechanical wave properties decrease with the weight of the particle. This is the reason why we do not have carbon bridges, oxygen bridges, etc., in the same sense, since they are too heavy to display such an effect.

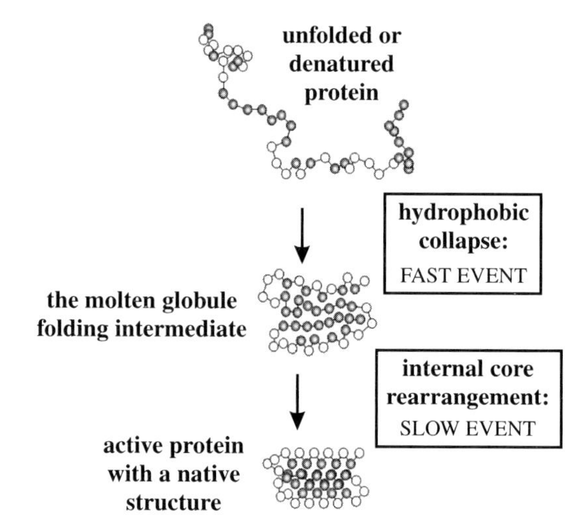

Fig. 5.1. The two major steps of in vitro protein folding on a hypothetical example of a small protein. *Grey* and *white circles* represent hydrophobic and hydrophilic amino acids of the protein, respectively. Conformational states of the target protein were adapted from Thirumalai and Guo (1994)

folded protein. If the protein is larger than 30 kDa, the molten globule is usually fairly stable. In the partially folded state of molten globules, the α-helices and β-sheets have not found their correct, tightly fitting relative position, which means that the protein does not yet have a stable tertiary structure. Molten globules still have large unburied hydrophobic surfaces. This makes them vulnerable to unspecific attachments, a process known as aggregation.

The second major step of protein folding is the slow, rate-limiting step. Here the inner, hydrophobic core of the protein is reorganized. In parallel with this, unique high-energy bonds are formed, such as disulfide bridges and ion pairs, and the peptide bonds are isomerized beside proline amino acids. The free energy gain due to these processes enables the formation of local, thermodynamically unstable, 'high-energy' protein structures, which are stabilized by thermodynamically favorable conformation of the rest of the protein. These high-energy segments of proteins can stabilize themselves by forming complexes with another molecule. They thus often serve as active centers of enzymes or as contact surfaces between the various proteins involved, e.g., in signal transduction.

Protein folding is not a straightforward process. Dead-ends, reverse reactions, futile cycles are all characteristic features of this process. A minor amount of fully folded, native protein always coexists with various forms of molten globule and the remaining traces of unfolded protein. Aggregation of unfolded protein molecules and molten globules is a major threat that would drive the majority of folding intermediates into unproductive side reactions, long before they could reach their fully folded, competent state. Moreover, unaided protein folding often leads to folding traps. In one of these traps, the rearrangement of the hydrophobic core is often prevented. Nascent proteins have the additional problem that they have to fold before they are even ready. The first protein segment which leaves the ribosome[6] will certainly have a different energy minimum to the whole protein. Therefore, in many cases, in vivo protein folding has to be delayed, and the premature folding of the first synthesized part of the protein has to be delayed until the rest of the protein has been synthesized (Bryngelson et al., 1995; Dill et al., 1995; Dobson et al., 1994; Kim and Baldwin, 1990).

In the next two sections, I use the network approach to see how proteins are designed and assisted in accomplishing the enormous task of protein folding described above.

5.2 Energy Landscapes

In the last section we saw that protein folding is governed by a search for the conformational state of the polypeptide chain with the lowest energy. We may understand this process better if we draw all the possible conformations with their corresponding energy levels. This is called an energy landscape. Landscapes were first introduced to science in 1932 (Wright, 1932) and fifty years later the powerful concept was applied to understand protein folding (Bryngelson and Wolynes, 1987; Bryngelson et al., 1995; Dill, 1985; 1999). The landscape of protein structures is shown in Fig. 5.2. To explain it, let us go through step by step.

To make the energy landscape understandable visually, the various conformational states of the protein are viewed in two dimensions only. Thus the protein conformations occupy the (x, y) plane. The energy corresponding to each of these conformations is plotted on the vertical or z axis. The large cavity at the front represents the absolute energy minimum of all conformations. In proteins, this is usually called the

---

[6]The ribosome is a macromolecular complex formed by a large number of RNA and protein molecules. The ribosomes are responsible for protein synthesis in our cells.

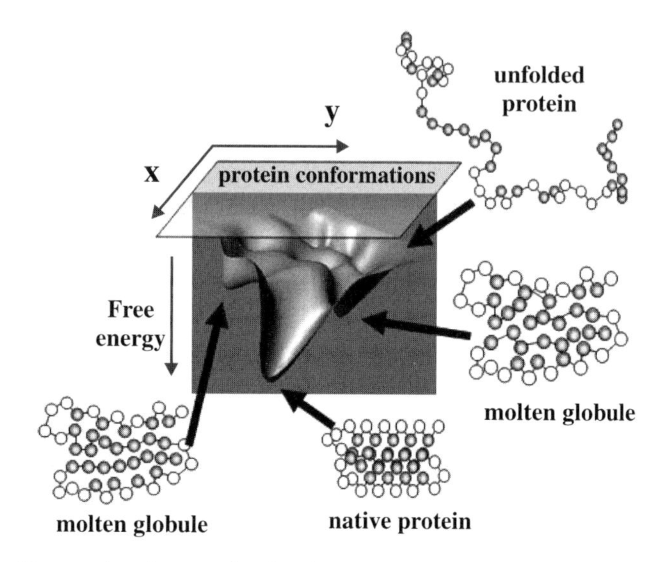

Fig. 5.2. Energy landscape. In this hypothetical energy landscape, the (x, y) plane serves to locate the various conformational states of the folding protein. The free energy of the given conformation is shown on the *vertical axis*. The approximate positions corresponding to the unfolded protein, the various molten globules, and the native protein are indicated

native conformational state of the protein. All other depressions in the landscape represent local energy minima corresponding to folding intermediaries, e.g., molten globules. The saddles between the depressions show the activation energies that the protein has to collect to get from one stable state (local energy minimum) to another. Similar landscapes can be constructed to describe the stability conditions of complex systems other than proteins, e.g., ecosystems, social networks, etc. The general term 'stability landscape' denotes all these landscapes and will be detailed in Sect. 12.2.

The energy landscape can also be visualized as a contour plot similar to those used on maps (see the top middle panel of Fig. 5.3). Here the various contour lines represent the energy levels and the actual position on the plane of the figure gives the conformation of the protein.

A network representation of the energy landscape was introduced and analyzed by Doye (2002). To illustrate this approach, the top right-hand panel of Fig. 5.3 shows the network representation of the energy landscape in the top left-hand panel. In this network, nodes

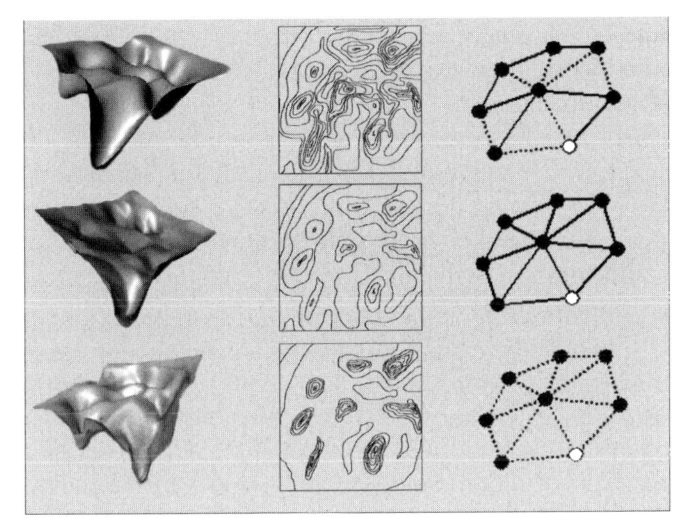

Fig. 5.3. A two-dimensional conformational space of a hypothetically simple protein is depicted to illustrate the importance of having the right amount of weak energy links in order to stabilize energy networks. The *horizontal plane* shows protein conformations and the *vertical axis* shows their energy levels. All eight energy minima remain the same in the 'normal' (*top row*), 'all-strong' (*middle row*) and 'all-weak' (*bottom row*) energy landscapes. However, the activation energies between the minima are variable, minimal and maximal in the normal, all-strong and all-weak energy landscapes, respectively. The *left-hand panels* show a 3D representation of the energy surface, whilst the *central panels* are contour plots of the energy levels around the eight minima, and the *right-hand panels* show the network representation of the transitions between the minima drawn using the concept due to Doye (2002). On the energy nets, the *white dot* represents the absolute energy minimum, while strong and weak energy links are marked with *continuous* and *dotted lines*, respectively (Csermely, 2004; 2005). Figure courtesy of Máté Szalay of the LINK group

represent energy minima and energy links[7] correspond to activation energies. The energy landscape of proteins has both a small-world and a scale-free character (Doye, 2002; Doye and Massen, 2005; Rao and Caflish, 2004; Scala et al., 2001).

Protein folding takes fractions of seconds or minutes, depending on the protein and the circumstances. The small-worldness of the energy landscape network gives us another explanation as to why our

[7]The expression 'energy link' is used to discriminate the links of the energy network – which actually denote transitions from one conformational state to another – from the links of the conventional, topological networks.

proteins fold so efficiently. Protein folding is often guided by the modularity of the energy landscape. In these cases, the landscape has a set of high activation energy barriers (weak energy links in the energy net), which exclude a major part of the possible conformational states from the folding process (Otzen and Oliveberg, 1999; Plotkin and Wolynes, 2003). Here again, we may observe a helpful mixture of global connectivity and confined relaxation. Global connectivity ensures that most starting conformations will find the single native conformation. Confined relaxation restricts further changes of the conformation once it has reached a certain level of development towards the native state.

Energy nets may provide further insights to extend our knowledge of network stability. In the energy net, an energy link can be defined as strong if the corresponding transition state in the energy landscape has low activation energy. An energy link is weak if the corresponding transition state has high activation energy. Weak energy links mean that the conformational change between the two corresponding nodes (i.e., protein conformations having a local energy minimum) has a low probability due to the high activation energy between the two energy states. If we imagine the extreme situation in which there are only strong energy links between the local energy minima, all activation energies will be low and all transitions will occur easily (middle panels of Fig. 5.3). We have an undefined (unstable) system, because the protein shifts to the global energy minimum without any appreciable transition time in local energy minima, and this is characteristic of unrestricted, fast protein folding.[8] If we have only weak energy links between the energy minima (bottom panels of Fig. 5.3), all local energy wells will behave as folding traps. The system becomes superstable, but the network itself is not defined, because we can hardly find any folding pathways in which the conformation gradually changes until it reaches the global energy minimum. The folding pathway of regular proteins corresponds to the 'normal' network, shown in the top panels of Fig. 5.3, where the distribution of energy link strength is balanced between stronger and weaker energy links (Csermely, 2004).

As my last example of energy nets I have chosen a pair of similar energy landscape networks, where the stability of two smaller systems were analyzed and compared. One of them had 19 argon atoms and the other was an assembly of 32 potassium chloride molecules (Ball et al., 1996). The argon network was grown by accommodating one or at most two atoms at a time, while the potassium chloride network was

[8] These landscapes were called buffed energy landscapes by Plotkin and Wolynes (2003).

grown by absorbing larger clumps of ions. In other words, in contrast to the argon network, the potassium chloride network was modular. An even more important difference occurred in the energy levels. The argon network showed monotonic sequences of energies, while minima of the potassium chloride network fell on multiple sequences which varied a lot, and arose from various locally stable potassium chloride crystal structures. In summary, the argon network was rather uniform, while the potassium chloride network was more diverse.

The topological differences between the networks provides a further, convincing explanation as to why potassium chloride can be condensed much more easily than argon. *"Something is not right here. Argon is a noble gas. I learned that it has very weak interactions between its atoms. As a consequence of this it does not condense well. You do not need this network hocus-pocus to explain the difference between potassium chloride and argon!"* Yes, Spite, this is true. However, argon condenses even less well than you would expect from the energy difference. This additional difference is explained by the differences between the network structures. Argon tends to stack in disordered, glass-like forms, while potassium chloride has a funnel-type energy landscape, similar to proteins, and also 'folds' (condenses) quite efficiently, like proteins and other similar macromolecules.

Weak topological links of a strong ion. I cannot resist making a comment here, in the form of two questions. What type of topological links will be more prevalent in the diverse, module-like topological network of potassium chloride than in the uniform network of argon? What may contribute to the stabilization of potassium chloride condensates? I leave the answers to you.

5.3 Weak Bonds in Protein and RNA Folding

What helps proteins to achieve their final structure so efficiently? The network approach has already given clues to find the answer. In the last section, we learned that the energy landscape of proteins simultaneously provides a confined relaxation and constitutes a small world, where the absolute energy minimum can be reached easily from any local energy minima (Doye, 2002). Small-worldness is a typical feature not only of the energy, but also of the topological networks of both globular and fibrous proteins (Bagler and Sinha, 2005; Greene and Higman, 2003; Scala et al., 2001). Interestingly, key amino acids (nucleation centers), which were shown to govern the folding process, form highly connected hubs in this small-world topological network (Vendruscolo et al., 2002). Moreover, the small-world-type connectivity increases further during the folding process (Dokholyan et al., 2002).

In contrast to many self-organized networks, the degree distribution of protein topological networks seems to be Poissonian and not scale-free (Bagler and Sinha, 2005). Protein topological networks often have modules, which we call domains. Domains usually fold separately, have a function, and are conserved during evolution. The distribution of the folds of various domains follows a scale-free pattern (Koonin et al., 2002) implying that there is a small number of very 'popular', stable folds and that we have a relatively large number of unique, orphan folds. This is not a miracle; it is selection. Evolutionary selection preferred those structures which were both stable and folded easily. These structures are the ones which have the common feature of small-worldness.

If domains behave like network modules, it is not surprising that interdomain contacts are weak. This helps to stabilize the protein structure and gives an advantage to the complex formation of different proteins, characteristic of higher organisms. Weak interdomain contacts are often provided by fluctuating water molecules (Csermely, 2001a). Water-mediated interactions play an important role not only in binding interfaces, but also in the folding of monomer proteins. Moreover, most water-mediated interactions are long-range interactions, implying that an organized set of water molecules mediates long-range interactions between proteins that are approaching one another (Kovacs et al., 2005; Levy and Onuchic, 2004; Liu et al., 2005; Papoian et al., 2004; Pertsemlidis et al., 1999).

Molecular 'washing machines' help protein folding. In the next section, I introduce molecular chaperones. Chaperones are special proteins which help protein folding. Here I only mention them to note an important element of the mechanism whereby they fullfil their task. Indeed, the key word is water. At least one class of chaperones works like a washing machine. Using the energy gained from ATP hydrolysis,[9] they periodically stretch the misfolded protein (Chan and Dill, 1996) and then release it again to let it collapse to a more compact state. During this cyclic motion, they wash the internal, hydrophobic core of the protein through with water, again

[9]The energy of ATP hydrolysis most probably does not cause the stretch, but– like a Maxwell demon – just organizes the Brownian motion of water molecules into a 'macroscopic' motion by preventing their contribution in any other directions than the desired one.

and again, thereby accelerating its folding (Csermely, 1999; Kovacs et al., 2005).

An important mechanism whereby water helps protein folding and complex formation occurs when it acts as a lubricant, assisting protein motions (Barron et al., 1997). Although several proteins can withstand a transfer to non-aqueous media, most enzymatic functions are stopped in the complete absence of water. Lyophilized food and dried meat last longer not only because the bacteria which have invaded them are then doomed or go into a non-metabolic spore state, but also because most digesting enzymes are inhibited. Moreover, dry proteins have a 'memory'. They preserve enzyme activity, if their structure has been previously stabilized. Dry proteins 'remember' the active state because their conformational changes are frozen in the absence of water (Klibanov, 1995).

Can proteins move without water? There are quite a few, sometimes contradictory observations concerning residual protein mobility in the absence of water. On the one hand, a 'monolayer' of water molecules is needed on the protein surface to restore the dynamics of biomolecules. The dynamics emerges when individual water molecules establish the percolation of their hydrogen-bonded network (Oleinikova et al., 2005; Rupley and Careri, 1991). On the other hand, in many enzymes, a residual enzyme activity can still be observed at very low hydration levels (Kurkal et al., 2005). Detailed investigations were able to discriminate protein movements, called slaved processes, which require the contribution of water as solvent, and movements which are independent of the solvent, called nonslaved processes (Fenimore et al., 2002).

In summary, water helps protein mobility. But how does it affect protein folding? A recent paper by Peter Wolynes and coworkers (Papoian et al., 2004) showed nicely that water efficiently lowers the saddles (activation energies) of the energy landscapes and makes previously forbidden conformational transitions possible.[10,11] Water molecules form

[10] The other 'weakly linked protein folding assistants', molecular chaperones, have also been proposed to smooth the energy landscape of folding proteins by Ulrich Hartl and coworkers (Brinker et al, 2001).

[11] In agreement with the in vivo situation, in silico protein folding is also helped by lowering the saddles of the energy landscape, as has been shown in the protein models of Zhang et al. (2002).

weak links with each other and with all atoms of proteins they interact with. Therefore, we can formulate the above statement in the following way: *weak links lower the saddles (activation energies) of energy landscapes* (Kovacs et al., 2005).[12] This statement can be generalized to all stability landscapes, as I will show in Sect. 12.2.

The effect of water accords with the functional definition of weak links. Let me repeat here the functional definition of weak links from Sect. 4.2: a link is defined as weak when its addition or removal does not change the mean value of a target measure in a statistically discernible way. When he gave this definition, Berlow (1999) also stated that weak links would decrease the variation and noise in the system. Individual water molecules do not change the minima of the energy landscape. They cause no change in the stable protein structure either. However, by the addition of water molecules, several previously inaccessible conformational transitions will become much easier, and as a consequence a larger convergence to the most stable protein conformation will be possible. The final assembly of available protein conformations will not have as many trapped conformers as before and will be much more uniform, with decreased variation and noise. This process gives a rather good match with the functional definition of weak links.

I hope I am not over-explaining this phenomenon. If so, I have an excuse: this statement (weak links make the energy landscape smoother) is an extremely important statement. This will explain why weak links decrease diversity, and how can they regulate the 'punctuatedness' of the punctuated equilibrium (Gould and Eldredge, 1993; Kovacs et al., 2005). I will return to these thoughts in Sect.6.3 and give a final synthesis in Sect. 12.4.

But let us return to water for the time being. Water helps the conformational transitions of proteins. What happens if these transitions are not smooth? An avalanche will occur. Early work by Ansari et al. (1985) already showed the existence of protein quakes, i.e., the cascading relaxation of myoglobin after the photodissociation of carbon monoxide. The protein quake implies the same scale-free statistics for

[12]The content of Fig. 5.3 is in an apparent contradiction with this statement. However, the similar words conceal a great difference. In the energy net representation, weak energy links were those saddles (activation energies) which were high between two energy minima. In the topological network of protein-related physical interactions, water-mediated weak links actually transform the weak energy links of the former energy net into strong energy links, implying an easy transition between two energy minima.

protein dynamics as we have seen for other self-organized systems in Sect. 3.3.[13]

A number of protein kinetics, including the above-mentioned carbon monoxide dissociation, enzyme actions, exchange of protein protons with those of water and protein folding, are all similar to Levy flights and obey scale-free statistics (Dewey and Bann, 1992; Flomenbom et al., 2005; Metzler et al., 1998).[14,15] The scale-free statistics reflects the scale-free and hierarchical structure of the energy landscape (Yang et al., 2003), as will be described in Chap. 12. These findings mean that most protein movements are restricted and rather small. However, proteins might also make a big jump. Fortunately this only happens very rarely. This scale-free kinetics is strongly related to the scale-free structure of protein surfaces (Goetze and Brickmann, 1992) and to the presumed scale-free transport 'channels' inside the proteins (West et al., 2002).

Proteins are stabilized by weak bonds. Both secondary and tertiary protein structures are stabilized by weak, rather dispersive forces: hydrophobic bonds, hydrogen bonds and van der Waals forces. Strong bonds, such as disulfide bridges or salt bridges, may exert more influence but are used rather sparingly to achieve structural stability. Why are strong bonds missing? The network approach may explain this apparent discrepancy. A greater involvement of strong bonds would induce a lower cooperativity in the folding process and would make the isolation of the local energy minima in the energy landscape much stronger.

Ribonucleic acids (RNAs) are stabilized by weak bonds. Similarly to proteins, RNA structure can also be perceived as a mixture of strong and weak bonds, where strong bonds are represented by the hydrogen bonds between base pairs and weak bonds are the van der Waals forces between various segments of the RNA molecule. It is the latter which gives the

[13]It would be interesting to check whether protein quakes are really bigger in the absence of water than in its presence, as expected.

[14]The scale-free temporal pattern resembles that of pink noise, which was highly correlated and had a 'memory' (see Sect. 3.1). In agreement with this, a memory landscape of single enzyme molecules has been suggested (Lu et al., 1998; Edman and Rigler, 2000).

[15]It should be noted that the scale-free distribution becomes more complex in most cases, since the underlying protein structure is hierarchical and modular.

3D RNA structure its unique shape and stability (Csermely, 1997; Sclavi et al., 1998).

Protein complexes are also stabilized by weak links. Due to recent advances in the detection limits of experimental techniques, we have more and more examples showing that protein–protein (and most probably, protein–RNA, protein–DNA) complexes are indeed stabilized by weak links between the complex-forming molecules (Smith et al., 2004; Swanson et al., 2003).

We have reached the end of the first trip into Netland. It is time to have a look at the souvenirs among our luggage. What have we learnt? Simple, chemical systems like condensed potassium chloride molecules or larger entities like proteins gave the first examples of the stabilizing role of weak links. Weak links were forces or chemical bonds in these cases. Moreover, weak links were shown to be efficient at smoothing the energy landscape and therefore facilitating the transitions of the whole network from one state to another. Our next journey will take us one level higher. Proteins, which were the networks in this chapter, will form the elements of the networks in the next. What is the network then? The first system which reached the complexity we call life: the cell.

6 Weak Links and Cellular Stability

This will be our second journey into Netland. We are going to visit cellular networks. I first introduce various cellular networks and continue with a wide variety of proteins which may all modulate the stability of our cells. In the last two sections, I also explain the role of stabilizing proteins in evolution, cancer and other diseases, and aging.

6.1 Cellular Networks

Cells are built from a wide variety of molecules. All these molecules build networks. Some of their interactions are largely unspecific and can be described in general terms. An example is the self-association of lipids to membranes. However, most of the interactions between cellular molecules are unique and specific, and require a network approach for a detailed description. One of the best examples for the description of unique cellular interactions between molecules is the protein–protein interaction network (see Fig. 6.1), where the elements of the network are proteins and the links between them are permanent or transient bonds. At a higher level of complexity, we have networks of protein complexes, where individual elements may be regarded as modules of the large protein–protein network. The cytoskeletal network and the membrane–organelle network are good examples of these larger networks. In the cytoskeletal network, the elements of the network are individual cytoskeletal filaments, like actin, tubulin filaments, or their junctions, and the links between them are the bonds. In the membrane–organelle network, various membrane segments (membrane vesicles, domains, rafts, of cellular membranes) and cellular organelles (mitochondria, lysosomes, segments of the endoplasmic reticulum, etc.) are the elements, and protein complexes usually link them together.

Both the membranes and the organelles contain large protein–protein interaction networks (Sőti et al, 2005).[1]

"Something is not clear to me here. Are biochemists playing around when defining such overlapping networks? Why is a protein–protein interaction network not good enough to describe the whole cell?" The easy answer would be that we have other molecules than proteins in the cell. However, the real reason is a bit more complex. When we speak about protein–protein interaction networks, we never mean the interaction of this little green protein A on the right side of the cell with that larger pear-like protein B sitting beside it. The interaction of protein A and protein B in the network sense means that there is a good chance that we will find their complex in the given type of cell. In networks, the notions 'protein A' and 'protein B' refer to a population of these proteins, and not to individual proteins A or B. In contrast, many of the other networks which have larger protein complexes as elements are defined as networks of unique partners, where the interaction of mitochondrium A-245 with mitochondrium A-312 can be traced and monitored continuously by modern techniques of cell biology.

I am sorry to have to tell you, Spite, that the situation is even more complex with cellular networks than was suggested in our previous discussion. There are a rather large number of these networks in which the elements and links are functionally defined (see Fig. 6.1). As an example, in signaling networks the elements are proteins or protein complexes and the links are highly specific interactions between them which undergo a profound change (either activation or inhibition) when a specific signal reaches the cell. In the metabolic networks, the network elements are metabolites, such as glucose, or adenine, and the links between them are the enzyme reactions which make one metabolite from another. Finally, the gene transcription network has two types of element, namely, transcriptional factor complexes and the DNA gene sequences which they regulate. Here the transcriptional factor complexes may initiate or block the transcription of the gene's messenger RNA. The links between these elements are the functional (and physical) interactions between the proteins (sometimes RNAs) and various parts of the gene sequences in the cellular DNA. As we begin to learn more about the molecular composition and regulation

[1] Here the examples (actin, tubulin, vesicles, rafts, lysosomes, etc.) are just given for those who are familiar with cell biology. It is not necessary to understand them here, since the basic concepts of this chapter can still be followed. I therefore ask the reader coming from a different background to skip these examples, or consult cell biology textbooks or publicly available glossaries for their explanation.

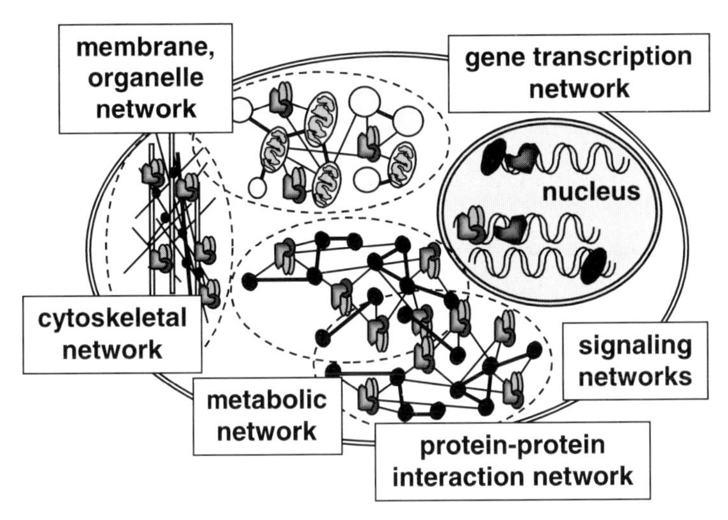

Fig. 6.1. Cellular networks. The figure illustrates the most important networks in our cells. Many of these networks, like the transcriptional or metabolic networks, are functionally defined. In the metabolic network, the various metabolites are the elements and the enzyme reactions are the links. In the gene transcription network, the transcription factors and genes are the elements and functional interactions between them are the connecting links. Other networks, like the protein–protein interaction network or the cytoskeletal network, have proteins as elements, and permanent or transient bonds between these proteins as links. All these networks overlap one another considerably, and some of them (like the membrane–organelle network) contain modules of other networks (e.g., the organelle specific modules of the protein–protein interaction networks)

of our cells, the definition of more and more networks becomes feasible. A sign of this is the fact that, recently, more and more specific transcriptional networks have been defined for various physiological events, such as development or aging, and for various human diseases.

Cellular networks have all the major network properties described in Chap. 2. They are often small worlds, with a scale-free degree distribution and a motif-rich, modular and hierarchical structure (Almaas et al., 2004; Bergmann et al., 2004; Bortoluzzi et al., 2003; Chen et al., 2003; Jeong et al., 2000; Park et al., 2005a; Ravasz et al., 2002; Stuart et al., 2003). However, when examining the above features in the context of cellular networks, it is important to scrutinize the validity of the dataset use a suitable sampling procedure and data analysis, as already described in Chap. 2 (Arita, 2004; Ma and Zeng, 2003; Tanaka, 2005).

Can we make an exact copy of ourselves? *"As you were listing all the interesting features of cellular networks, I kept asking myself: when we know the position of all the proteins in the cell, will we just be able to put them together and get another cell which is identical? Going even one step further, if it was a nerve cell and we copied this procedure with many of them, could we arrive at a cloned brain, which has the very same thoughts as the original?"* This is very good question, Spite. Congratulations! I seriously doubt whether we will ever be able to do this. The first objection is that we cannot yet isolate all the protein molecules, set their modification patterns and position them exactly as they were in the 'sample' cell. However, there is a much more serious objection than this. All these networks are self-organized networks. This means that they have a memory. A cellular protein may have several possible neighbors, but it will only bind to some of them. What determines which of the possible neighbors will be the real one? All the past events between these network elements determine their current interactions. Moreover, the events which changed the interactions of these elements with the rest of the network also influence complex formation between the proteins. Hence, for a correct copy we have to repeat many of the past events in the process of network self-organization, which makes the whole adventure quite impossible, not only in technical terms, but even theoretically.[2]

6.2 Stability of the Cellular Net

After a brief introduction to the various cellular networks, I will describe a landmark experiment which gave insight into the emergent properties of cellular networks. In 1998, Suzanne L. Rutherford and Susan Lindquist published ground-breaking results in Nature, showing that compromised chaperone function leads to the appearance of silent mutations in *Drosophila*. Molecular chaperones, or in other words, heat shock proteins or stress proteins are highly abundant proteins that have been conserved throughout evolution. Why are these proteins known by so many different names? The name 'chaperone' refers to their function. Chaperones are the physicians of the cell. They heal other proteins. As mentioned in the Preface and in the last chapter, chaperones form the most ancient defense system of our cells. They recognize both half-ready and damaged proteins, prevent their aggregation, and help them to complete their folding process or to re-fold. This is why chaperones are also called stress proteins. If our cells suffer any damage, the cellular proteins also become

[2] I am grateful to Gergely Hojdák for this question.

damaged. Therefore, we need more chaperones after stress. In agreement with this, chaperone synthesis is up-regulated in damaged cells. This happens in parallel with the inhibition of most other protein-synthetic events. The reason for the general no-go sign to protein synthesis is that it is very costly. If the cell is in danger, its energy reserves are low. A stressed cell is a very utilitarian hospital, where only doctors and nurses get their breakfast, lunch and dinner.[3] After stress, more or less only chaperones are synthesized in the cell. This is why they are called stress proteins. Heat shock is the archetypal stress, and heat shock was the damage under investigation when Ferruccio Ritossa discovered stress protein synthesis in 1962 (Hartl, 1996; Bukau and Horwich, 1998; Csermely et al., 1998; Ritossa, 1962).

What is stress? The word 'stress' was coined by Hans Selye (1955; 1956). In this book, I use a definition from the context of the cellular net. Stress is any unexpected, large and sudden perturbation of the cellular network, to which the network (1) does not have a prepared adaptive response or (2) does not have time to mobilize the adaptive response. In this book stress is used differently from stress in the usual sense in physics, where it is a force that produces strain in a physical body. I should note that this definition is very close to the relaxation problems mentioned in Sect. 3.2. If the perturbations are of unusual type, or if they are too big, or arrive one after the other in too great a number, the network may be in trouble. If the network has time, it will be able to remodel itself. This is called an adaptive response. However, if the perturbations come too fast, the network has to mobilize its general defense, the stress response.[4]

[3]In the hospital of the stressed cell, porters may also get some food, since the unequal ion balance must be maintained at the plasma membrane.

[4]Cells behave quite similarly to families, when they put grandma's silver, Tante Sissi's china and any other family treasures into a safe place in the hope of preserving a few resources for recovery after damage. Tante Sissi – Aunt Sissi – was our neighbor when I was a child. She was already unthinkably old when I was born, and as the widow of a high-ranking officer in the Austro-Hungarian monarchy, which had become obsolete half a century before, taught me German with a genuine Austrian accent every Wednesday. The highlight of these sessions was an original Meinl tea, which I had to welcome with the appropriate enthusiasm for such a rarity. Tante Sissi will appear on several occasions in later sections of the book to illustrate social behavior.

More on the name 'chaperone'. The word 'chaperone' comes from French and was coined by John Ellis (Hemmingsen et al., 1988) after Ronald Laskey used this expression for a histone chaperone, nucleoplasmin (Earnshaw et al., 1980). Originally, chaperones were old ladies who accompanied beautiful young girls to the grand ball many years ago. As mentioned above, cellular chaperones prevent the aggregation of proteins. Aggregation is an unplanned, rather tight interaction of two partners in which they stick together. Similarly to the old ladies who went to the ball to protect young girls from this kind of unplanned, tight interaction, their cellular counterparts do the same with the inexperienced and naïve proteins when they run into trouble.

Let me return to 1998 when Suzanne Rutherford bred her 10 400 fruit flies. All of them were sons and daughters of funny couples. One of the parents was normal. However, the other had a mutation in Hsp90 which compromised the function of this chaperone.[5] 10 226 flies looked normal. However, 174 of them were miniature Frankenstein monsters. Eyes were missing, distorted or repositioned, wings grew deformed, legs were transformed, bristles were duplicated and halteres were crippled. Altogether 23 types of malformation were catalogued with a minimum occurrence in 3 and a maximum in 48 flies (Rutherford and Lindquist, 1998).

What can be the reason for these malformations? Similarly to many other chaperones, Hsp90 is known to participate in embryogenesis (Csermely et al., 1998). The first explanation which comes to mind is rather straightforward: damaged Hsp90 derailed embryonic development. However, such an effect should derail the development of far more than just 174 fruit flies out of the 10 400. Moreover, many of these malformations were inheritable even after a transient inhibition of Hsp90. Exactly the same type of malformation was observed in the grandchildren of 9 monster types. This suggested a genetic background, which brings us to the next possible explanation: Hsp90 inhibition increases the mutation rate. The mutation rate was carefully checked and many other experiments were also performed. All suggested that the mutations causing the distortions were originally present in the *Drosophilas*. However, they were not visible.

These silent mutations affected the phenotype only if Hsp90 was inhibited. Once silent mutations got exposed, they destabilized the cells

[5] In the expression Hsp90, Hsp stands for heat shock protein and 90 refers to the molecular weight of this chaperone, which is 90 kDa.

Fig. 6.2. Mutations causing the distortions were originally present in the *Drosophilas*. However, they were not visible

involved in development and led to their increased diversity, inducing a larger number of unexpected developmental phenotypes. A new conclusion was born: the Hsp90 chaperone buffers the effects of silent mutations to induce diversity in developmental morphology (Rutherford and Lindquist, 1998).

Mutations keep their silence in many different ways. In describing the findings of Rutherford and Lindquist (1998), I used the expression 'silent mutation'. Mutations can be silent in many ways. Here, silence means that, although the mutation causes a phenotype change at the level of the organism, this change is concealed by the buffering effect of chaperones. There are mutations which are more silent than those concealed here, since their presence does not lead to any change in the phenotype. This second type of silence is a permanent silence, as opposed to the conditional silence of chaperone-buffered mutations. Permanent silence is rather widespread due to gene duplications and degenerate pathways in various networks, which efficiently substitute the diminished or missing function.

"Now I understand the chaperones, and even the sudden outbreak of the monstrous Drosophilas. But you still owe me something. Where is the cellular net?" It is coming soon, Spite! We are almost there. First I would like to make a few extensions of the original statement:

1. Is *Drosophila* the only species where chaperones stabilize the effects of silent mutations? No, *Drosophila* is not unique. Chaperone-induced buffering has been found in other species, such as the plant

Arabidopsis thaliana or the bacterium *Escherichia coli* (Fares et al., 2002; Queitsch et al., 2002).

2. Is Hsp90 the only chaperone to stabilize the effects of silent mutations? Hsp90 is not alone in buffering developmental diversity. Over the past few years the findings of Rutherford and Lindquist (1998) have been extended to other chaperones like Hsp60 and Hsp70 (Fares et al., 2002; Roberts and Feder, 1999).
3. Are chaperones the only proteins to stabilize the effects of silent mutations? Chaperones are not unique in buffering developmental diversity. There are several other proteins which increase the morphological diversity of the *Drosophila* phenotype (de Visser et al., 2003; Gibson and van Helden, 1997; Gibson and Wagner, 2000; Scharloo, 1991). The generality of the effect has been addressed by Wilkins (1997). Moreover, Bergman and Siegal (2003) showed by modeling experiments that the deletion of many proteins both from a model network and from yeast cells may cause an increase in phenotypic diversity. These data warn us that there are many more proteins than just chaperones buffering phenotypic diversity.

Here we have to stop for a while. An exciting question arises: how do all these proteins buffer phenotypic diversity? While it was just a chaperone, we believed we knew the answer: the chaperone deals with the mutant proteins, preventing them from causing trouble. When stress comes, all types of mutants are released and they start to cause various unexpected effects and interactions leading to increased diversity in the next generation. However, this very same mechanism clearly cannot work with all the proteins uncovered. While some of the non-chaperone proteins like those participating in various cellular signaling steps may affect developmental diversity in a direct and rather specific way, there may also be a general effect here which should not be linked to the specific function of the individual proteins. Spite, listen! Here comes the cellular net. The network approach seems to be rather effective in keeping the specific features of the network elements (here the various proteins) in the background and concentrating on their position in the broader context, throughout the whole network, i.e., the cell.

Here is one more thought before really launching into the cellular network.[6] As mentioned in Sect. 3.1, the development of diversity is

[6] I know, this chapter is beginning to look like a dummy version of a Beethoven symphony: whenever you start to think: "Now he's really getting to the point of it!", a new development is inserted. But please bear with me. The coda will be quite good, I promise.

linked to noise. Higher noise brings greater diversity. Those mutations made the poor *Drosophilas* not only monstrous, but also noisier.[7]

Cellular noise and diversity. Kuznetsov et al. (2002) found that the messenger RNAs of at least 45% of yeast genes are present in less than 1 (less than one!) copy per yeast cell on average. Many genes are switched on sporadically, producing messenger RNAs in occasional pulses present in only a few cells. An individual cell may contain only a few molecules of several key proteins, which are distributed in various segments of the cell. Low reactant numbers can lead to significant statistical fluctuations in molecule numbers and reaction rates, producing a rather high level of noise and diversity (Rao et al., 2002). Moreover, in both prokaryotes (Elowitz et al., 2002; Ozbudak et al., 2002) and eukaryotes (Blake et al., 2003), it has been shown that an increase in the noise of gene transcription leads to an increased diversity of final transcripts, and hence to an increased diversity in phenotype (McAdams and Arkin, 1997; Levin, 2003).

We have finally arrived at the cellular net. We will use it to see if there might be a common mechanism behind the role of all those proteins inducing more noise and diversity. What do we know about the network properties of chaperones, which have the best established and most general effects on phenotypic diversity among the wide variety of proteins? Chaperones are typical weak hubs in the cellular protein net since they form a multitude of low affinity interactions with a large number of other proteins and with each other (Kovacs et al., 2005). Chaperones are forced to leave their weak links during stress. What happens if the system starts to lack weak links? Is there any connection between weak links and noise? Yes, there is: long-range or intermodular weak links make the system less noisy. Weak links help relaxation and increase system integrity. Let me sum up all of this: if we disturb the weak links of chaperones, the cellular integrity is decreased, the relaxation of perturbations also decreases, the cellular noise increases, and local tensions develop, which may all significantly contribute to the development of the observed diversity. This brings me to state the following hypothesis:

[7] The noise is cellular noise here, but one never knows: even their mutant buzz may be noisier, repelling the Beethoven-loving *Drosophila* members of the other sex.

Weak links of the cellular protein network buffer noise and diversity. Besides their specific effects, chaperones and other proteins may also buffer developmental diversity by forming a large number of weak links in the cellular protein network. The increase in diversity goes in parallel with an increase in the cellular noise level. As an additional support to these ideas, Tsigelny and Nigam (2004) recently demonstrated that chaperones decrease the noise of protein folding in a model system.

The effect of chaperones agrees with the functional definition of weak links. Let me reiterate here the functional definition of weak links from Sect. 4.2: a link is defined as weak when its addition or removal does not change the mean value of a target measure in a statistically discernible way. When he gave this definition, Berlow (1999) also stated that the removal of these weak links would increase the variation and noise in the system. We do have an agreement here. The incapacitation of the Hsp90 chaperone did not change the *Drosophila* population in a statistically discernible way, since 174 divergent monsters did not make any significant changes in the mean parameters of 10 400 *Drosophilas*. However, the variation did increase. Similarly, the increased diversity caused by the non-chaperone proteins listed above was usually also confined to a smaller segment of the population. Moreover, the above definition of weak links is also in agreement with the data of Bergman and Siegal (2003), who selected a gene set which had no significant change in its expression after knocking out 53 yeast genes (Hughes et al., 2000). Examining the variability of gene expression of the same selected gene set, Bergman and Siegal (2003) found that it was increased after the deletion of the 53 yeast genes, in agreement with the experiments with molecular chaperones.

Are there any other proteins which may be good candidates both for making a large number of weak links and showing a capacity to buffer cellular diversity?

- **The p53 tumor suppressor as a buffer of diversity.** One of the possible candidates to buffer cellular noise and diversity is p53, a transcription factor involved in the proper control of the cell cycle, which suffers various debilitating mutations in many types of cancer. It has long been regarded as a highly-connected node in the cellular network (Vogelstein et al., 2000) and as a buffer for developmental noise as well (Aranda-Anzaldo and Dent, 2003). The potential role for p53 to buffer cellular noise is in agreement with the presumably increased noise during malignant transformation.

- **Prions as diversity generators.** Prions are peculiar proteins. They have two conformations, the normal and the infectious. The energy levels of the two prion conformations are separated by a high activation energy. This allows the conversion of normal prions to infectious prions only under specific conditions. One of these conditions can be the presence of an infectious prion, which catalyzes its own formation from normal prions. The conformational difference gives prions a molecular memory and serves as a source of epigenetic inheritance (Uptain and Lindquist, 2002).[8] In the case of yeast prions, the infectious [PSI$^+$] prion form makes ribosomes to skip the termination of protein synthesis at the stop codon of the messenger RNA in approximately 0.2 to 16 percent of cases. This will give rise to proteins extended at their C-terminus, which may bear a new function. The normal prion form is spontaneously converted to the infectious [PSI$^+$] form in approximately one of a million yeast cells. Therefore in a regular yeast population some of the members 'automatically' acquire a new phenotype. This phenotype might be eliminated from the population or fixed by mutations in the skipped stop codons (True and Lindquist, 2000; True et al., 2004; Wilson et al., 2005).

Protein aggregates as noise generators in neurodegenerative diseases. Infectious prions form protein aggregates (Uptain and Lindquist, 2002).[9] Prion aggregates display great diversity (DePace and Weissman, 2002), making them a good candidate for a noise generator. Moreover, aggregation itself increases noise and diversity, since chaperones and other proteins co-aggregate with prions and other aggregates. This may preferentially break the original weak links of the protein network of the cell, since proteins involved in strong links cannot easily be removed from their original position and so cannot easily be captured by the growing cellular aggregates. Besides infectious prions, there are numerous other forms of ex-

[8]We call inheritance epigenetic if the inheritable property is not transmitted via a DNA sequence, but is inherited with the help of other molecular mechanisms. Such a mechanism may use the modulation of DNA accessibility by DNA methylation or histone modification. Epigenetic molecular mechanisms also include RNA- and protein-based inheritance.

[9]The enhanced aggregation is what makes infectious prions dangerous for nerve cells in the sheep disease scrapie or in its human forms like the Creutzfeldt–Jakob disease. Aggregation may capture essential proteins and it disturbs cell life in many ways as described here.

tensive protein aggregation which are most characteristic of neural cells and cause severe neurodegenerative diseases, like Alzheimer's and Parkinson's diseases. Cellular noise may also be increased in these disease states, preventing the proper function of the affected nerve cells.

Ways of optimizing noise and diversity. As already mentioned in Sect. 3.1, we require an optimal level of noise to obtain stochastic resonance. In this chapter we have seen that we need an optimal level of diversity to survive. The optimization has to be finely tuned, since if buffering were too low or too high, silent mutations would never accumulate or get released, respectively. How are noise and diversity optimized at the cellular level? As discussed, chaperones help damaged proteins to refold, and hence promote cell survival. Consequently, a wise cell wants to be soaked in chaperones to obtain her bonus for life. But this would be a bad strategy. Noise and diversity cannot be dampened beyond a certain level. In other words, a lot of weak links are bad for your health! (Ask your physician or pharmacist!) In agreement with this, chaperone levels are tightly regulated, not only from the bottom, but also from the top. An increased chaperone capacity leads to a decrease in chaperone levels (Dressel et al., 2003; Feder et al., 1992; Gülow et al., 2002; Rubenstein and Zeitlin, 2000). In summary, it seems that if noise becomes suboptimal, then buffering gets reduced. We might have another solution to restore the balance between noise and buffering. If buffering cannot be decreased, increased noise generation might also help. As a possible example of this, yeast prions are much more frequent in laboratory strains than in natural or industrial isolates (Pal, 2001). Laboratory strains of the yeast fungi possibly enjoy stress levels several orders of magnitude smaller than those observed by wild-type yeast strains. Noise becomes suboptimal, and the noise-generating prions become over-expressed.

Aggregation as a friend: molecular crowding reduces noise. To show the complexity of life, let me describe another thought concerning aggregation. All the aggregation phenomena mentioned so far have been extensive, avalanche-type, irreversible processes, which produce strong links and capture various regular members of the cellular protein net, forcing them to leave their original weak links. These aggregation processes are most probably noise generators. However, we might have another type of unplanned protein association. If the association of proteins is a finely tuned, reversible process, than it results in numerous novel weak links which may actually decrease cellular diversity. If the cell can slightly modify its conditions, preferring the slow, self-organizing development of weak links, this would serve as an ideal tool for noise regulation. This is a typical slow adaptive response, as opposed to the stress response which is always produced in a hurry. What are the cellular tools for the regulated development of weak links? In this chapter I have mentioned quite a few ways, such as chaperones, prions and other proteins, to fulfill this requirement. An additional mechanism is related to water. Water-mediated weak links are highly efficient at stabilizing protein structure (Kovacs et al., 2005). Water may act at the cellular level as well. However, its action here is completely different from that mentioned before. Cellular water content cannot be reduced to an extent that would seriously affect the weak links of protein structures. If cellular water is just slightly decreased, molecular crowding occurs (Hall and Minton, 2003). Here the 'free water' between proteins gets reduced, creating better conditions for low affinity protein association which develops weak links. It is quite likely that the cell learned to regulate its water content and, consequently, the level of molecular crowding as a finely tuned adaptation to varying conditions. [After writing this remark, I learned about the modeling results of Morishita and Aihara (2004), showing that molecular crowding may indeed reduce the noise of gene expression.]

Protein surface: a fractal for attachment optimization. *"In spite of your prediction of the Morishima and Aihara (2004) paper, something is not clear to me here. You said in Sect. 5.1 that proteins were like nano-LEGO. Their vastly different amino acids never fit together. And what happens now? You 'predict' that whenever they have a small chance of coming closer, they will just bind to each other. Do you not feel that there is a gap here?"* You are right, Spite. If proteins meet just by chance, they cannot make strong links. However their surface is just optimal to make low affinity interactions, i.e., weak links. Proteins have a self-similar, fractal surface. The fractal dimension which denotes the exponent of the scale-free distribution of self-similar elements of the protein surface is variable in individual protein regions. This means that protein surfaces are multifractal. This allows a good opportunity for the loose accommodation of a wide variety of partners (Lewis and Rees, 1985). Moreover, once two proteins get close to each other, water and chaperones help both their positioning and mutual conformational rearrangements (Kovacs et al., 2005; Liu et al., 2005).

Weak links can be fairly general in the protein net. In Sects. 2.4 and 4.2, I mentioned that most links are weak in a scale-free network. However, the link strength distribution of cellular protein–protein interactions has not yet been analyzed. Are there any data to support a large amount of weak links in the cellular protein net? Annotated protein–protein interaction databases, e.g., that of von Mering et al. (2002), contain relatively few 'highly reliable' interactions and a vast number of 'unreliable' interactions. While many of these 'unreliable' interactions may indeed be artifacts, quite a number of them may represent low-affinity, weak links between proteins. Kinetic measurements of seemingly stable cellular complexes, such as membrane rafts, the cell nucleus, or clathrin skeleton, showed the surprising flexibility of these complexes. As an example, a core histone stays for only a few minutes in the tightly packed DNA nucleosome. This crazy flip-flopping may only occur if a large amount of the protein–protein links are in fact weak (Kenworthy et al., 2004; Misteli, 2001; Wakeham et al., 2003).

Are unstructured proteins noise buffers? Recently, a large number of proteins have been identified which contain shorter or longer sequences with no secondary structure. These sequences participate in many low affinity protein–protein interactions (Dunker et al., 2002; Tompa, 2002; Uversky, 2002; Wright and Dyson, 1999). Chaperones are unstructured proteins (Tompa and Csermely, 2004) which buffer noise and diversity. Whether unstructured proteins generally buffer cellular noise and diversity is an exciting question for future study.

Let me summarize what we have said so far. Cellular diversity is regulated by a complex array of noise buffers and noise generators. Chaperones and p53 help to decrease, while prions and other protein aggregates probably increase cellular noise and diversity. Both mechanisms might involve large changes in the configuration and the level of weak links in the cellular protein network. Are protein–protein interactions the only way for cells to stabilize themselves? Obviously not. There are many thousands of specific regulatory elements, like the negative feedbacks in genetic networks (Becskei and Serrano, 2000) described earlier. Besides these, modules of protein networks like cellular organelles may also play an important role in cellular stabilization.

Organelle diversity as a stabilizer of eukaryotic cells. Recent data indicate that mitochondria form a network which is able to show collective phenomena, such as synchronization, topological phase transition, etc. This significantly increases the speed with which a signal, like that of reactive oxygen species, can travel through a large cell, such as a cardiomyocyte spanning 0.1 mm (Aon et al., 2004a). The existence of a well-coordinated mitochondrial network and the established links between other cellular organelles, such as mitochondria and the endoplasmic reticulum, the endoplasmic reticulum and the cell nucleus, etc., makes the presence of a cell organelle network possible inside eukaryotic cells. Subcellular organelles are not identical. They have qualitatively and quantitatively different components, different environment, different damage, different ages, etc. However, they do have a similar function, and they are therefore also linked to similar elements of the cellular net. This complex similarity and diversity pattern enables the emergence of several weak links between the individual organelle and other elements of the cellular net. Most probably, these weak links also help to stabilize cellular functions.

As we get close to the end of the first part of our trip amongst the cellular networks, I must make two things explicit. I always mentioned cellular stability, noise and diversity in general terms instead of saying, for example, that chaperones buffer developmental noise by acting on the protein network of the cell. This implies two generalizations:

- Chaperones and all the proteins and organelles mentioned above act on all cellular networks including protein, genetic, metabolic, cytoskeletal and organelle networks (Sőti et al, 2005).
- The second generalization means that chaperones and all the above mechanisms regulate not only developmental but presumably all types of diversity.

However, this leads us to the next section, since we need stress (Rao et al., 2002) to provoke the diversity of diversities.

6.3 Stress, Diversity and Jumps in Evolution

The story of our second trip into Netland is almost complete. We have got the following message: weak links may provide stability to the cellular net. It was also shown that a decrease in weak links brings greater noise and diversity to our cells. We have seen a number of potential molecular mechanisms. However, two elements are still missing:

- How and when are weak links decreased in natural conditions?
- What are the consequences of sudden increases in diversity?

Chaperone function may be compromised in several ways. Chaperones can be inhibited either pharmacologically or by introducing debilitating mutations into them. However, the most important inhibition is competitive and occurs during stress. Stress produces a large number

Fig. 6.3. Stress is any unexpected, large and sudden change in the life of the network, to which the network does not have a prepared adaptive response or does not have time to mobilize its adaptive response

of damaged proteins which all compete for the same set of chaperones. Chaperone levels do increase under stress, but damaged proteins may easily outnumber the available chaperones and cause a chaperone overload (Csermely, 2001b).

What happens in the cellular net if stress occurs? The cellular resources become exceedingly sparse and the cell has to concentrate all its energy to maintain the most important pathways. Everything which is not absolutely necessary is stopped. Weak links are released and do not reform again. However, the disappearance of weak links makes the system unstable. As a specific point, stress induces the reconfiguration of protein networks around chaperones by cutting their existing weak links. How is this done? Chaperones become overwhelmed with damaged proteins and release the silent mutations they protected before. As a consequence the phenotype will display profound changes.

Disintegration of the cellular net during stress. As mentioned above, stress deciphers the original weak links of the cellular protein network leading to cellular instability. The stress-induced reconfiguration of the cellular net resembles the topological phase transitions described in Sect. 3.4. From the random to scale-free topology of our protein network under normal conditions, stress may shift the net to a star phase and further to a disintegration to subgraphs. This latter phenomenon, which breaks the netsistance (see Sect. 4.3) and implies the death of the cell, may actually occur during apoptosis. Apoptosis often accompanies severe cellular stress (Sőti et al., 2003). Apoptosis is programmed cell death where, in the final executive phase, special proteases called caspases destroy several key proteins of the cell (Sreedhar and Csermely, 2004). It is an exciting question as to whether caspase substrates were 'selected' to destroy hubs of the cellular net in such a way as to ensure a fast and lethal disintegration of the cellular networks.

Stress increases population diversity, not only in development, but also in a surprisingly large number of cellular features, as summarized in the seminal review by Rao et al. (2002). A summary of the various forms of stress-induced diversity is given in Table 6.1.

How does stress induce all this diversity? The exposure of silent mutations described in detail in the last section is just one of the mechanisms which have been developed to increase diversity. Cell cycle progression, mitochondrial activity, oxidation and epigenetic regulation are all among the large number of events which will undergo variable changes during stress. Stress mobilizes several pathways to increase variability in the genotype, such as decreased fidelity in DNA replication, mutations, recombination, etc. (Radman et al., 2000). As an additional mechanism, an over-representation of repeats in stress-response genes has also been observed, which may induce phenotypic variability due to genetic recombination or slip-induced mispairing of bases (Rocha et al., 2002).

We have reached another point on our trip where we may take a rest. We have learned that stress induces a great diversity in an astonishingly large number of system properties. We have also learned that there are many mechanisms behind this. When we sit down for a while, we may start to think: what is missing? Just the essence is missing. Why is this good?

To give the answer let me return to the studies by Rutherford and Lindquist (1998) and repeat their conclusion here: Hsp90 buffers the phenotype of silent mutations. As I mentioned, chaperones (including Hsp90) get saturated by damaged proteins during stress, silent mutations became exposed, and diversity develops. This connects the results of Rutherford and Lindquist (1998) with evolution. If a larger stress comes, the *Drosophila* destiny has two ways to go:

- **Genome cleansing.** The normal *Drosophila* population survives. Silent mutations cause malformations and the crippled *Drosophilas* die out from the population, either directly or by natural selection due to their mating disadvantage. In both cases genome cleansing will occur and the *Drosophila* population may start to collect the

Table 6.1. Stress-induced diversity in various complex systems. As already mentioned in a general context in Sect. 4.1, due to the deductionist mainstream scientific approach, individual responses under stress have received little attention until recently. The list here is therefore far from complete. Most of the data, including those where no direct reference is given, are from the excellent reviews by Booth (2002), Himmelstein et al. (1990) and Rao et al. (2002)

Source of diversity	Reference
Prokaryotic cells	
Bacterial stress unpredictably activates lysis of bacteria by their pathogen, the lambda phage	Arkin et al., 1998; Vohradsky, 2001
Stress induces the death of individual bacteria at various times	Lewis, 2000
The doubling of different bacterial cultures varies under stress	Plank and Harvey, 1979
Stressed bacteria move towards their food with highly variable speeds and mechanisms	Alon et al., 1999; Levin, 2003; Spudich and Koshland, 1976
Stressed bacteria develop various numbers of small DNA fragments, called plasmids	
Pathogenic bacteria show a great variability in the different segments of their life cycle when stressed	
Stressed bacteria switch from one survival strategy like spore formation to quite another, rather unpredictably	
Eukaryotic cells	
Stress induces a large variation in the speed and selectivity of protein transport	Simon et al., 1992
The doubling of different cell cultures varies upon stress, together with the length of different parts of the cell cycle	Brooks, 1985
Stressed stem cells differentiate to a much more diverse set of differentiated cells and the speed of differentiation also varies greatly	Mayani et al., 1993
Heat shock induces the death of individual cells at various times	Yashin et al., 2002
The regularity and speed of the polymerization of actin, the major constituent of the cellular cytoskeleton, varies a lot in stressed cells	van Oudenaarden et al., 1999
Stress brings a great variability to coordinated gene transcription; both the onset and the extent of transcription varies	

Table 6.1. Cont. Stress-induced diversity in various complex systems

Source of diversity	Reference
Eukaryotic tissues, organisms	
Stress induces a great variability in the phenotype of the next generation	Rutherford and Lindquist, 1998; Queitsch et al., 2002
Stressed plants show a great variability in their regeneration from tissue culture, both in speed and in final shape	Finnegan, 2001
The flowering time of stressed plants shows a great variation	Finnegan, 2001
Certain tumors show a highly variable speed of development if the host becomes stressed, both at the primary tumor level and in metastasis development	Cook et al., 1998; Kemkemer et al., 2002
Artificially induced genes (transgenes) show a large variability in their expression and final effects in stressed mice	Elliott et al., 1995
Blood vessel formation shows a great variability both in speed and in the structure of the vessels in stressed animals	
The immune response of stressed animals becomes much less predictable than that of control animals	

new silent mutations again after the stressful event. This is the fate of the *Drosophila* gang in your orchard on an unusually hot summer's afternoon. However, on certain rare occasions, they may undergo a big enough stress for another outcome.

- **Evolutionary jump.** The *Drosophila* cannot survive the stress. If they had no diversity reserves in the form of silent mutations, all of them would die. This is THE END, folks. The last *Drosophila* may buzz a requiem, and after a decent decline, bring the whole species to the graveyard. However, if the effect of silent mutations can engender unprecedented diversity, one (some) of the stress-induced phenotypes may be able to survive under the drastically changed conditions. A charming *Drosophila* lady, e.g., with one eye on her hind leg and an excellent view of the hideously dangerous *Drosophila* back-catcher as it seeks its dinner for the day, will mate with a handsome guy, also with an eye on its hind leg, and become the Founding Mother of the new *Drosophila* population, to be remembered forever. I have to make the point explicit because you might not have got it yet: an evolutionary jump[10] has happened. (Actually, the eye jumped from the front to the hind leg, but this is not the important part of the story.)

Since the time of Charles Darwin (1859), the mechanism behind evolutionary jumps has remained a rather unexplained phenomenon. Wings do not develop gradually: first the forelimb is extended; then it turns a bit backwards; then it grows even longer. In the next step it begins to be covered by longer epidermal scales than usual. Then the scales are gradually transformed to feathers. After 150 years of debate we still do not know the exact chain of events explaining how wings actually developed (Dial, 2003). But we know something for sure: it was not as gradual as described above. Life simply does not work this way. Gould and Eldredge (1993) formulated the concept of punctuated equilibrium in 1972 to outline the theoretical background for evolutionary jumps.[11] However, the molecular mechanism was missing. Hsp90 gave the first clue to solve the puzzle as to how evolutionary jumps and punctuated equilibrium develop at the molecular level.

It is not buffering but the molecular mechanism itself that was new in the work by Rutherford and Lindquist (1998). The concept that developmental diversity can be buffered was introduced by the classic studies of Schmalhausen (1949) and Waddington (1942; 1953; 1959) who called it canalization. Hsp90 gave the first molecular explanation for the mechanisms behind canalization. Is the concept of cellular weak links novel? Sangster et al. (2004) already raised the idea that Hsp90 participates in the remodeling of protein networks. However, the assumption that Hsp90 may do this as part of the weakly linked elements of this network, and that other modulators may also behave as capacitors of evolution using this molecular mechanism was suggested later (Csermely, 2004; 2005).

Knowing the molecular mechanisms, it is not so surprising that the probability of evolutionary jumps can be regulated. Increased buffering causes fewer jumps. Conversely, less buffering leads to more jumps. Earl and Deem (2004) showed that the propensity for faster or slower evolution, which is called evolvability, can be selected. Their findings

[10] In strict terms, the expression 'evolutionary jump' is restricted to really big changes in evolution which obviously did not occur from one generation to the other.

[11] For speciation, however, the change requires a reproductive isolation in space (Gould and Eldredge, 1993).

Fig. 6.4. An evolutionary jump had occurred

probably imply the regulation and selection of a number of molecular mechanisms during evolution.

1001 ways to adjust evolvability. Evolvability can be regulated by a number of mechanisms. Using the excellent summary by Molnar (2002), let me list some of the possible options here:

- **Weak-link-induced buffering.** As discussed above, weak links of the cellular network can adjust the level of buffering. A lower ratio of weak links may lead to a higher evolvability.
- **Redundancy and degeneracy.** As shown in Sect. 4.5, redundancy and degeneracy both stabilize networks and increase the number of weak links.

Decreased redundancy and degeneracy leads to higher evolvability (Willmore et al., 2005).

- **Noise generators.** On the other side of the coin, noise generation can also be achieved at the level of cellular networks. The higher the noise, the higher the evolvability of the system.
- **Error-prone replication.** When the DNA is copied, the evolvability can be regulated by regulating the fidelity of the process. More errors lead to a higher evolvability of the system.
- **Recombination.** An increased recombination rate leads to a more frequent repositioning of DNA segments in chromosomes. This also increases evolvability.
- **Level and fidelity of repair.** Repair of random damage is a key point for information stability. However, a carefully tuned regulation of the level and the fidelity of this repair can be a key mechanism for the regulation of evolvability. Lower and poorer repair leads to higher evolvability.
- **Replacement fidelity.** Repair can affect not only parts, but whole modules, or elements of the bottom network. This repair is better called replacement. A good example is the replacement of damaged cells by the proliferation of stem cells. If the fidelity of the replacement is lowered, evolvability may grow.
- **Redundancy.** Genetic information is often highly redundant. Several organisms contain duplicates of the whole genome, multiple copies of genes are even more prevalent. Decreased information redundancy or blocked release of spare copies may accelerate evolvability.
- **Storage.** Fluctuating resource density may be counteracted by efficient storage and proper mobilization. Although seemingly less efficient than the other mechanisms, inhibition of storage and release may also increase evolvability.
- **Multilevel selection.** The various sources of multilevel selection (Lewontin, 1970) may also regulate evolvability. As an example, a less tightly regulated spontaneous abortion during human pregnancy may increase evolvability.
- **Feedback.** Inhibition of regulating feedback mechanisms may significantly help the development of higher evolvability. Many other network motifs, such as balanced activation–inhibition circuits, may also regulate evolvability.
- **Network topology.** Besides weak links, cellular networks offer a number of other features which may all regulate evolvability. Jordan and Molnar (1999) described the bridge structure of cellular networks as a 'reliability factor', i.e., a factor reducing evolvability. Indeed, if we delete bridges which may act as stabilizing 'cross-links' of the network skeleton, the underlying fractal pattern of cellular networks (Song et al., 2005a), we may increase evolvability.

The 1002nd way: stress may induce evolvability via the topological phase transition of cellular networks. Liberman et al. (2005) explored various evolutionary scenarios as a function of the underlying graph structure. While random graphs suppressed evolutionary selection, scale-free and star structures were potent selection amplifiers. If we compare these findings with the topological phase transitions of network structure detailed in Sect. 3.4, we may summarize that under low levels of stress, a random network structure is preferred, and this has an inherently low evolvability. As resources diminish and stress grows, the network is transformed into a scale-free and a star topology, which increasingly speed up the evolvability of the system.

Laboratory strains evolve faster. *"A recent report by Gu et al. (2005) showed that the usual laboratory yeast strain which was isolated approximately 70 years ago from a rotten fig has elevated evolutionary rates, when compared to a wild strain isolated recently from the lungs of an AIDS patient. As far as I know, laboratory strains should enjoy a low-stress environment. They receive plenty of food, the nutrient composition is usually the same, the environment is more or less sterile, and in most cases there is plenty of room for daily life. If it is a high level of stress which speeds up evolution, how could the laboratory strain evolve more quickly?"* This is a nice reference, Spite. I looked at it, too. My first remark is that the difference was only 15%. However, the difference was significant and does indeed demand an explanation. The laboratory conditions provide an extreme example. Due to the lack of stress, the selection pressure here is usually low, and this allows 'experimentation' with a number of changes which would die out in normal, stressful conditions. Laboratory yeast strains have had time to grow a large number of silent mutations.[12] You will be surprised, Spite, but notwithstanding the long stress-free periods from time to time, the laboratory strains are under greater stress than the wild-type strains. When no experiments are performed, yeast strains are often set aside and the number of surviving yeast cells becomes extremely reduced. These periods result in extreme population bottlenecks which are known to accelerate evolution. The earlier data on the higher prion content of laboratory yeast strains (Pal, 2001) may provide an additional explanation. The lab strains may have made an 'overshoot', subsequently producing an overcompensating increase in their inherent noise which led to faster evolvability.

[12] It is worth noting that, in the original Rutherford and Lindquist (1998) experiment, the laboratory strains had more than twice as many silent mutations as the wild-type strains.

Evolutionary continuity at the molecular level. The requirement for continuity in evolution has also been formulated at the molecular level. Maynard-Smith (1970) emphasized the functional continuity of consecutive mutations in protein evolution by introducing the concept of protein space. Chaperones like Hsp90 serve perfectly to bring this concept into question as well, since they may smooth the transition between the original wild-type protein and a stable final mutant by helping the unstable mutant proteins to fold. At the protein level, the phenotype (a new enzyme activity, a novel site to get new binding partners, etc.) may also be revealed in stress, when chaperones are forced to leave the crippled protein mutants alone. Chaperones help jumps in protein evolution and actually make punctuated equilibrium (Gould and Eldredge, 1993) of the protein evolutionary landscape possible.[13]

For optimal evolvability, we require stress to expose the silent mutations. However, we also need peaceful periods to let the silent mutations develop. We need a certain level of noise to help the cellular network to reach the next equilibrium in the stability landscape. However, if noise grows too big, the system cannot dissipate the flow perturbations, and becomes continuously unbalanced.

Stress-management helps evolution. Major transitions in evolution (Maynard-Smith and Szathmary, 1995) may have required a finely tuned series of gross perturbations to induce these real evolutionary jumps, and peaceful periods to allow relative stability for the gradual development of network symbiosis and complexity.

Noise management helps the evolution of multicellular organisms. Noise management may have helped the development of the complex transcriptional control of multicellular organisms. Prokaryotes need a high translation rate for a high growth rate, which is a prerequisite for their fitness to compete efficiently for available food and outgrow other bacteria that might want to eat the same food. High translation invokes high noise. In contrast, for multicellular eukaryotes, a fast cellular growth rate is a ma-

[13]The dual role of chaperones here, i.e., their weak-link-induced stabilizing effect at the same network level, allowing unstable proteins to survive, and their weak-link-induced stabilizing effect at the cellular level, resembles the effects of weak links in society, which will be described in Chap. 8. We will see that weak links stabilize *both* the society and its members who participate in these weak links.

lignant suicide. Cellular association requires a low translation rate, which is less noisy, and provides a 'noise space' for the development of complex transcriptional control, which gives an advantage to form multicellular organisms (Levine and Tijan, 2003).

We have almost reached the end of our trip through the cellular net. Time to look again at our luggage and ask what we have learnt. Stress-induced development of diversity may help population survival and may be an important element of evolutionary jumps. Part of the mechanism behind this is a decrease in weak links in the cellular net, which leads in turn to increased noise and diversity. Both noise and diversity have to be kept at an optimal level. Noise and diversity management is an important element of long-term survival. As an introduction to our own involvement in the worlds of noise and diversity, let me end this section with a few examples illustrating the importance of optimal noise management.

Mona Lisa had an optimal level of weak links. We need an optimal amount of weak links. Too many of them would make us rather insensitive. Too few would mean a dangerously high level of potential instability. Developmental asymmetry shows the level of stabilization at sensitive and important points during embryonic development (Kowner, 2001; Willmore et al., 2005). A large asymmetry means a devastatingly high noise. On the other hand, perfect symmetry leads to over-stabilization and little room for evolutionary flexibility. A highly asymmetrical face is perceived as a monster. A fully symmetrical face will raise suspicions. Such a face is a baby face, which is not the best. If you doubt this, try to copy the left half of your beloved's face with your computer and see if you like the result. If asymmetry is present but not too dominant, we consider the face to be beautiful (Perrett et al., 1994; Swaddle and Cuthill, 1995). Eye development is another well-known example of developmental instability, behaving like litmus paper for the weak-link optimum. I have sobering news. When you are immersed in the beautiful eyes of your sweetheart, what is happening is no more and no less than an assessment of the weak-link optimum. We are not alone. Female birds select their mates by the beauty of their songs. Song-learning excellence shows the presence of the same stabilization (Nowicki et al., 2002) as the appearance of the most beautiful face.

Successful politicians may have an optimal level of weak links. Mate selection by birds and humans may not be the only example where we use an assessment of weak-link content. The recent report by Todorov et al. (2005) shows that facial appearance may have an unexpectedly high impact on the final outcome of US elections. Congressional elections between 2000 and 2004 were mostly won by candidates who had a mature face as opposed to candidates with a baby face. In the light of the above findings, it would be very interesting to reassess the facial images and compare the level of symmetry. Personal and community preferences may have a common root.

6.4 Cancer, Disease and Aging

"This has indeed been quite an adventurous trip. I have seen crippled Drosophilas, networking mitochondria, a molecular crowd, even jumping evolution, but tell me, besides Mona Lisa and successful politicians, where does this leave us humans?" This section will be about us. As mentioned, chaperones are conserved throughout evolution. Chaperone-induced buffering is at work in you as you read this text (it may become somewhat compromised by the stress I have caused with some of my claims, but let me hope it is still functioning). I will review three situations when buffering becomes compromised and diversity develops in humans.

6.4.1 Cancer

Cancer cells live in constant stress. The tumor tissue in the suffering patient has gone through an evolution of months or years at best. Other organs are slightly better off, being products of evolutionary selection for times that are orders of magnitudes longer. Consequently, angiogenesis (blood vessel formation) is not part of the blueprint for most tumors. Hypoxia and nutrient deprivation of tumor cells are prevalent. These lead to increased glycolysis, a subsequent production of lactic acid, and the acidification of the cellular environment. Moreover, all these conditions vary abruptly due to the instability of microcirculation (Baish and Jain, 2000). In addition to all these factors, the immune system also attacks tumor cells and many types of cancer arise in a background of chronic inflammation, with permanently high immune activity (Loeb et al., 2003). Cancer cells do indeed live under constant stress. Besides the stepwise mutations of critical nodes in the protein network and the developing genetic instability (Hanahan and Weinberg, 2000), stress further enhances the instability, noise and diversity of malignant cells. Reish et al. (2003) have given a good example of the instability of cancer cells showing a decoupled replication of several cancer genes in the two strands of DNA. These genes would normally be copied synchronously to the new DNA strands.

Weak-link-induced stabilization may play an important role in protecting us from malignant transformation and from the development of the more aggressive, metastatic forms of tumors (Csermely, 2001b). However, the instability of cancer cells gives us an additional tool to fight against them, namely, the error catastrophe (Eigen, 2002; Orgel, 1963). If we diminish the residual stabilizing weak links in tumor cells, the instability may go to such an extreme that the malignant cell is killed. Not surprisingly, inhibitors of the typical weak linkers, molecular chaperones, have been successfully introduced as anticancer agents (Neckers, 2003). Destruction of weak links may be one of the rich mechanisms triggered by these multitarget drugs.

Pink noise against cancer. Pink noise may be especially good at provoking the error catastrophe in tumor cells. As mentioned in Sect. 3.1, pink noise occasionally involves very large fluctuations. These fluctuations may push cancer cells beyond the instability threshold, without necessarily breaking the stability of healthy cells in an irreversible manner, as white noise of similar magnitude would do. Therefore in each of the chemotherapeutic, radiation and hyperthermic protocols, the use of a pink-noise-modulated flux of the noxious agent may prove to be highly beneficial.

6.4.2 Disease

Stress and cancer lead to destabilization of our cells. In modern medicine, Hans Selye (1955; 1956) was the first to note that, in spite of the various causes and final outcomes of diseases, most patients have many similar symptoms. In fact, they are simply sick. What if cellular instability in terms of disturbed relaxation and increased noise is a general phenomenon of the 'sick state' (West and Deering, 1994)? In agreement with these assumptions, heartbeats, circadian and sleep-phase variations all become noisier in sick patients (Goldberger et al., 2002; West and Deering, 1994).

Weak link therapy. Combination therapy is a success story in recent medicine. Combination therapy leads to smaller side effects. Moreover, with combined medicines, less resistance develops against the treatment. These benefits are usually regarded as an effect of the smaller doses of each of the drugs one might use, compared with the dose of a single drug administered alone (Borisy et al., 2003; Huang, 2002; Sharom et al., 2004). This is obviously a great advantage. But what happens if we consider combination therapy, multitarget drugs (Morphy et al., 2004; Youdim and Buccafusco, 2005) and natural remedies in the context of the cellular net? These drugs all interact with their targets with low affinity. If they inhibit their target, they turn a strong link into a weak one. Weak activation also makes weak links (Agoston et al., 2005). Multitarget drugs stabilize the cell, besides their multiple effects. The smaller noise leads to fewer unexpected side effects and, since the whole system gets more stable, it is not forced to shift to a new equilibrium, the resistant state. Weak link therapy may be a winning strategy for future medicine (Csermely et al., 2005).

Hospitals dispensing Western-style medicine look more and more like automobile repair shops. We bring in our sick relative with the secret expectation that we will get back a repaired, shining, polished 'original'. In the hospital it is often only the 'sick part' of the person that is treated, whilst the process leaves the real disease of the complex living person untouched. In contrast, traditional Chinese, Indian (aryuvedic) medicine and alternative (complementary) medical treatment concentrate on the whole, and do not mobilize the vast knowledge mankind has collected about the molecular background of our body over the past two centuries. Hopefully, with the advent of the fast-propagating network and system biology approach, these two treasures will soon be reconciled.

6.4.3 Aging

We all have a sickness that cannot be cured: aging. There are three major theories of aging:

- **The mutation accumulation theory** says that the deleterious mutations that take effect at an advanced age are not cleansed by evolutionary selection, since aging occurs after the peak of reproduction. Therefore, these mutations can accumulate through generations.
- **The pleiotropy (or antagonistic pleiotropy)** theory suggests that pleiotropic genes with good early effects are favored by selection even if the same genes later show deleterious effects.
- **The disposable soma theory** states that the prevention of late defects (like better scavenging of free radicals) would require a lot of intrinsic resources. Large investments in these costly mechanisms

would help longevity but, in parallel, would decrease survival at a younger age (Kirkwood and Austad, 2000).

Aging is accompanied by a general increase in noise (Carney et al., 1991; Goldberger et al., 2002; Hayflick, 2000; Herndon et al., 2003), in parallel with a decrease in complexity (Goldberger et al., 2002). As an interesting example showing the similarities between stress and aging, fasting stress undergone by young individuals reproduces the irregularities of cortisol secretion in elderly people under normal, non-stressed conditions. The additive effects of stress and aging are shown by the fact that cortisol secretion in elderly people becomes even more irregular after fasting (Bergendahl et al., 2000).[14]

The age-induced destabilization of cellular nets is perceivable in everyday life, too. Sudden tears and running noses are both signs of the same phenomenon: the increased noise of aging. Synchronization goes downhill too. When we grow old, first jet-lags become more of a nuisance, and then suddenly you find yourself sleeping during a colleague's lecture and lying awake the whole night afterwards (Weinert, 2000). The seminal and unfortunately rather unnoticed paper by Himmelstein et al. (1990) gives a good summary of the erosion in homeostatic capacity during aging, which is more prevalent if elderly people have lived a life in poverty, under stress or under racist pressure. Not surprisingly, many of the longevity genes are hubs which control somatic maintenance and repair (Ferrarini et al., 2005; Kirkwood and Austad, 2000), most probably damping the increased disorder and noise of the aged organism.

The increased noise of aging and the wisdom of the youngest child. It is rather common in the fairy tales to find that the youngest child is the wisest. Elderly parents often have rather talented children. If our cells become noisier as we grow older, more cells will be unusual in an old woman or man than in a young person. Noisier maternal and epigenetic effects may give rise to a larger proportion of unusual gifts – and unusual problems.

Aging of cellular networks. Deterioration of complex structures is seen all around our body during aging. Loss of the fractal structure of the dendritic arbor from neural cells of the motor cortex may lead to the

[14]The take-home message is eat well, if you grow old. But life is complicated! You should not eat too much, since it will oxidize your proteins, and accelerate your aging process.

Fig. 6.5. The erosion in homeostatic capacity during aging becomes especially prevalent if elderly people have lived a life of poverty, stress or racist pressure

increased frequency of devastating falls by elderly people (Scheibel, 1985). The trabecular network of our bones and the network of bone-remodeling osteocytes have a scale-free and small-world pattern. The deterioration of these networks occurs if physical exercise is not sufficient, during osteoporosis and other age-related diseases. The analysis of bone network integrity can be used for the prediction of bone fractures, often having fatal consequences in advanced age (Benhamou et al., 2001; Bourrin et al., 1995; da Fontoura Costa and Palhares Viana, 2005; Gross et al., 2004; Hruza and Wachtlova, 1969; Mosekilde, 2000).

Do other complex systems also age? Do systems like the World Wide Web, the world economy, social networks, ecosystems, or Gaia get more unstable as they develop?[15]

[15] I am grateful to Bálint Pató for these questions.

It is difficult to establish a direct link between increased noise and loss of complexity. A loss of complexity (system integrity) will lead to increased noise and, conversely, increased noise with decreased repair and remodeling processes causes a loss of complexity. These two phenomena are different sides of the same coin and may form a vicious circle, aggravating the status of the aging body.

The weak-link theory of aging. Let me make an 'in neuro' experiment here. As we learned in Sect. 2.4, most of our cellular networks have a scale-free link strength distribution. Aging induces random damage in networks. At the molecular level, this means that mutations and free radicals affect various proteins in a random fashion. Most of this random damage will affect weak links. What is the result? The loss of weak links leads to a similar, but unbalanced system. What do we see in aging? I think you have the answer.[16]

Chaperone overload: a possible cause of civilization diseases. A special case of the loss of weak links is the chaperone overload in the cells of the aging body (Csermely, 2001b). Chaperones do not only buffer silent mutations in bacteria, fruit flies and plants. Obviously, the human being is no exception. We also harbor mutant genes in our cells, which remain silent because their damaged proteins are constantly repaired by chaperones. Due to the success of medicine and changes in our Western lifestyle, natural selection had been practically switched off by the 20th century. Moreover, we successfully avoid devastatingly large stresses in our life (life threatening diseases, etc.). These lead to the accumulation of silent mutations in the human genome. Even if there is any exposure of silent mutations, medical care saves us and no selection occurs. In principle this is not a problem. We have chaperones, and they will take care of our silent mutations. However, aging incapacitates and overloads chaperones. As one sign of this, 50% of our proteins become oxidized by the time we reach the age of 70 to 80. Chaperone overload leads to the exposure of silent mutations in aged people. This may contribute[17] to the development of age-related polygenic diseases, such as

[16]As I mentioned above, we see the same noisy, unbalanced system in aging as after the destruction of weak links. Actually, the loss of weak links may also account for the disintegration of responses (Goldberger et al., 2002), since in most cases the long-range, bridging links are also weak.

[17]The extent of contribution is not known. Currently, it may be negligible, but it will grow with each human generation. We may have a further 600 years until it becomes a serious threat. However, doctors should watch out today: with each

cancer, diabetes and others. As a rather surprising coincidence (or perhaps more than a coincidence), Azbel (1999) showed that human life-expectancy follows different rules before and after 75 years, which may signal that "very unusual latent mutations [...] are switched on [...] by some kind of evolvability" (Azbel, 1999). Chaperone overload may give an efficient and very general mechanism for this 'switch', which is applied to all silent mutations and not only the 'very unusual' ones. Thus chaperone overload may be a quite general reason behind age-induced disorganization of cellular responses.

We have come to the end of our trip into the cellular net. Before going a level higher, let me list the precious stones we have collected along the way:

- an optimal level of diversity and noise is important for survival;
- the cell uses a large number of noise buffers and noise generators to do this task well;
- weak links may provide an important element of noise and diversity control at the cellular level;
- noise management[18] may provide excellent opportunities for evolutionary jumps;
- diseases like cancer and aging are accompanied by an increased level of noise and diversity to the extent that the damage of weak links may be a cause of aging.

Finally and most importantly, all these mechanisms are astonishingly general and conserved.

---

coming generation the aging diseases will become more and more unpredictable due to the increasing exposure of random, silent mutations.

[18]The best recipe for a high evolution rate involves intermittent long periods of low noise, and short periods of high noise, or stress.

7 Weak Links and the Stability of Organisms

With this part of our trip, we discover more familiar parts of Netland. This chapter will be about ourselves. Our body is a network of cells and organs. Unfortunately, this complex world is largely unexplored from the standpoint of network behavior. This is partly because elements and modules are mixed up here. Sometimes an organ may be regarded as an element of the body net, whilst in other cases it has to be treated as a module of its cells. Moreover, the complexity of the interactions makes exploration and description even more difficult. This is why I will sometimes use examples from the animal world to bring out the network properties of this exciting network: our own body.

7.1 Immunological Networks

The immune system has to solve four problems:

- the self/non-self recognition problem,
- the signal-to-noise problem,
- the context problem,
- the response problem.[1]

This complex task can only be solved by a network (Cohen 1992a). Immune cells provide perhaps the most sophisticated network in our body. Not only is each cell different from any other and capable of interacting with any other, but the system topology changes continuously and long-range interactions are also possible via cytokines and other mediators. This system offers perhaps the largest repertoire of events which call for stabilization. The idea of immunological networks was first introduced by Niels K. Jerne as the idiotype network, i.e., an

[1]These four basic problems of the immune system can be formulated with the following three questions: For the first two, what should the immune system attack? For the third, when should the immune system attack? For the fourth, how should the immune system attack?

inter-reactive network, where antibodies of one lymphocyte serve as an antigen for another[2] (Jerne, 1974; 1984). Later this concept was extended to networks of natural autoantibodies, which has been called the immunological homunculus, or immunculus by Irun Cohen (1992a; 1992b). Cytokines and other participants of the immune response also form networks.

Immunological networks possess all the properties forming the signature of a network presented in Chap. 2. The reactivity pattern of the repertoires of both T and B lymphocytes, which denotes a set of positive and negative actions on a series of antigens, has a scale-free distribution (Burgos, 1996). The scale-free pattern invokes the self-similarity of fractal structures. Indeed, Jerne (1984) noted that the immune system is complete in the sense that, if we were able to remove 90% of B-cell-derived antibodies randomly, the remaining ones would still represent a complete repertoire.

Among the major network properties, the highly flexible modularity of the immune network seems to be especially important. Modularity helps the dissection of the network of natural autoantibodies, the immunculus, from the networks of acquired immune response to various antigens. Modularity also gives rise to stabilized, localized immune activation circuits as well as to immunological memory (Cohen and Young, 1991; Varela and Coutinho, 1991; Weisbuch et al., 1990). The segregation of a module within the immune network is similar to the segregation of synchronized oscillators within a larger network described by Winfree (1967) and the behavior of the Boolean networks of Kauffman (1969). Utilizing the generality of the immune response, distortions of the immune network have been proposed as highly sensitive potential markers of pathogenic metabolic changes (Poletaev and Osipenko, 2003).

The dynamical properties of the immune network start from its very development. The development of B cell clones in newborns leads to a general decrease in the connectivity of network members from several dozen to around 6 to 9. Low connectivity comes in parallel with high affinity interactions and massive antibody production (De Boer and Perelson, 1991). In agreement with these changes, the B cell repertoire becomes more restricted as mice grow from 1 to 2 weeks old (Burgos, 1996). These changes are reminiscent of the random to scale-free topological phase transition described in Sect. 3.4. In that transition, the highly connected, low complexity network was transformed to a more

---

[2]The sequence of the idiotype network antibodies may be continued: the antibodies of the second lymphocyte serve as antigen for the third, etc.

sparsely connected, but complex network. The change was provoked by a decrease in the available resources.

During these changes, the immune network can recruit and lose elements, change topology, and show restricted percolation phenomena (Brede and Behn, 2002; DeBoer and Perelson, 1991). The connection density has a profound effect on immune properties. If there are too few connections, the network cannot function. This seems rather obvious. However, if there are too many connections, the network will not function properly either (DeBoer and Perelson, 1991; Varela et al., 1991). This resembles the dilution effect which occurs when a node has too many neighbors and their influence is diminished. As we saw in Sect. 3.3, the dilution effect gives rise to unpredictable cascading failures (Watts, 2002). The immune network displays scale-free behavior in time, with $1/f$ noise-type fluctuations[3] in the idiotypic network of natural antibodies (Lundkvist et al., 1989), which is disturbed in autoimmune disease (Varela et al., 1991).

Antigen exposure can be regarded as a disturbance of the original network status which provokes a relaxation phenomenon (Varela and Coutinho, 1991). This assumption makes it much easier to understand why the Watts condition (Watts, 2002) mentioned above may appear deleterious for the overconnected immune net. In the immunological network, relaxation is achieved by a massive change in the elements, connections and topology of the network as a whole. In agreement with this, the B cell repertoire becomes wider after antigen exposure (Burgos, 1996).

Antigen hunger: A potential basis for immunological diseases. Let us reconsider the above statements for a moment. If antigens provoke the relaxation of the immunological network (Varela and Coutinho, 1991), meaning that antigens serve as a resource for the immune net, and not long after birth the immunological network is likely to undergo a random to scale-free network transition, than in our early childhood our immunological network could be said to have indulged in a plethora of antigens. This should not be considered as a tragedy. Just as our brain feels well only when it is bombarded by information, our immune system feels well only if it is challenged by antigens. The young immune system is not fighting with the antigens of the neighboring world. In fact, quite the opposite. Repeated exposure to antigens keeps the young immune system perfectly resourceful and relaxed, allowing it to remain in the random phase, and prepared for

[3]This is the same as $1/t$ noise, a form of pink noise (see Sect.2.1).

Fig. 7.1. Just as our brain feels well only if it is bombarded by information, our immune system feels well only if it is challenged by antigens

any future challenge and change. So the situation may actually be reversed. When we moved into the first caves, got our first clothes and grew up, our immune system was deprived of the joyful source of relaxation and an increasing antigen hunger developed. From this point of view, the increasing sterility of the current epoch is a true disaster, since it deprives the immune system even further of the much-awaited antigen relaxation. No wonder that increasing tensions develop in the immune system, which may lead to bursts of inflammation, allergy (Yazdanbakhsh et al., 2002), autoimmunity or immune deficiency.[4] Many of these bursts may follow a scale-free pattern, resembling the avalanches of self-organized criticality described in Sect. 3.3. It is quite likely that we will soon have to administer regular 'dirt injections' of carefully selected antigens, to ensure the proper relaxation of our immune system. In a few years from now, this immune repair may be as natural a behavior as the other regular relaxation repair we indulge in, e.g., eating, sleeping, and laughter.

After the clone selection theory of Burnet (1959), the later network concepts of immune recognition (Jerne, 1974; 1984; Cohen 1992a) hy-

[4]In extreme cases, chronic and severe antigen deprivation may even produce a series of topological phase transitions leading eventually to a transition from the star phase to the isolated subgraph phase, which can in fact be identified with a total collapse of the immune system.

pothesized a lot more weak links in the immune network.[5] The degenerate pathways whereby various packages of antigens interact with the immune system and provoke an immune response certainly give rise to numerous weak links, as reflected in the models (Brede and Behn, 2002). In spite of this increasing knowledge of the immunological network, consideration of the role of weak links in the stabilization of the immune response are premature, since we do not yet have enough data to assess the effect of the affinity repertoire on the stability of an immune net containing millions of differing elements. However, as a final remark, let me demonstrate the strength of the concept with a hypothetical example.

Autoimmune vaccination focuses and probably stabilizes the immune response. Vaccination is no longer an intervention used exclusively against infections. It is now becoming accepted as an efficient treatment against autoimmune diseases as well. Here vaccination modulates weak links by rearranging regulatory connections of the immunculus (Cohen 1992b; 2002). What happens if the vaccination not only redirects but also stabilizes the immune response by introducing additional weak links? *"Something is wrong here. A direct immune response for a new antigen should introduce strong links if any, not weak ones."* This treatment indeed introduces strong links in the form of the immune response against the newly added antigen. However, in comparison with this strong response, all the residual responses including the autoimmune responses become weak, even if they retain their original strength on an absolute scale. Acquiring the status of weak links, the former strong links of the autoimmune disease now contribute to the stabilization of the immune response. Having smaller fluctuations, the autoimmune response will not reach the activation threshold very often and the patient's symptoms will improve. We have a little data to support this model. The loss of scale-free characteristics of autoantibody fluctuations in autoimmune disease (Varela et al., 1991) may reflect a distortion in the topology and link strength distribution of the underlying immune network structure.[6] In agreement with the model by Stewart et al. (1989), which suggested that autoimmunity develops due to an inadequate connectivity of self-reactive lymphocyte clones, vaccination may help to shift the immune network link strength pattern back to the scale-free distribution, where stabilizing weak links predominate.

[5] Jerne himself marked the third generation of anti-idiotypic antibodies as 'buffering sets' in his original publication in 1974.

[6] Self-organized networks tend to be scale-free in multiple dimensions (see Sect. 2.2 for more details).

7.2 Transport Systems

Transport systems were introduced in Sect. 2.2, where I described the allometric scaling laws. These laws are empirical laws concerning the scale-free behavior of metabolic rates, heartbeats and lifespan, to name but a few contexts where they are relevant. All these measures behave as a linear function of the 1/4 or 3/4 power of the body mass.[7] As described in Sect. 2.2, the 3/4 power law dependence of the metabolic rate is difficult to explain if one starts from simple geometrical considerations. One of the examples above, namely, the metabolism, depends on area. Mass obviously depends on volume, which would give a 2/3 power law function rather than 3/4. However, if the geometry of the blood vessels is taken into consideration, a 3/4 power law function arises. The geometry of the branching transport network does indeed show a scale-free, fractal distribution both in our blood vessels and lungs (Banavar et al., 1999; McNamee, 1991; West et al., 1997). Allometric scaling can be extended to individual cells (West et al., 2002), and is in good agreement with the assumptions concerning the scale-free, fractal geometry of the cytoplasm (Aon et al., 2004b). Three-dimensional, scale-free networks display an optimized degree of efficiency, suggesting that evolution shaped them to minimize the costs of transportation (Garlaschelli et al., 2003).

The blood vessel network is degenerate. The branching network of vessels represents an arterial tree and is connected to a similar venous network. However, this tree is not regular. A small difference between vessel lengths and radii has to be introduced and can be experimentally verified to explain the observed heterogeneity of blood flow (van Beek et al., 1989). Collateral blood vessels do not duplicate the transport function. They perform the same function slightly differently. They are degenerate. Thinking back to Sect. 4.5, we may guess the type of links that arise when degeneracy occurs.

Once again, we find an interesting coexistence of scale-free behavior in space and time. Heartbeat dynamics in an undisturbed state

[7] The allometric scaling laws follow the equation $P = cM^{\alpha}$, where P is the property, c is a constant, M is the mass of the organism or organelle, and α is a scaling exponent which depends on the nature of P. The most thoroughly studied of these is the basal metabolic rate, which obeys Kleiber's law (1932). Here P is the basal metabolic rate, i.e., the amount of energy per unit time required by a living organism to remain alive, and the scaling exponent is 3/4. The value of the exponent is different in the other examples. Hence, the dependence of heartbeat and lifespan have an α exponent of $-1/4$ and $+1/4$, respectively. As a word of caution, note that this simple rule is sometimes overinterpreted (Dodds et al., 2001).

has multifractality, which means that the scale-free behavior can be described by more than one exponent.[8] However, this is true only for a healthy person. In patients suffering from congestive heart failure, multifractality is lost and gives rise to much simpler scale-free behavior. This behavior can be characterized by a very narrow range of scale-free exponents and can be regarded as monofractal (Ivanov et al., 1999). The simplification of the pattern in a disease state resembles the loss of scale-free characteristics of autoantibody fluctuations in autoimmune disease (Varela et al., 1991), described in the last section.

7.3 Muscle Net

The stability of our movements requires a well-orchestrated interplay of various agonist and antagonist muscles and their governing neural functions. The generation of muscle force requires the dynamic recruitment of a large number of motor units with tens of thousands of individual neuronal commands. This process would certainly make the motor function very suitable for network analysis. However, the network approach is still largely absent from this field. As an example of the sporadic data indicating network behavior, Duarte and Zatsiorsky (2000) measured pressure fluctuations in experimental subjects standing on a force plate. The fluctuations during prolonged unconstrained standing showed a scale-free distribution.

Analyzing muscle control from the output, the movements and the generated force provide a good measure of system stability. The complexity of movements, as well as the variability of maximal forces and discharge speeds between motor units all make muscle control a noisy process. The behavior of noise becomes greater with advanced age (Enoka et al., 2003). The endpoint variability of the movement depends on the signal-to-noise ratio, which can be adjusted by the stiffness of the affected muscles (van Galen and Huygevoort, 2000). In other words, if we need higher precision, we flex our antagonistic muscles together with a greater force. Precision comes with struggle. (Many of my colleagues probably agree that more PhD students should take a course on muscle control to learn about this.)

[8] For a more detailed description of multifractality, see Sect. 2.2 or the glossary in Appendix B.

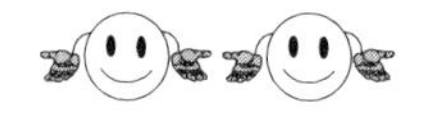 **An alternative explanation of struggle benefits.** A balance of agonist and antagonist muscles at a higher force level invokes higher signals, which may increase the threshold and make the background noise less effective (van Galen and Huygevoort, 2000). However, another explanation may also be given. Yes, weak links again. Similarly to the idea outlined in Sect. 7.1, the muscle control system may use the increased strength of its key elements to make the strength distribution more uneven, and the stabilizing effect of weak links more pronounced. We should never forget: a weak interaction can be weak only in the presence of some really strong interactions. The strength increase may come only by the simultaneous flexing of antagonistic muscles, if we want to preserve the main character of the original movement. We probably need a struggle of opposing strengths to achieve stabilizing weakness.

The analysis of the output of muscle control already gave some hints about the potential involvement of network stability and weak links in the regulation of movement precision. However, an analysis of the input, the concerted activation of the numerous motor units, seems to be even more promising. Motor units are governed by the activation of motor neurons, which is often a synchronized process. However, the extent and level of synchronization are finely tuned in both the time and the frequency domains. Increased synchrony reduces the time required to reach maximal force (Semmler, 2002). However, the same increase induces larger fluctuations in muscle strength, decreasing the steadiness of the exerted force (Semmler et al., 2003; Yao et al., 2000). We have agreement with the output analysis above, since a smaller synchrony between motor units increases the chances that alternative, canceling forces will act against each other, which is the case for the increased muscle stiffness mentioned above. Table 7.1 summarizes a large number of examples to show that a decrease in motor unit synchrony leads to an increase in precision; and vice versa, a synchrony increase makes the final movement noisier.

From the exciting and highly coherent list of experimental data in the table, let me highlight some of the results of Semmler and Nordstrom (1998). They found that the synchrony of the index finger motor units increased as follows:

$$\begin{aligned}&\text{musician's right (dominant) hand} = \text{musician's left hand}\\&= \text{control person's right hand} < \text{control person's left hand}\\&= \text{weight lifter's right hand} = \text{weight lifter's left hand}\,.\end{aligned}$$

Table 7.1. An intermediate level of motor unit synchronization is needed to achieve maximal movement precision

Condition	Motor unit synchrony	References
Aging	Coherence increases	Semmler et al., 2003
Holding versus moving of object by macaque monkeys	Increased synchrony in motion	Baker et al., 2001
Musicians versus weight lifters	Decreased synchrony in skill training versus strength training	Semmler and Nordstrom, 1998
Skilled hand versus other	Decreased synchrony of skilled hand	Semmler and Nordstrom, 1998
Thumb and index finger versus other fingers	Decreased synchrony of thumb and index (skilled) fingers	Bremner et al., 1991
Stroke patients versus healthy subjects	Increased long-term synchrony of stroke patients	Datta et al., 1991
Parkinson's disease patients versus healthy subjects	Increased long-term synchrony in Parkinsonism	Datta et al., 1991

Thus, in terms of precision and motor unit synchrony, piano or flute training gives you two right hands instead of one, whereas weight lifting leaves you with two left hands. In contrast to the struggle mentioned above, brute force is not really helpful to achieve precision. Here is another lesson for PhD students!

The need for increased muscle sync: birth. Motor unit synchronization and the subsequent destabilization of the muscle net can be helpful. And not just helpful, but essential. Childbirth requires the contraction of the uterus. Approaching term, the incoherent contractions become more and more synchronized leading to an increased susceptibility of the fetus–mother system to external perturbations (Sornette, 2002). This may actually work like an integrative computer so that, when the increasingly synchronized contractions exceed a threshold with the addition of the integrated perturbations, the baby is born.

In summary, if we have a predictable task, requiring a well-practised, fast action, like weight lifting, a large synchronization develops. (From this point of view, birth may be regarded as a special form of weight lifting.) This is achieved by strong links between muscle units. On the other hand, if we have an unpredictable, complex task requiring movement precision, like music performance, we need a submaximal, optimal level of motor unit synchronization. Partial synchronization, which causes a divergence in the action of individual motor units, is a typical case of weak link generation at the muscle network level.[9] Strong links are required for well-defined, predictable tasks, which already have an adaptive response. Weak links are necessary for changing and carrying out unknown tasks of high complexity. I return to the more general consequences of these statements in Sect. 9.5, where I redefine the differences between engineers and evolution as a tinkerer. Engineering usually solves a well-defined, predictable task, in contrast to evolutionary tinkering, which solves unpredictable, changing tasks of high complexity.

The emergence of weak links at least coincides with the stabilization of our movements. Where can we find a causative link? Movement of muscle units is governed by motor neurons residing in the central nervous system. Thus the final source of synchronization and network regulation is a level higher. This level, the network of neurons and associated glial cells, will be the subject of the next section.

7.4 The Neuro-Glial Network

The maximal information processing capacity of the human cortex is around a terabit per second, which is comparable to the backbone capacity of the world Internet in 2002. The human cerebral cortex contains approximately 8.3 billion neurons and 67 trillion connections. The length of these connections in a single human brain is between 8 and 800 times the diameter of the Earth. It is not therefore surprising that cortical neurons form a rather complex network. But since information processing is very costly,[10] it is no more surprising to find that

[9] Once technical advances permit, this important lesson may be applied in the construction of artificial limbs.

[10] Besides skeletal muscle our brain is the other major energy consumer in our body: just try to sit calmly and read this book without eating for several hours and you will see how hungry you become – not for more science in this case, but for food.

only about 1 to 16% of neurons operate at any given time (Aiello and Wheeler, 1995; Laughlin and Sejnowski, 2003; Sporns, 2003).

Our brains seem to have read Chap. 2 rather well, learning the lesson and displaying all the usual features of a typical network. The neural net is a small world which is ensured by dense local connectivity and relatively few long-range connections. In agreement with this, neural connection length displays a scale-free distribution. The neural network is modular and the modules are highly interconnected. At intermodular boundaries, special overlaps, the so-called fringe areas, can be found, which may lead to either facilitation or occlusion of the neighboring neuronal modules. Finally, the neural net is nested. This means that the network of neurons contains local circuits, which are multiple synaptic junctions with complex electronic and chemical interactions. These local circuits further contain a number of molecular networks in the individual synapses. All these features serve a compromise between maximization of computational power and complexity and minimization of the costly wiring and maintenance needed to achieve this (Agnati et al., 2004; Buzsáki et al., 2004; Eguiluz et al., 2005; Kniffki et al., 1993; Laughlin and Sejnowski, 2003; Sporns, 2003).

Unless you are familiar with neuro-anatomy, or you are involved in one of the rapidly developing neuronal imaging techniques (Freund, 2003; Gulyas, 2001), it will be rather difficult to obtain a direct view on the network properties of the neural net above. However, there is a related feature which you definitely will have observed. In agreement with the above structure, our mistakes also follow scale-free statistics (Gilden et al., 1995). This means that when we try to make a replica of a target length in space or time, most of the time we make only a small error. However, we sometimes definitely make a bigger mistake. Is this bad news? It will become even worse. Sooner or later we will make a real blunder, too. (In another way, this is also good news: I have at least one excuse for a big mistake in this book.) Scale-free statistics is characteristic of error in a large variety of cognitive tasks[11] (Gilden, 2001).

Higher brain functions emerge from a transient synchrony between neural cells. Synchronization has already been postulated by Berstein in his classic paper of 1945, where he stated that (Sporns and Edelman, 1998): "There has to be an inevitable synchronization, both by frequency and by phase of numerous low-frequency oscillators in the

---

[11] The scale-free error distribution is characteristic of response latency, repetition and discrimination accuracy.

cortex." Indeed, it has been demonstrated that temporarily synchronized neurons ensure successful memory formation (see Sect. 3.5) (Fell et al., 2001). Neural networks demonstrate several oscillatory bands between 0.05 Hz and 500 Hz. Slow oscillations require very large networks and efficiently modulate faster oscillations confined to smaller neuronal space (Buzsaki and Draguhn, 2004). Recent data showed the existence of two dynamic, diametrically opposed, widely distributed and anticorrelated networks of brain oscillators (Fox et al., 2005). In a number of diseases, such as Alzheimer's disease, autism, schizophrenia and attention-deficit disorder, a disturbed neural synchrony was observed (Stam et al., 2005; Stelt et al., 2004).

Saints and geniuses: Exceptional cases of extensive synchronization and coherence. Extensive synchronization and coherence of electroencephalograms (EEG) reflect an increase in long-range connectivity of the neuronal network. Rather interestingly these states were found to be characteristic of emotionally intensive states, tasks requiring efficient information processing or creativity, giftedness and meditation (Aftanas and Golocheikine, 2001; Jausovec and Jausovec, 2000; Orme-Johnson and Waynes, 1981; Petsche, 1996). (I just note in parentheses that the nested sync I suggested as a possible explanation of the Jungian synchronies of fulfilled prayers and dreams also invokes similar mental states.)

As described in Sect. 3.5, synchronization is helped by small-worldness and scale-freeness, typical features of neural networks. Weak coupling has also been suggested to facilitate synchronization (Sect. 3.5). What are the weak links between neurons? Many of the indirect neural regulatory mechanisms (Paton and Vizi, 1969; Vizi, 1979; 1984) and the so-called volume transmission (Agnati and Fuxe, 2000; Agnati et al., 1986) of diffusible messengers, such as nitric oxide, or carbon monoxide, almost certainly play a role in accomplishing this important task. However, we do not know enough about the strength or the sophisticated regulation of these interactions to assess their contribution in a general manner.

Besides direct neural transmission, or volume transmission, astrocytes provide another candidate for weak interneural links.[12] Astrocytes demarcate gray matter regions from gliovascular units (Nedergaard et al., 2003) and could easily be one of the weak linkers between the modules of the neural net. The ratio of astrocytes to neurons in-

[12] I am grateful to Péter Száraz for this idea.

creases steadily as brain function develops. They form an electrical syncytium (Nedegaard et al., 2003) which ensures a weak link with neighboring neurons. Astrocytes have coordinated calcium signals with neighboring neurons and regulate synaptic transmission, synaptogenesis, synaptic maturation and elimination (Hirase et al., 2004; Newman, 2003d; Slezak and Pfrieger, 2003), which predispose them as an ideal candidate for a stabilizing weak link in neural networks.

We may have quite an elaborate network of weak links in our brain. These links may help synchronization and the integration of brain function. What are the consequences of their contributions? Dreams, learning and even our consciousness may all require weak links in the neural net. These complex brain functions lead us to human behavior, which will be the subject of the next section after the following comments on dreams, learning and various levels of consciousness.

Weak links may help us to dream. The longest and most intense dreams occur during the rapid eye movement (REM) phase of sleep and can be characterized by (a) memory selection, (b) bizarre and unrecognizable representation of memories, and (c) high emotional content. Cognitive tests on associative memory suggest that REM dreams are modulated primarily by weak neocortical associations (Stickgold et al., 2001).

Weak links may help learning. The aim of learning is to make strong links in neural networks. However, most of our memory is helped by associations with weakly defined subsystems like colors, emotions, smells, etc. This is called the context effect. If divers learn words under water, they remember them much better under water than on the shore. If we learn words under the influence of alcohol, we will remember them better if we drink again. Similarly, alcoholics find hidden reserves of money or liquor when they get intoxicated again (Goodwin et al., 1969; Smith and Vela, 2001). Recent data indicate that the slow-wave, non-REM phases of sleep may be involved in remodeling of the neural network, thus assisting the consolidation of a previous learning process (Huber et al., 2004). It is currently unknown how this network remodeling affects weak versus strong links between neurons and neuronal modules.

Weak links and consciousness. Consciousness facilitates widespread connections between otherwise independent modules of the neural net increasing both the complexity and synchronization of brain function.

Both the loss of neural synchronization or over-synchronization, e.g., in slow-wave sleep or in epileptic seizures, respectively, reduce complexity and may lead to unconsciousness. The simultaneous presence of functional integration and segregation, parallel and recursive signaling (called reentry), as well as the large repertoire of differentiated neural states are key features of neural complexity and the conscious state. These all require an optimal amount of ever-changing synchronization (Baars, 2002; Tononi and Edelman, 1998; Tononi et al., 1992; 1998). Synchronization and connection of divergent brain areas and functions are all helped by weak links.

Reflections of the outside world. We have quite numerous mechanisms in which we either construct internal images of the outside world or allow the triggering of pre-set responses by the outside world. Internal images of the outside world are formed when we imagine other people's mental state by adopting their perspective with the help of a special set of neurons called mirror neurons (Gallese and Goldman, 1998). Triggering of pre-set responses by the outside world occurs when we become exposed to unconscious information like primes, which are images too short to be recognized consciously (Kunde et al., 2003). In all these mechanisms, a broad coherence of the neuronal network is expected to operate. This coherence also requires a broad synchronization and connection of divergent brain areas, which is once again helped by weak links. These weak links not only allow the various levels of our consciousness to develop, but also lead us to the next section: the human psyche.

7.5 Psycho Net

The network properties of our brain invoke a network approach to rationalize and understand the human psyche. The Freudian revolution broke the transparency of the 'Cartesian theatre' of our inner world, showing that it cannot be discovered even by its owner (Pleh, 1988). This broken world was calling for a more complex approach. The basis of psycho nets was established by Carl Gustav Jung in 1921, when he published his landmark book on psychological types. Stern's differential psychology (Stern, 1911), typology (Jung, 1969; Kretschmer, 1921) and factor analysis (Cattell 1978; Eysenck, 1970; Spearman, 1931) are some of the major schools and methods of psychology which are inherently related to network studies.

In spite of all these psychological networks, rather surprisingly, neither the four major network properties mentioned in Chap. 2, small-worldness, scale-freeness, nestedness and weak-linkness, nor the dynamic network characteristics such as the presence of a giant component, percolation, self-organized criticality, topological phase transitions, synchronization, etc., have been systematically addressed in order to understand the structure and dynamism of our psychology. In the rest of this chapter, I outline a few starting ideas to illustrate the wealth of potential applications of the network paradigm in psychological studies.

As one of the rare examples of the network-related dynamical aspects of psychological events, I have already mentioned the scale-free statistics of ticks in Tourette-syndrome patients (Sect. 3.2) (Peterson and Leckman, 1998). The variation of the time interval between two ticks can be interpreted as a self-organized criticality event, a tick quake. During a tick quake a gradual tension develops, and the relaxation comes in the form of the tick.

Adolescence as a psycho quake. The development of the psyche is not monotonic. Evolutionary periods are interspersed with sudden bursts of development, as in the punctuated equilibrium of evolution (Gould and Eldredge, 1993). One of these bursts occurs in adolescence. Here self-accomplishment and partner contacts are often felt to be delayed and a tension gradually develops. On the other hand, the teenager leaves the safe circles of the family and school and becomes increasingly exposed to completely different environments, where she has to cope with the new situations alone. From time to time, these occasions may trigger an avalanche-type development of the psyche, or psycho quake.

A healthy psyche needs a balance of strong and weak links. We need a healthy balance of strong and weak links to maintain our psychological stability. Unbalanced links characterize a number of psychic disorders:

- Monomaniacs and multi-talented people. Both monomaniacs and multi-talented people tend to be emotionally unstable. Monomaniacs may have an emotional network resembling a star network, where all the links are strong. On the other hand, multi-talented people are closer to an all-weak, random network. To help their stabilization, monomaniacs require various secondary activities, while multi-talented people need strong guidance, to provide their missing weak or strong links, respectively.
- Autism. Another potential example of a strong-link surplus to various, and not necessarily personal elements of the environment may be autism,

where novel situations, ambiguity, and the appearance of weak links may cause a panic reaction revealing a high level of instability.
- Emotional traumas. Emotional traumas often result in a closed personality, where the former personal network is reduced to a few strong links.
- Borderline personality disorder. Undeveloped, broken or unbalanced emotional links of early childhood may cause a borderline personality disorder, where the link imbalance is extended during adolescence and leads to a general instability of the personality.

Linville (1987) showed that the complexity of the self protects against the destabilizing effects of stress. A higher self-complexity here may mean multiple social dimensions or friends from a grossly different social context. All of these result in an increased number of weak links between the various self-aspects and cognitive modules, which counteract the destabilizing effects of stress. A balance between strong and weak links is also a prerequisite of the 'flow', a state of harmonious creativity (Csikszentmihalyi, 1990).[13]

The dangers of being talented. According to a recent report by Beilock and Carr (2005), individuals with a high working memory capacity showed a decrease in capacity for accurately solving difficult mathematical problems when they were in a situation leading to high emotional pressure.[14] So far the results are rather obvious: stress provokes an attention deficit. The interesting part was that individuals with a low working memory capacity showed a much smaller decrease under pressure. In fact, the talented group lost all its advantage in the stressful situation. As an explanation for these observations, the authors suggest that performance pressure may consume the working memory capacity surplus of talented individuals. The network approach may provide an alternative explanation. As described in Chap. 6, extreme abilities may correlate with a generally more unbalanced status of the underlying cellular networks. If the talented person is already less well balanced from the beginning, the larger intrinsic noise may provoke a stochastic resonance and may help the stress to exceed an 'inability threshold'. In the average person, the intrinsic noise is small, and stress thus has no chance of reaching this inability threshold.

STRONGLINKERS and WEAKLINKERS: A possible pair of new personality traits. A recent hypothesis by Bateson et

[13] I am grateful to Mária Herskovits, Cleopatra Ormos and András Szabó for many of the ideas in this paragraph.

[14] The high emotional pressure involved a combination of monetary incentives, peer pressure and social evaluation.

al. (2004) raises the possibility that animals and humans can be divided into two basic phenotypes: the SMALLS who adapted to survival under harsh conditions, and the BIGS who adapted to proliferation using rich resources.[15,16] Thinking further about the consequences of this division, I would like to put forward the idea that SMALLS are correlated with a 'thrifty', rather introverted psycho-type, while BIGS are correlated with a 'spending', rather extraverted psycho-type.[17]

BIGS do not fear that they will not be able to reciprocate the goods they receive. BIGS are often altruistic, hoping for a later return. BIGS care about possessing as much of the rich resources as possible, explore new possibilities, and make all sorts of alliances to achieve these goals. BIGS expand, explore and make long-range contacts, which are weak links. In contrast, SMALLS try to spend as little as possible. SMALLS minimize their contacts, which are costly to build and maintain. SMALLS withdraw, restrict, and maintain only the safe, long-term contacts, which are strong links.

Bateson et al. (2004) also added that these phenotypes cannot change from one moment to another. For a complete change, one must wait 2 to 3 generations.[18] Consequently, if Bateson et al. (2004) are right – and their arguments are rather convincing – then we have fairly distinct STRONGLINKER and WEAKLINKER phenotypes, in both animal and human behavior. A few features of the extreme versions of both STRONGLINKER and WEAKLINKER behavior are listed in Table 7.2.

STRONGLINKERS and WEAKLINKERS are equally good. I would like to stress that both phenotypes are equally important and valuable. Without either of them, humankind would have been wiped out

[15]The SMALL phenotype would correspond to a modified version of the 'spore' and 'dauer' states of more primitive organisms, where these local energy minima in the phenotype landscape are even more widely separated. Formerly, a similar typology was called the thrifty geno- or phenotype, referring to the genetic consequences of the lifestyle change after the excessive hunting period, which made big game extinct approximately 11 000 years ago in the Paleolithic (Neel, 1962), or to the effects of poor nutrition in fetal or early postnatal life, resulting in a sensitivity to diabetes as well as obesity, high blood pressure and other elements of the metabolic syndrome (Hales and Barker, 1992).

[16]The names are a bit misleading. The SMALL and the BIG phenotypes are complex traits and cannot be reduced simply to someone's height.

[17]The dichotomy of the SMALL and the BIG phenotypes can be extended even further. The same distinct response pair is a valid description of plants and ecosystems, which usually display two different and widespread evolutionary responses (D'Odorico et al., 2005).

[18]As an indirect proof of this, certain spending habits, e.g., charity donations, decrease in lawful tax-evasion tactics, should show a roughly two-generation lag-time after the general appearance of discretionary income in a population.

by natural selection a very long time ago. STRONGLINKERS are necessary to build the core of our networks. Even more importantly, STRONGLINKERS are the only solution during hard times and stress. WEAKLINKERS are useful in periods of expansion. However, they have to be controlled by STRONGLINKERS, otherwise the network will overspend its resources, and become inefficient, overconnected and unstable.

STRONGLINKERS and WEAKLINKERS are not always different. Bateson et al. (2004) argued that the phenotypes corresponding to STRONGLINKERS and WEAKLINKERS are a kind of inherited 'fate', and may change only after 2 to 3 generations, in which members experience the opposite environment to an earlier one. If our grandparents and parents lived in relative poverty, we may have a winning lottery ticket, but this does not mean that we will transform from a STRONGLINKER to a WEAKLINKER from one day to the next. The following three notes aim to give a more detailed picture of this rather deterministic statement to help the reader to understand it better.

- STRONGLINKERS and WEAKLINKERS both have strong *and* weak links. Both STRONGLINKERS and WEAKLINKERS represent only trends, which mean that WEAKLINKERS will definitely have a few strong links, and STRONGLINKERS may easily end up building a weak link.
- A mixed strategy wins in uncertainty. As circumstances evolve, we change from STRONGLINKERS to WEAKLINKERS or vice versa. This does not happen like a kind of miracle. After decades of hidden development, the former WEAKLINKER does not suddenly wake up one morning as a STRONGLINKER. A mixed psycho-type may easily occur for generations. Usually a mixed strategy is not a winning strategy. However, the mixed strategy might become a winning strategy in times of uncertainty, when outside signals suggest the development of both psycho-types. As a rather far-removed but suggestive piece of evidence for this behavior, a mixed strategy becomes the winner in dryland plant ecosystems which have been exposed to random interannual fluctuations in precipitation (D'Odorico et al., 2005).
- Social dimensions. The parallel development of STRONGLINKER and WEAKLINKER behavior may appear in different social dimensions. We may behave like a STRONGLINKER at home with our family and manifest a WEAKLINKER character with our friends, lover, and colleagues, or vice versa.[19]

[19] I am grateful to András Szabó for these ideas.

Table 7.2. Differences between the STRONGLINKER and WEAKLINKER personality traits. The table lists typical features of extreme STRONGLINKER and WEAKLINKER personality traits for the sake of easy discrimination between the two. In reality, various mixtures of the above extremes are usually present

Feature	STRONGLINKER behavior	WEAKLINKER behavior
Friendship network	Relies more on family links, usually has only a few very good friends, and masters reliable, life-long contacts	Has friends from a rather broad circle, including different backgrounds and lifestyles, but may lack close friends; new friends often become more important than old ones
Cognitive and emotional hierarchy	Centered on a few ideas and emotions; change of cognitive sets is difficult; highly hierarchical; self-disciplined	Many competing ideas and emotions; low, conflicting and ambiguous hierarchy;[a] playfulness, instability, ineffectiveness
Internal images of the outside world	Rigid, match-seeking usually occurs with pre-set ideas	Flexible, highly adaptive to new information and environment
Cognitive dimension	Low; can imagine only a few attitudes; maximal capacity to understand interacting people may be easily saturated	High; can imagine numerous attitudes and assess a large number of interacting people; may become overcomplicated
Exploration	Very efficient; prefers short range search with very rare jumps; exponent of Levy flight[b] is big	Inefficient; enjoys longer jumps, unexpected explorations; the exponent of the Levy flight is small; may shift towards random walk
Relaxation	Disturbed to keep rare energy packages in the system	Undisturbed to release surplus energy as fast as possible
Ambiguity tolerance	Low	High
Efficiency in well-defined tasks	High	Low

Table 7.2. Cont. Differences between the STRONGLINKER and WEAKLINKER personality traits

Feature	STRONGLINKER behavior	WEAKLINKER behavior
Creative behavior	High in structured, hierarchical schemes, like engineering, mathematics, religion	High in diffuse structures, like social sciences, humanities, interdisciplinary research, or art
Attitude towards resources	Spending-conscious; thrifty; minimizes excess expenditure	Achievement-conscious; generous; maximizes income
Major problems	Loneliness, depression, too rigid, too logical lifestyle, instability due to absence of weak links	Unfocused life, variable motivations, lack of endurance, ineffectiveness; instability due to overconnectedness
Network structure	Scale-free network gets closer to a star network	Scale-free network gets closer to a random network

[a] I am grateful to Cleopatra Ormos for this idea.
[b] An efficient search pattern (Levy, 1937) with a scale-free distance distribution (see Sect. 2.2).

STRONGLINKERS may favor STRONGLINKERS while WEAKLINKERS may favor WEAKLINKERS. Pettijohn (1999) has shown that, in stagnant and pessimistic social conditions, American movie actresses with mature facial features have been more popular. However, when social and economic conditions were prosperous and optimistic, actresses with neonate facial features were popular. This preference was not valid for male actors and the preference may be derived from many sources including perceived competence, perceived nurturing abilities, etc. In spite of these restrictions and explanations, let me play with the idea that facial differences also reflect differences in facial symmetry. The neonatal face is more symmetrical than most mature faces. As discussed in Sect. 6.3, increased symmetry reflects lower stress during embryonic development and is probably related to more weak links in cellular networks providing a greater stability. A neonatal face may be more typical of a WEAKLINKER than a STRONGLINKER. During prosperous and optimistic social conditions, society presumably contains a greater number of WEAKLINKERS than STRONGLINKERS. As a grossly simplified explanation for the above findings, during prosperous and optimistic times, the WEAKLINKER social majority selected a WEAKLINKER actress as a preference. In connection with these remarks, it would be very interesting to analyze the results of

Fig. 7.2. BIGS expand, explore and make long-range contacts which are weak links. SMALLS withdraw, restrict and maintain only the safe, long-lasting contacts, which are strong links

recent US congressional elections (Todorov et al., 2005), where in 70% of the districts a mature-faced candidate won, as opposed to 30% of the districts, where a baby-faced candidate was the winner. The election results should correlate with the US population density. The reason for this expectation is that the US population density is the only population measure which correlates with the average height of the current US population (Komlos, 2005). Population density and average height correspond to crowding stress and richness in resources, respectively. Hence, where population density is low, a WEAKLINKER social majority is expected, which may select a WEAKLINKER, baby-faced candidate in the elections.[20]

The benefits of linker-symbiosis: Are STRONGLINKERS necessary to protect WEAKLINKERS? I have described the help

[20]This expectation may be spoilt if sparse populations have no chance of building social links and become STRONGLINKERS due to their isolation. One might argue that, at the current level of technical development in the USA, this may no longer be a typical situation. However, data collected by Liben-Nowell et al. (2005) showed the extreme importance of geographical distribution on the development of virtual, Internet-based friendship networks, which means that people have a very strong sense of their geographical status and population density even today.

given by low achievers to high achievers in the cellular metabolic net in Sect. 3.3. Low achievers are bad enzymes with a stable structure, whilst high achievers are good enzymes with a labile structure (Shoichet et al., 1995). Perturbations might remain longer with the low achievers. In this way low achievers may protect their more vulnerable, high-achieving counterparts, thus facilitating the record-breaking enzyme activities of the high achievers. A rather loose analogy with a network two levels higher, the social net, raises the following question: are STRONGLINKERS necessary to protect WEAKLINKERS in society? Is trouble dissipation by STRONGLINKERS a mandatory 'trouble sink' in society to keep the vital bridges of the sensitive WEAKLINKERS alive? Can one say that STRONGLINKERS build local circles, helping and ensuring the local dissipation of trouble, while WEAKLINKERS provide global coupling to ensure the emergent properties of the social network as a whole? Do WEAKLINKERS build a rich club, where 'club members' have a large number of connections? Where can we position the rich clubs of WEAKLINKERS in society? Are WEAKLINKERS mostly members of the middle, or upper-middle class? Do the really rich become STRONGLINKERS again, and withdraw to a VIP club building only a small number of strong links, and directing a few influential WEAKLINKERS to go out and cause trouble (Masuda and Konno, 2005)? It would be interesting to examine assortative social networks from this point of view.

It is a basic rule of psychotherapy that the psychoanalyst cannot establish strong personal contact with the patient. *"I have not participated in any psychotherapy yet, but I cannot believe that you are right. How can any therapy be successful without a strong contact?"* Intensity and strength are not the same, Spite. Psychotherapeutic contact is a very intensive contact, but it is both purpose-oriented and limited to the sessions. This is why it remains a weak link between the two multidimensional personalities. The task-oriented approach is an element of the success here. Besides the ethical norm which prevents the therapist from exploiting the patient, many examples show that a sufficient distance is needed to give real psychological help (Degenne and Forsé, 1999; Freud, 1915; Kawachi and Berkman, 2001; Veiel, 1993). Weak links seems to help our stabilization better than strong bonds.

Weak links stabilize partnerships. It is rather commonplace but still worth noting here that long-term partnerships and the health of their participants are stabilized by the weak links of affection rather than the strong links of burning love. As a modern form of this, care can be expressed more by frequent SMS messages than by a limited number of lengthy phone calls. Frequent SMS messages here correspond to multiple weak links as opposed to lengthy calls corresponding to occasional strong links.[21]

Social networks may influence the psyche in a number of ways. However, here the interpersonal links are much more important than the intrapersonal network structure. I will therefore list the remaining examples of these interactions in Sect. 10.3, under the heading of social capital.

Leaving our body (do not worry, I neither drank too much nor invite you to come with me for a splendid session of levitation; we just finished our third trip into Netland), let me summarize what has been gathered together here. We have gleaned quite a bit of evidence according to which various networks in our body, such as the immunological networks, the blood flow system, the muscle net, the neuro-glial network, and our psyche, all involve a large number of weak links. Moreover, this part of Netland is found to be full of smoking guns. In all these networks, there were quite suggestive hints that they might be stabilized by weak links. However, most of the formal proofs are still lacking. Perhaps it is time to start work!

[21] I am grateful to István Kovács for this idea.

8 Social Nets

This is not only the time to get down to work, as I noted at the end of the last chapter, but also a time to thank you for your patience in coming along with me on this trip to Netland. We have reached an important point. We are just about to rise above ourselves. In the last chapter, we surveyed some of the networks in our body, and in this chapter the same body will be an element of a larger network, the social net. The current chapter will give me a good opportunity to understand my obsession with building social networks. The first sign of this came as rather a shock. It was a great surprise to my 17 year old soul when a sociological survey showed that I was the only person in the class who was in rich contact with two competing groups (a typical weak hub in a bridging position, I would say now). Later I found myself organizing scientific societies, political protests, NGOs, firms and world congresses. I had better spare you from the whole list. It is high time to discover at last what were the properties of the networks I spent several decades building. So, many thanks again for staying with me in this self-organizing task. However, you will have to wait a while yet for the 'real thing'. I begin by introducing the ancestors of our own nets: the animal communities.

8.1 Animal Communities

Have you ever tried to interview a honey bee? In spite of the low information content of the potentially fatal encounter (the interrogator will be beaten and, as a rather unfair exchange, the bee dies), we know quite a lot about insect social nets. The most fascinating features of insect colonies are the self-sacrificing altruism expressed by colony members, the complex division of labor, and the highly efficient adaptation to environmental changes. As a part of the adaptational plasticity, insect social nets have dense contacts, making the propagation of information an easy matter within the colony. The insect net seems to be a small world possessing a scale-free degree distribution. The network is hierarchical with the queen at the top, but many decisions are made locally in the various modules led by key figures such as 'scouts' or 'dancers'. Intermodular contacts are made by weak links, which are key regulators of task redistribution in the eventuality of any change in the environment (Fewell, 2003). The very similar ant networks display percolation thresholds and transient synchronization events, and their random to scale-free topological phase transition has already been mentioned in Sect. 3.4 (Bonabeau et al., 1998a; Karsai and Wenzel, 1998; Le Comber et al., 2002; Theraulaz et al., 2002). Animal communities display an almost complete list of the most important network properties. However, some exciting points are still missing. It would be nice to see whether ant or bee colonies display avalanches and additional phase-transitions.[1]

Queens of honey bee colonies mate with several males. This results in a high genetic diversity among their workers. Consequently, thresholds for different tasks differ and change as bees grow older. Younger or older bees have a low threshold for tasks inside or outside the nest, respectively, which means that younger bees work mostly inside, while their elder peers go out more often. Different honey bee thresholds for foraging allow a more efficient response to fluctuations in the availability and need for resources. As a result, not all bees immediately switch the search from one type of food or direction to another. Moreover, surplus bees can be redistributed from one task to another (Bonabeau et al., 1998b; Page and Erber, 2002). As another example, brood temperature is an important factor for honey bee pupae development. A difference of 1.5°C in brood temperature may turn the affected bee from a good learner to a mentally handicapped bee (a zom-bee!). Thus, it is no small advantage that the brood nest temperature is three times more stable around the optimal 35°C in genetically diverse colonies than in genetically more uniform ones. The explanation is similar to the differences above. Diversity in temperature response thresholds prevents all the bees from starting to fan or instigate metabolic heating at once, to decrease or increase brood temperature, respectively. Such behavior would cause rather big overshoots and fluctuations in brood temperatures, especially after a change provoked by the environment (Jones et al., 2004; Tautz et al., 2003).

[1] Possible experiments may include the examination of a sudden burst of behavior change during a gradual repopulation after removing ants or bees from the nest, or the test of a more hierarchical organization in the case of food shortage.

In summary, bee diversity allows the development of degenerate responses, increasing the number of weak links in the bee net. We should listen to honey bees and learn that difference pays off. So does tolerance of difference. (The other take-home message, i.e., that queen-promiscuity is beneficial for the stability of the society, may be more honey-bee specific!)

Several features of insect community networks can be extended to other animals. The scale-free statistics of group size distribution seems to be a general phenomenon, valid not only for insects, but also for tuna fish, sardinellas and African buffaloes (Bonabeau et al., 1999). Dolphins also form groups with a scale-free distribution. These groups are, however, much more structured. A highly flexible, multilevel super-alliance of bottlenose dolphins helps their stability and survival. The dolphin network is a small world, characterized by a high redundancy of contacts. Changing alliances in the super-alliance and redundant contacts both make the emergence of multiple and interchangeable weak dolphin links very likely (Connor et al., 1999; Lusseau, 2003). Hierarchy levels may vary. An interesting topological phase transition has been demonstrated with macaques, where food scarcity provoked a scale-free to star transition giving rise to a despotic society (Hemelrijk, 2002). Baboons also form complex, multilevel social nets. The human brain has developed to keep an inventory of our contacts (Dunbar, 1998). However, each of our acquaintances is multivalent.[2] Baboons have at least a part of this rich inventory. They can discriminate between kinship and rank hierarchies (Bergman et al., 2003). Multiplication of the perceived social net by diverse role-attributions raises the number of weak links.[3] Baboons have a female-dominated society, since baboon males disperse from their natal groups, while baboon females stay there. Social support among female baboons extends to a large segment of the whole community and helps infant survival. Grooming becomes very intensive and reciprocal. Isolation from this tight grooming net of weak links significantly lessens reproduction success (Noe, 1994; Silk et al., 2003).

---

[2]Tante Sissi was not only a neighbor of ours in my childhood, but also an excellent teacher of the German language every Wednesday, and the best living relic of tea ceremonies in the whole town.

[3]I had to meet Tante Sissi every Wednesday, because her German was excellent, but then she had got used to the habit of brewing each tea three times – the king's tea, the citizen's tea, and the proletarian tea as she called them, and for some reason I always got the last version – Wednesday certainly became the only possible day of the week when I could be forced to meet her.

As a take-home message, I conclude that the emergence of weak links grows in parallel with the stability of all the animal societies mentioned. Bee colonies provide excellent examples of the stabilizing power of diversity in response thresholds. These thresholds are genetically determined but can be modulated by age, experience, nutrition and pheromones (Pankiw et al., 1998; Page and Erber, 2002). Bee–bee contacts not only transmit chemical signals but also knowledge through the 'dancing' mentioned earlier. Weak bee–bee links, such as an irregularity or divergence of pheromone and dance instructions, may further increase the stabilizing diversity of bee responses. The take-home message here is that – fortunately for us – absolutely perfect dictatorships cannot be efficient in the long run.

The stability conditions of bee colonies strongly suggest the beneficial effects of weak links. Baboons, however, provide direct proof for weak-link-induced stability. Grooming of female baboons has been demonstrated as an especially useful weak link to facilitate community coherence. I will return to this in Sect. 8.6, where I will list several pseudo-grooming elements of our own (post)modern society.

8.2 A Novel Explanation of the Menopause

The menopause, reproductive cessation at an advanced age, occurs in a number of species, such as non-human primates, rodents, whales, dogs, rabbits, elephants, and domestic livestock, and is especially long and pronounced in humans. In spite of this lengthy list, the menopause is unusual. Most animals die when they lose fertility. In the summer of 2004 I was rather perplexed to witness hundreds of bombyx mothers dying right on the spot where they laid their thousands of light-brown eggs. Fortunately our own mothers are luckier. Why did our great grandmothers develop a menopause? Menopause may be (a) a cultural artifact, (b) a senescence-related phenomenon, (c) a protection against the propagation of genetic damage; and lastly and perhaps most importantly, (d) a 'grandmother effect' in the form of a combined adaptive response to prolonged infant dependency (Hawkes et al., 1998; Shanley and Kirkwood, 2001; Sherman, 1998; Peccei, 2001). In these explanations, the cultural artifact refers to the fact that we now live long enough to experience the menopause. If menopause is related to senescence, then fertility just decays, like eyesight. If the menopause has developed as a protection against the propagation of genetic damage, then it prevents late pregnancies which may give rise to various genetic malformations, such as the Down syndrome. Lastly, the grandmother effect means that grandmothers can assist their adult daughters in raising their grandchildren.

Many of the above theories, such as the cultural artifact theory have received quite a bit of criticism (Peccei, 2001). Recently, it is the grandmother effect that has gained the most attention. Packer et al. (1998) showed that the menopause of olive baboons and African lions is not associated with an improved reproductive performance of children or better survival of grandchildren. Lahdenperä et al. (2004) analyzed multi-generational life data sets of Finnish and Canadian families living in the 18th and 19th centuries. They concluded that in both societies long-living mothers increased the reproductive success of their children, allowing them to 'breed' (scientific language can be quite interesting sometimes) earlier, more frequently and more successfully. However, when children reached the end of their fertile periods, rates of female mortality started to accelerate. A few centuries ago, Finnish and Canadian great-grandmothers were not very well tolerated. The difference between baboon and lion versus Finnish and Canadian mothers might arise from the extent of the menopause. The shorter menopause may restrict intergenerational help by a parallel own parenthood in the animal species (Shanley and Kirkwood, 2001). Indeed, the post-menopausal periods were roughly twice as long for Finnish and Canadian mothers as for baboon and lion females.

The grandmother effect seems to be rather well established. It certainly works. Grandmas do indeed help their daughters. This is undoubtedly a reason, but are we sure that it is the only one? Have you seen a grandma who had nothing else to do but help her daughter 24 hours a day? I would bet that this level of intensity would not boost 'breeding' rates, but would be more likely to have an adverse effect, e.g., the young couple must at least have some privacy to breed more frequently. So I would propose that the menopause is also beneficial in providing more weak links within the animal or human community. Females after the menopause are not the subject of fights or attempts at seduction. Their connection structure shifts significantly to contain far more weak links than before and provides additional stability for the whole community. In agreement with this proposal, the cohesive force in both dolphin and baboon networks is mostly maintained by adult females (Connor et al., 1999; Lusseau, 2003; Silk et al., 2003). *"I have not had the opportunity to see family videos from the 18th and 19th century, but I bet that Finnish and Canadian women of the time were not all as beautiful as the wife of Menelaos, and were not often abducted by Paris to Troy!"* Spite, have you ever seen a family video of Paris and Helen? Regardless of any debate about abduction, younger women certainly have a tighter control and daily schedule than grandmothers (see, e.g., that peculiar breeding duty above). It would be very exciting to see whether the grandmother effect increases with the coherence or wealth of the community, allowing more free time and opportunities in the form of chats, visits to the market, etc., to establish weak links.[4] It may be that the contradictory appearance of the grandmother effect is due to the fact that researchers did not include the resources of the society in their analysis. If weak links do indeed play an important role in the grandmother effect, then this effect should be strongest in communities where resources are not sparse, but also not too plentiful.

Women survive stress better than men. I hope the above examples have convinced you that we need women. *"Yes, I definitely agree!!!"* Spite, let me finish! We need women for the stabilization of our social groups. And although we need them in prosperity, we need them even more in stress. Men tend to 'cope' with stress by displaying greater competitiveness, hostility, social withdrawal, and substance misuse, whereas women do so by seeking out social support. Male behavior is not only self-destructive, but also undermines group stability, while female behavior is both stabilizing and contributes to female longevity (Skrabski et al., 2004; Taylor, 2002). Spite, if we want to live longer, we should start to copy this.

After this comparison of the baboon, the lion, and Finnish and Canadian mothers, I will continue with the most studied networks, social nets. I begin by describing their general features, highlighting those which are specific to ourselves. Then my pet idea comes along (you may have guessed what it is: weak links), and finally, I show how to use social nets for information management, one of the most important reasons for their development besides the division of labor.

8.3 Stability of Human Societies

In the last section, grandmothers undoubtedly emerged as a high point of evolution on Earth. In spite of this, societies contain a lot more actors than just grandmothers. Social communities provide a rich and

[4] I have to make the remark here that the wealth-induced promotion of weak links is not unlimited. Extreme wealth reverses the trend: really rich people tend to build strong links again, as summarized by Granovetter (1983).

very promising field of network studies. This was recognized a long time ago (Lotka, 1926; Moreno, 1934). In fact, with a little simplification, I may say that the network theory was born as an attempt to categorize and understand the bewilderingly enriched halo of human relationships emerging in a systematic way with the birth of our modern societies.

We are not exceptional in this respect. Social networks have almost the same properties as any other networks, including ant and bee communities. You may recall here our own herding behavior in panic. Before describing the special features of social nets, let me first list their general features. I gave quite a few examples of the small-worldness of social nets in Sect. 2.1, so let me restrict myself here to a Hungarian study. I was proud to read that the ten million people of my little country form a truly well-connected small world. In a survey by Utasi (2002), one third of Hungarians reported that they knew a national celebrity personally. *"This result depends heavily on what these guys call a national celebrity. We now have real-life shows 24 hours a day. Sooner or later one of your neighbors will ask for a camera in the bedroom and suddenly become a national celebrity in the field of nighttime acrobatics!"* Spite, 'these guys' as you call them are average Hungarians like yourself. They named real celebrities who had achieved something truly important and valuable.

Having saved my national pride from Spite, I started to read Mark Newman (2003c): "Consider two (fictitious) individuals. Individual A is a hermit with a lousy attitude and a bad breath to the point where it interferes with satellite broadcasts. He has only 10 acquaintances. Individual B is erudite, witty, charming, and a professional politician. She has 1 000 acquaintances. Is the average person likely to know A and B? Absolutely not. The average person is 100 times more likely to know B than A, since B knows 100 times as many people." My illusions were shattered. Mark had proved to me that my beloved Hungary is nothing special. The take-home message is simpler: well-known people are well-known. Let me make a final note on the small-worldness of social networks using another quotation from Mark Newman (2003c): "There are a number of morals to this story. Perhaps the most important of them is that your friends just aren't normal. No one's friends are. By the very fact of being someone's friend, friends select themselves. Friends are by definition friendly people, and your circle of friends will be a biased sample of the population because of it." Life is demanding. If you want to have a non-biased sample of friends, you should start to look for the hermit with the lousy attitude and the astronomically bad breath. NOW! This is the bad part of it. What is the good part?

You may delete the long row of numbers of charming politicians from your mobile.

Social networks are scale free. Liljeros et al. (2001) showed that the majority of the 2 810 Swedes who filled in a questionnaire on the most private aspects of their lives in 1996 had had zero or one sexual partners in the preceding year. However, a few of them had had more than ten. If cumulative numbers were assessed, the most active male had had almost a thousand partners already. However, this 'achievement' was almost unique. The same statistics appear again. You always have a chance of picking up a partner with an order of magnitude higher promiscuity (you may call it experience, depending on your own habits – Spite! Please do not grin!), but you have exactly an order of magnitude lesser chance for this. Good news for HIV prevention efforts. It is enough to warn, protect, educate and vaccinate the rather tiny, most active segment in order to make an efficient campaign for the whole. How can this tiny segment of the society be identified in an efficient, lawful and decent manner? The answer is not easy and requires a good measure of wisdom.

Social networks are modular and hierarchical. Modules, especially smaller ones, are bound together by strong links, while intermodular contacts are given by weak links (Girvan and Newman, 2002; White and Houseman, 2003). Modular structure is especially important for the efficient functioning of the social net and will be described further a bit later, where the special properties of these networks will be listed.

Social nets display similar dynamism to other networks. They do have a giant component and subsequent percolation, self-organized criticality, topological phase transitions and synchronization. Here the giant component means that most members of our social networks are connected. For this reason, percolation occurs, which means that news and effects can potentially reach almost all members of the network. Self-organized criticality is revealed by the avalanches I described in Sect. 3.2, and topological phase transitions have been listed in Sect. 3.4 and will be detailed further in Sect. 10.2. For the synchrony of social groups, audiences clapping together (Néda et al., 2000) or stadium visitors performing a Mexican wave (Farkas et al., 2002) are very good examples.

Besides the general network properties, social networks do display some key differences from other nets. A human baby has to learn a great deal. The human brain has quite remarkable plasticity. Rather a large amount of brain capacity has been developed to keep a complete record of our contacts.[5] Whom did I speak to, what did I tell her, how did she react, what am I supposed to do when I meet her next time, and so on. This is a LOT of information. Not surprisingly, we have to restrict friendship circles quite drastically to cope with the abundance of characters, situations, expectations and memories. We live in concentric circles of approximately 5, 15, 35, 80 and 150 people (Dunbar, 1998; Hill and Dunbar, 2003). This may correspond to our family and best friends (5 people), close friends (15 people), colleagues and pals (35 people), club and congregation members, fellow fathers and mothers at our kids' school, etc. (80 people), and our 'village' (150 people). Interestingly, people living in a megalopolis like New York also demarcate their 'village' in their neighborhood. Why do they do this? The answer is simple. We simply cannot put more faces into our cache memory. Why am I so sure that the size of our current cache memory is about 150 people? First of all I trust Dunbar (1998). Secondly, if you are a scientist, you agree that we have a good personal example of this. Imagine a conference. If it has 35 participants, it is a rather private discussion, where you will have a chance (and in most cases you dare) to discuss all the details. If it has 80 participants, it is a workshop, where you will still know most of the people by the end of the 2 or 3 days. If it has more than 150 participants you are lost. You cannot even remember all the faces. You feel lonely and tend to cling to the colleagues you knew from before.

The conference behavior mentioned above is a highly typical feature of social networks. You tend to associate with people who are similar to you. Since they also associate with people who are similar to them, you will in most cases find their friends friendly. Their friends will be your friends, too. This process, which is called clustering, is further helped by those occasions when you go out together to have a beer or make up a bowling party. Another rather characteristic feature of social networks is assortativity. It stems from the same habit: similar people like each other. Well-known people, who are usually hubs of the social network, attract other well-known people. Hermits make friends

[5] Let me give a personal example to demonstrate this. If you experience a big stress, or you have received a lot of information and got very tired, one of the first brain functions to switch off is your politeness. Exhaustion makes you inconsiderate. This is not because you like doing this, but rather because taking into account all the preferences, expectations, and game rules of others requires the handling of an astronomically large amount of data. This consumes a LOT of energy which can no longer be supplied by your exhausted brain, and therefore the whole function will be shut off and you will spend an afternoon when you will unintentionally but systematically hurt every single person who comes close to you.

Fig. 8.1. People living in a megalopolis like New York like to demarcate their 'village'

with hermits. (If they have that astronomically bad breath described by Mark Newman, they might use the Internet. However, if both of them have the same bad breath, then the similarity is complete and assortativity may develop without gas masks.)

As mentioned with regard to baboon society, our recognition is multivalent.[6] This complex recognition limits our capacity for the small circles mentioned before, bringing a highly developed modularity to the

[6] A good example of this multiplicity is Tante Sissi's German course, proletarian tea ceremony, and unforgettable role as a neighbor-interpreter, whereby she

hierarchy of human networks. However, the same complex recognition gives us the chance to form long-range contacts, and this provides an extraordinary flexibility and effectiveness to our networks.[7] We may be proud of ourselves. Our net is richer than that of any other animal society (Newman, 2003b; Newman and Park, 2003). Moreover, our network analysis and networking is a major part of those features which make us human. You may have been unaware of this fact up to now. However, it is high time you recognized your inherent commitment to network research!

The richness of our social dimensions is easily handled if the underlying social network is unchanging (stable). Small villages and other closed communities may master rules and attitudes with century-long practices and traditions. However, the large cities and our rushing, globalized world change our social net from one minute to the next. Expectations can no longer be foreseen and actions cannot be adjusted. We need a new vehicle for stability, and we have found it. Hammurapi, Mohammed, Moses and many other great personalities of human history proclaimed rules. Indeed, besides dogs, humans are the only creatures which follow rules, these being formed as laws in larger or changing societies. Our obedience to the law becomes a much more important principle than the actual benefit or righteousness of the law. Law-abidance is a highly important element of the stability of modern societies (Csányi, 2005). Law-abidance works as a strong link, providing a backbone of behavioral codes and restricting ambiguity.

Strong links are important in fresh democracies. When a sudden change occurs in your society, as happened in Central Eastern Europe in 1989, you think the world has been opened to you. It is very difficult to learn that even an extremely bad law is a thousand times better than the absolutism of the best and most understanding person on Earth. Moreover, it is difficult to accept that, without keeping even the most idiotic rules, democracy does not work. (For you, Spite, I have to add here that, without achieving consent to change these idiotic rules, democracy does not work either.) We have to keep this in mind when trying to export democracy to other regions of the world. A careful balance of strong and weak links is required, and we need highly detailed plans to facilitate the development of both.

---

uninterruptedly repeated the German text to the Hungarians in German, and the Hungarian text to the Germans in Hungarian for hours on end.

[7]I will return to the explanation of clustering and assortativity in the synthesis of Sect. 12.2.

We have learned that strong links are important because they substitute the lost continuity of modern societies. What other stabilizing forces do we have? You know what comes now, don't you? Allow me to jog your memory: weak . . . ? *"Weak what?"* Spite!!! Have you just started to read this book? Now go to the kitchen, make a strong coffee, and come back again.

Did you get that strong coffee? Good! Now you will get your weak links too, to get yourself stabilized. Let me remind you of Chap. 1 and the landmark studies by Mark Granovetter (1973). He discovered that useful information often comes from far away,[8] via long-range links. These long-range links are weak links (Onnela et al., 2005). Extending Granovetter's studies, Lin (1999) generalized the statement, saying that strong links are status-preserving, connecting similar elements, while weak links are status-enriching, connecting different elements.

However, Granovetter took a further great step. Comparing his primary findings on job-finding efficiency, he also stated that "weak ties play a role in effecting social cohesion" (Granovetter, 1973). Indeed, intermodular contacts are also weak. If social modules are bound together, conflicts can be settled peacefully. On the other hand, segmented societies have high conflict levels. Later, I will provide examples to illustrate the consequences of a lack of intermodular contacts.

If intersegmental weak links are important, how can we enrich them? We should learn from the bee and ant communities. We need divergence between the modules and between the individuals forming these modules to develop weak links. Different sensitivity thresholds are important examples of these divergences, since they develop the division of labor (Page and Erber, 2002). For the different sensitivity thresholds, Fewell (2003) gives the following excellent example: "Used dishes pile up in the sink, producing a continuously increasing stimulus. The dishes go unnoticed until the threshold of the most sensitive to them is met, and he or she washes them. This removes the dishes as a stimulus, further reducing the likelihood that the other group members will ever wash them. The result is a dishwashing specialist (much to his/her dismay), and a set of non-dishwashers. Similar interactions across other chores, from cleaning the bathroom to taking out the garbage, generate a division of labor for the household."

[8] The word 'far' demands some explanation here. In social nets distance is not physical distance in space, but distance along the social dimensions mentioned before. We can define distance in social networks via their assortativity and clustering. Due to the assortative and clustered network, our neighborhood is highly similar to us, and the 'further' we go, the more different are the people we reach.

The division of labor generates weak links and becomes a part of cohesion. However, I must warn here that the division of labor is a double-edged sword. On the one hand, the greater heterogeneity we have, the greater is the number of intersegmental relationships, and this gives greater chances that they will be based on weak links. On the other hand, a high drive for efficiency may overspecialize and alienate the members of various groups and act just in reverse, causing segmentation and instability (Degenne and Forsé, 1999; Durkheim, 1933; Utasi, 2002).

Diversity stabilizes only if tolerated. The very same diversity destabilizes if it is not tolerated. Continuing my remarks on the double-edged sword characteristic of labor division, let me add here that its cause, diversity itself, is also a double-edged sword. In Hungary, we had a great king, St. Stephen, roughly a thousand years ago. Before he died, he wrote a set of Admonitions to his son, St. Emeric. One of the most important pieces of advice he gave was this: "A country with one language and one custom is weak and perishable. Therefore, I order you, my son, to show goodwill to our visitors, and protect and cherish them, so that they will prefer to stay with you rather than to live elsewhere." Indeed, St. Stephen allowed Czech, German and Italian priests, as well as many refugees to come to Hungary, and strongly enforced this idea that the local Hungarians should recognize and tolerate them. This thousand year old counsel is valid today: all forms of diversity (Afro-Americans, Chinese, gays, gypsies, handicapped, Islamites, Jews, Kurds, lesbians, religious sects, talented people, Turks, to name but a few) will develop an abundance of weak links and will thereby increase the stability of the society – if they are tolerated by the majority and remain open themselves. But this very same diversity instantly becomes destabilizing if it is not tolerated by the majority. Then the minority becomes segregated and closeted, and will develop strong links. Consequently, the society will be segmented and unstable. From the middle of our own homogeneous group, the labeling of the diversities is rather easy. On the other hand, tolerance might be dangerous, since we have to fight with our own friends to get them to accept aliens. Is all this effort worth making, or would we do better to go back to the Middle Ages, before St. Stephen? I leave the answer to you.

Besides intermodular contacts, division of labor, and diversity, are there other means to develop weak links? Please pay attention! Our next animal lecture will now follow. We shall leave the bees and ants, and try to learn something from baboons. Grooming developed

Fig. 8.2. Grooming is a hygienic action but developed as a way of stabilizing society

amongst our monkey ancestors as a highly efficient way of stabilizing the monkey society. In principle, grooming is a hygienic action. However, monkeys spend an enormously large segment of their lives in grooming, in a way that goes far beyond the most extreme requirements of hygiene. We simply need to get groomed. However, to do it day and night in the 21st century would be very tiresome. Instead of grooming, we smile and chat. These are the most traditional pseudo-grooming actions (Dunbar, 1998). Friends, gossipmakers, hairdressers, madams, priests, psychologists, (to name but a few in alphabetical order) may all contribute to the weakly linked, background networks of human society. Part of their job is to be our pseudo-groomers.

Grooming or pseudo-grooming is our evolutionary duty. *"What if I do not have time for it?"* Spite! You always have time for something when you really want it. This is no excuse. However, pseudo-grooming may be an annoying action to some, generally male members of our society. (This is one of the reasons why they die earlier, but I will explain this only in Sect. 10.3.) Very inadequate persons, the VIPs of our society, are usually hubs with the same 1 000 acquaintances, like the erudite, witty, charming professional politician. VIPs develop a pseudo-circle around them for pseudo-grooming. (We can be quite sophisticated sometimes!) Let me explain this. They are always helped by a number of people, like babysitters, chauffeurs, janitors, and gardeners to name but a few. These helpers are not helping the VIP because the VIP has already forgotten how to take care of his baby, how to drive, how to clean the toilet, cut the grass, etc. As their most important task, these helpers create a net of weak links around the VIP, to do the pseudo-grooming in his place, thereby increasing the stability around him.[9]

From now on, we shall consider the strong links of the social network. White and Houseman (2003) described the societies of the Middle East and Afghanistan (parts of the Balkan region can be added to their description as well). These societies form highly clustered units with extremely strong internal links. Clustering causes high segmentation. In such societies conflicts escalate when the clusters split into opposing factions along highly predictable fault lines (Otterbein, 1968). It is time to remember the baboons and dolphins again. There, aged females maintained numerous weak links in the society and stabilized the groups (Connor et al., 1999; Lusseau, 2003; Silk et al., 2003). In the clustered, highly traditional Afghan and Middle Eastern families, women stay at home. When they are allowed to go out, they must not make numerous contacts, since this is a sign of infidelity which may bring a serious penalty. The society thus loses one of its most important stabilizing forces: women. Segmented societies are unstable (Degenne and Forsé, 1999). A lack of women-induced weak links makes them dangerously unstable (White and Houseman, 2003).

At the end of this part, let me bring a well-known example: Romeo and Juliet. Just imagine, if the grandmas of the Capulets and Montagus had got together to have a good chat. Whilst knitting, they would certainly discuss the fate of those poor youngsters who have fallen in love with each other. A few tears, and the story is over! The end will be a happy marriage, and poor William may start to search for another silly segment of the world, where women are not allowed to do their evolutionary duty – get together and have a chat.

I think I have to make two remarks here:

- I am not saying that women's only role in life is to get together to have a chat. In the next section, I will discuss the idea that without them, without female-type behavior, no firms, no human endeavor can be successful. What I mean is that any male chauvinist who

[9] I am grateful to István Kovács for this idea.

despises a single woman for engaging in 'idle chat' for hours is endangering his own stability.
- My second remark is even more important. I hope you have noted that women's stabilizing role is strikingly missing in all the major conflict areas of the world. The fight for women's rights (chats heavily included) is not an altruistic act of the male-dominated society. It is an outright necessity to save the stability of our planet.

Weak links and general welfare. Let me repeat the idea of Sect. 7.5 concerning the WEAKLINKER and STRONGLINKER personality traits, in a different context this time. As I mentioned there, Bateson et al. (2004) suggested that we may have two phenotypes: the SMALLS adapted to survival and the BIGS adapted to proliferation. I have suggested that SMALLS might correlate with STRONGLINKERS, while BIGS can afford the costs of building more weak links. Bateson et al. (2004) warned that changes in these phenotypes take place by a slow process, requiring many generations. The lack of weak links in Afghanistan, parts of the Middle East, and Albania is probably not only a cultural trait. We need a significant increase in general welfare *for many generations* before we can expect these basic trends to change. I would like to stress that this is not a note against educational programs. We need them as well. This note warns against impatience. The destabilizing, STRONGLINKER behavior may be partly genetic (or at least epigenetic). It cannot change from one second to the next. We must wait. A hundred years perhaps.

You may lean back and say: "Afghanistan and the Middle East are both very far away. You have convinced me. I will write a check and support a women's rights movement, as well as a long-term project to increase food supply there." A noble act, indeed. However, you yourself have your own duty here. Granovetter wrote a sequel to his famous 1973 publication in 1983. In this paper he summarized findings according to which, in the job market, less well educated respondents were most likely to use strong links to find a job. However, he also gave an explanation: both upper and lower class individuals are embedded in strong links and suffer a lack of cognitive flexibility. As a result, they develop arrogance towards any other approach than their own (Granovetter, 1983).

Now take a deep breath, drink a glass of crystal clear water, relax, and most importantly: think. I want you to make an inventory. Write down a list of your 15 closest friends. (This is the second circle of Dunbar.) Could you finish your list? Do you have 15 close friends?

Is there anyone on your list, who has an extremely different cultural, educational or financial background to you? Is there anyone on your list, whose background, values and norms are radically different from yours? Is there anyone on your list who would not read this book under any circumstances? *If you do not have such friends, stand up, leave this book on the table and start to collect them.* How do you expect to understand your own country if you lack the cognitive flexibility to do so? Do you really think you are doing your fair share to preserve the stability of your nation? The outside dangers that threaten Western countries may be just a prelude to the real threat of the cognitive isolation and the subsequent instability that slowly arises from within.

Let me emphasize this again. We need both weak and strong links to make our society stable. Actually, we are quite good at expressing both needs in a balanced way. Our conservative parties (if they know what is meant by this name) favor the building of strong links. On the other hand, our liberal parties (if they know what is meant by this name) favor the building of weak links. Neither of them is wrong. We need both at all times. Sometimes we need one of them more. In a wise nation, this is the party which wins the elections.

The two sources of STRONGLINKERS. *"I have to stop you here. Referring to Bateson et al. (2004), you said before that STRONGLINKERS correspond more or less to the SMALL phenotype. Do you mean to imply that conservatives are the poorest, who did not eat enough either in the past or even now? I admit, I am a bit young to have gained extensive social experience, but this sounds very suspicious to me."* Spite, you have hit upon a very good point again. Let me quote Granovetter (1983), who summarized a large number of studies and showed that strong links are typical in both the top and bottom segments of society. The richest and the poorest are both segregated. Obviously, the traditional core of conservativism is the former, while the SMALL phenotype is mainly found in the latter. However, the occasional merger of traditional conservativism which aims at the top 20% with populism aiming at those in the bottom 20% may have its root in the common preference of these 'target populations' for strong links. In fact, this is a winning strategy, since 40% of votes are enough to win an election in most countries.

To end my meditations on strong and weak links, let me put them into perspective. Without strong links, society has no structure, and it falls apart. Strong links, traditions, and order are necessary for social peace and for the effective and reliable functioning of society. Without

Fig. 8.3. Weak links require and cause cohesion, solidarity and trust in the society

strong links, society has no resistance and will be destroyed by some bigger danger. On the other hand, affection, goodwill and tolerance of diversity all build up informal networks and weak links. Weak links require and cause cohesion, solidarity and trust in society. Weak links stabilize society, emerging as key elements of its fitness and forming a part of its social capital. I hope that my examples of Afghanistan and others will convince the reader that this is no longer a matter of a single society. It has become our problem, too.

It is time to pose the following question: if we do eventually achieve a stabilized society, what good is it? This may sound rather a silly question, but it is worth looking at it from the network point of view. Let us learn another lesson from bees and ants. One of the most important network-related society functions is information management. Bees have to look for pollen, water, and many other ingredients of their daily life. Not to mention the alarm reactions due to periodic attacks by the Big Honey Parasite, the beekeeper. How can we inform each other about the news? How can we get answers to our questions? Strong links provide a fast source of established and focused knowledge. If I have a precisely formulated question and I know that one of my close friends is aware of the answer, I should ask her, since she knows me and I can be sure that she will sacrifice some of her time to answer my question. However, if I have an ill-formulated question or have no idea who might know the answer, it is better to mobilize a weak link. Why will a weak link serve me better in this case? Weak links are long-range contacts and they provide access to non-redundant information. If my knowledge is completely inappropriate even to formulate my question well, then how could I dare to think that my best friend will know the answer? As Mark Newman said (2003c): "By the very fact of being someone's friend, friends select themselves." Friends are friends because they have redundant interests. *"Peter, you made such a great issue of my role in stabilizing my own nation a few pages before that I decided to extend my friendships. I now have friends who are completely different from me."* Spite, I am indeed touched by the influence my thoughts have had on you. But believe me, you have become quite unique. Most of our friends are similar to us. This is the reason for the high levels of clustering and assortativity in social nets.

Opinion is stabilized by weak links. An interesting example of information spreading is opinion formation in networks. If the population is only allowed to make local interactions, it may change its opinion abruptly, manifesting a transition after diffusion-type propagation. However, long-range contacts stabilize original opinion (Kuperman and Zanette, 2001), which may show the stabilizing effect of weak links in this system.

Let me put the same message into another form: several important studies have shown that successful social searches require intermediate to weak links (Granovetter, 1973; Dodds et al., 2003a). Moreover, multiple identities in the form of multiple social dimensions help social searches. First the search goes along the first identity and then it is followed by the second (Watts et al., 2002). However, there is a twist here. Weak links may only give you non-redundant information if this information is easy to access. Tacit knowledge, meaning information which is protected or difficult to formulate, requires more than a weak link. It also needs benevolence – and competence-based trust (Hansen, 1999; Levin and Cross, 2004). Exposure to tacit knowledge needs strong links. Strong links are not individual strong links here, but rather a common attitude across the whole community, ensuring an open, trustful and helpful atmosphere.

We need both weak and strong links to find important information in a social net. How many links do we need? Should we add more and more links? Overconnected networks have low innovation potential, because innovation then spreads only very slowly (Rogers and Shoemaker, 1971). Actually, this is not at all surprising, since it follows from the same dilution effect that we have seen in Sect. 3.2 (Fink, 1991; May, 1973; Siljak 1978; Watts, 2002). If most of the network members have 1 000 contacts like the erudite, witty, charming professional politician, my innovation will go round and round helplessly, and will reach distant elements of the network paradoxically slowly. Moreover, strong interconnections may mean increased coherence, which biases the whole network to strong links, and makes it rather closed to any change coming from outside. We need not only an optimal amount of weak links, but an optimal amount of links as well.

The secret of the 'Martians' – who actually came from Hungary. As an example of success and cohesion in social nets, it is worth mentioning the scientific collaborations of a few legendary Hungarians. They were called Martians due to the peculiar language they used, i.e., Hungarian, which no one else understands. János von Neumann and Leó Szilárd were multitalented people who introduced seminal thoughts in many areas of science. The weak interdisciplinary links these scientists provided certainly increased the speed of advance in all these areas. Pál Erdős should also be mentioned. He stayed 'only' in mathematics, but with an unprecedented multidisciplinarity, made a significant contribution in numerous mathematical subdisciplines. With a vast number of collaborators, he became a worldwide stabilizing institution of mathematical research for several decades.

Heresy as a network phenomenon. In the medieval Catholic Church the fight against heresy unwittingly utilized the lessons from the network approach. The initial general persecution was soon followed by the development of the Inquisition to target key individuals. The investigations carried out by the Inquisition tried to extract similar information to today's sociological surveys (albeit with a rather different outcome and methodology). Murder, imprisonment or stigmatization of heretics was actually a way to prevent their further contacts in the society. Several practices of the Inquisition resemble those used in current strategies to prevent the spread of viral (e.g., HIV) infections. Fortunately, 700 years of moral development, which reached their culmination after the inhuman and barbaric acts perpetrated during World War II, have excluded the death penalty, substituted hospitals for prisons, and given special treatment in the place of stigmatization (Liljeros et al., 2001; Ormerod and Roach, 2004).

With the Martians and medieval heretics (actually, these two types of people are sometimes quite close to each other; think of Giordano Bruno or Galileo Galilei), I have now reached the social networks of society modules. Some features of these module networks will be discussed in the next section.

8.4 Firms and Human Organisations

Organizations, firms, and governmental systems are social networks, and have the same properties. So why should I waste your time with this section? The fact is that these organizational networks have a few special features. I will concentrate on these specialties here. I am fortunate, because Rob Cross and Andrew Parker published an excellent book in 2004, called *The Hidden Power of Social Networks*. This book gives hundreds of good pieces of advice on how to make a business network more efficient. This long line of advice starts with a simple idea: networks are needed for the efficient life of a firm. Quite surprisingly, this trivial observation is not so trivial for many chief executive officers (CEO). Employees make networks even if they do not need to do so. A number of extensive studies have proved that highly sophisticated and astronomically expensive data-base behemoths remain rather underused, since employees turn to their colleagues with their questions. Moreover, networks are not only important for the firm. They will be important for you, Spite, whenever you start your Real Life (*"Too bad, this man has realized that I am only a fiction. Better to shut my mouth for a while."*). Distinguished high performers have a large and more diversified personal network. So the take-home message, Spite, if you do not want to end up in the bottom 20%, is to be a diligent student. However, if you want to end up in the top 20% and not somewhere in-between, be a diligent student and learn also how to make an efficient personal network (Cross and Parker, 2004).

The benefits of personal networks. *"What a silly title!"* Spite might add. (Where are you, Spite?) This is obvious. If you have connections, you can reach resources. You will have a job, solve your problems, and live a happy life. Personal networks mean even more than that. Dunbar (2005) describe the observation that the number of your close friends – the number of people whom you are emotionally attached to and dependent on – correlates with your perspective-taking ability. The more close friends you have, the better you can imagine another's position, motives, or future acts. This secures a great advantage to plan your reaction, as well as to prevent any foreseen damage. However, there is still another advantage. Now take a deep breath, drink a glass of crystal clear water, relax, and most importantly:

think. *The more perspectives you can imagine, the better you will understand the people around you.* A personal network with long-range contacts involving friends from different social circles not only stabilizes society by making crucial links between segments which would otherwise remain isolated, but also stabilizes the microenvironment around you. Stabilization of our microenvironment is one of the great innovations of evolution, as I mentioned in Sect. 4.3. Most of our technical achievements, like clothes, housing, fire, air-conditioning, world trade (to procure fresh strawberries at Christmas!) all serve to do just this. The more isolated you stay, the bigger evolutionary failure you are from the network point of view. Do you still wonder why such people remain unsuccessful? I will return to this problem in Sect. 12.4, where I try to give a synthesis of the most important thoughts in the book.

Most organizational networks are small worlds with numerous statistical features, such as their size or the standard deviation of their growth, and a scale-free distribution (Axtell, 2001; Stanley et al., 1996). Due to their rather sophisticated task distribution, organizations usually have a modular structure. Now here comes the key feature: the efficient working of an organizational network needs good bridges between modules. These bridges have to be redundant. If someone gets sick, there has to be at least one more person who remains as a bridge. It is even better if the bridges are degenerate, where the links have a backup not by simple parallelisms, but by the general connection scheme of the system (Cross and Parker, 2004). Here it is worth recalling those intermodular boundaries called fringe areas in neuronal networks in Sect. 7.4 (Agnati et al., 2004). Carefully developed fringe areas are of special importance in business organizations. Fringe areas are key points of innovation, too. Firms participating in new forms of industrial district develop an intermediate description of their problems called pidgin formalization (Sabel, 2002). These problem descriptions are sufficiently detailed to refer to the original problem, but sufficiently abstract to provide a large enough contact surface for experts with completely different expertise to assess and help to solve them. As we will see later in this section, pidgin formalization is helped by weak links and played a major role in solving the 1997 Aisin crisis of the Toyota complex (Watts, 2003).

If you want to work efficiently, you need hierarchy. Hierarchical organization is especially characteristic of firms. According to the model of firm growth by Stanley et al. (1996), on average 70 to 90% of company decisions are determined by hierarchical orders. Hierarchy leads to another special feature: leadership collaboration. If leaders of smaller units are just treated as parts of a larger hierarchy and their horizontal contacts are not promoted effectively, network fragmentation may occur (Cross and Parker, 2004). Remembering the definition of network stability in Sect. 4.3, you will agree that this is a particularly dangerous situation since it may lead to a breakup of the giant component of the firm's network. If the giant component is broken, ideas cannot go around the network and the firm is in fact dead. Leadership collaboration is a necessity for the netsistance (life) of organizational networks.

Firms also display the topological phase transition phenomenon. For instance, as a business group grows, the hierarchy may increase, and the network structure may be reorganized along the random $\longrightarrow$ scale-free $\longrightarrow$ star configuration axis (Stark and Vedres, 2002). Rather interestingly, I have not yet found studies of self-organized criticality and synchronization phenomena in business networks,[10] which remain subjects of future research.

Firm quakes. In spite of the lack of data, I am quite sure that firm reorganizations satisfy the criteria of self-organized criticality. Most firms have some degree of resistance against gradual change. The cycle thus starts with a continuously growing tension. Things get worse and worse, but remain latent. "If I say anything, my bonus will be cancelled." "I had better concentrate on my job." Then all of a sudden, something really bad happens. The firm loses a very lucrative bid. Or the CEO runs over a stray cat on the way home and cannot sleep all night. As he tosses and turns, his self-guilt gradually expands from the ghost of the cat to the whole firm. By the next morning, the reorganization avalanche hits. The outcome is rather unpredictable. However, both the periodicity and the extent of change might actually follow a scale-free pattern.

As I mentioned above, firms, governmental systems and most other organizations are typical formal networks, i.e., hierarchical structures where links are, by definition, strong. In the remaining part of this section, I will list some interesting examples to demonstrate the need for a careful balance between strong and weak links in business networks.

Dodds et al. (2003b) showed that the most stable firms (they are said to be ultrarobust) are those which have an informal, background

[10] The Schumpeterian avalanches of innovations mentioned in Sect. 3.2 and other studies on economics are not strictly concerned with business networks, so I do not take them into consideration here.

network in addition to the main-frame of the firm's network, the 'chain of command'. Why is this important? In an organizational network, the elements are individual people. They may get sick. Moreover, their mother may get sick. More commonly, they may get overloaded. (Consequently, they may get sick, but this is another aspect of system biology.) In a good firm the CEO talks to the janitor if they happen to share a ride in the elevator. (To start with, in a good firm the CEO and the janitor use the same elevator!) Moreover, a good firm has a specially designed sanctuary of information exchange, innovation and life-and-death decisions: the cafeteria.

An interesting example of system stability is described by Duncan Watts (2003) in his very informative book *The Six Degrees*. The story is about the Toyota group in Japan. By the end of the 1990s, the production of automobile parts had become highly distributed over the hundreds of Toyota-affiliated firms. Moreover, due to the highly efficient and reliable supply system, the inventory of all automobile parts was reduced to maintain the production for approximately two days. A crucial element in the proper control of the rear brakes, the P-valve, was exclusively manufactured by the Kariya plant of the company Aisin Seiki. No plant meant no P-valves. And no P-valves meant no brakes from the second day on. No brakes, zero cars instead of the 32 500 made daily in all Toyota factories. Early in the morning of Saturday 1 February 1997, the Kariya plant burnt down. Production came to a complete standstill. Such a staggering loss would cause disaster in most automobile companies. But not in Toyota. Within a few days, 62 firms had started to make P-valves. None of these firms had any prior expertise in P-valve production. One of them was actually a manufacturer of sewing machines, with no previous experience whatever in the automobile industry. In two weeks the production of P-valves (and Toyota cars) was back to normal again. How could this happen? Well, weak links had played their part once again. The stress was distributed to hundreds of firms instead of striking one. Moreover, Kariya plant engineers specified the problem in the broadest terms possible using the minimum level of details and concentrating on the function of the P-valves rather than the method they had used to optimize their production in their own plant. This is a very good example of pidgin formalization, mentioned earlier as a fringe area, i.e., a proper contact surface between firms (Agnati et al., 2004; Sabel, 2002). The P-valve substitutes produced after two weeks were not identical, but they were functional, and this degeneracy saved Toyota as well as a large fraction of Japanese automobile exports in 1997 (Nishiguchi and Beaudet, 2000).

Hong and Page (2004) gave an exact proof that if they hypothetically selected a problem-solving team from a diverse population of intelligent agents, a team of randomly selected people would outperform a team comprised of the best-performing agents. Diversity and weak links pay off. They not only increase group stability, but also raise the innovative potential of the group. In a 'real-world' example, MacDuffie (1997) compared the problem-solving of three automobile companies, General Motors, Ford, and Honda in the 1990s. It turned out that weak links played an increasing role in the conditions, organization and general policy for problem-solving in the three companies, in that order. Examples of these weak links were the free interactions cutting through departmental lines, bridging persons, ad hoc, problem-based groups, a low level of centralization, etc. Not surprisingly, the number of customer complaints was highest at GM and lowest at Honda at the time of the survey. The Honda philosophy that "a problem with our product is a problem for the whole company, not for any individual or department" clearly paid off. Needless to say, in the Honda plants, the cafeteria was a central place for discussions!

Other examples of system stability were revealed after the terrorist attack against the World Trade Center on 11 September. Cantor Fitzgerald lost 700 out of its roughly one thousand employees in the disastrous collapse of the twin towers. But business had to go on, and the remaining employees decided to continue. Fortunately, all computer data were saved in several backup copies in remote locations. However, a formidable problem arose: all those who knew the password to access the data had died. So what the remaining employees did was this (Watts, 2003): "They sat around in a group and recalled everything they knew about their colleagues, everything they had done, everywhere they had been, and everything that had ever happened between them. And they managed to guess the passwords." This outstanding achievement of collective memory shows the power of filtering mechanisms resulting in the emergence of the important message from the noisy information background. This emergence is based on weak links.

The benefits of women CEOs, and ambiguity. As is quite clear already from the above examples, stabilizing weak links cannot be achieved by instructions, reorganizations and CEO memos. The philosophy, the whole mentality of the firm has to allow and encourage the formation of ever-changing weak contacts. The mixed-type (feminine and masculine) communication and leadership style is very helpful in this. Such a style should mix visions and clear instructions with politeness, empathy, continuous appreciation and nurturing. Women in leadership positions (or men who have been able to learn from their female colleagues) are beneficial to promote weak links. Ambiguity is also helpful to promote weak links. Ambiguity is a form of degeneracy giving similar solutions to the same problem. I will detail the connection between ambiguous meaning and weak-link-induced stability in Sect. 9.1. On the face of it, ambiguity and firms do not seem to go together very well. A firm either has a goal or it does not. Neither benchmarks nor deadlines leave much space for ambiguous information. However, if benchmarks set the goals and not the method or the elementary steps required to achieve these goals, innovative solutions may develop better. Such a general policy requires decentralization and autonomy (Grabher, 1993). All these promote weak links, which are necessary to maintain the innovative potential of the firm.

My last example concerning the effect of weak links brings in another aspect of firms: their ownership. An unstable environment leads to the diversification of ownership (Stark and Vedres, 2002). This change can be regarded as the establishment of an array of stabilizing weaker links in society in times when extra stabilization is needed.

Hey, Spite, where are you? In the other sections, by this time you had already launched a heavy protest against my demagogic and one-sided advocacy of weak links. Well, wherever you are, I shall talk about the strong links from now on. A rather typical example of strong links was the coal, iron and steel complex of the Ruhr area in Germany a few decades ago. In this context, the work of successive generations built up a common language with regard to technical matters, contracting rules and various other aspects of business life. When this industrial sector was hit by the big changes that occurred during the 1970s, these strong links prevented recognition of the change. Instead of reorienting and downsizing, an enormous technological development took place, resulting in a 'sailing-ship effect'.[11] Parametric rationality won out over strategic rationality, not only in the coal, iron and steel complex, but in many other business sectors, including the automobile and high-tech industries, which were based on a culture of traditional, strong

[11]The sailing ship effect refers to an unnecessary overshoot in the technological development, using the example of the sailing ship, where the most important improvements occurred after the introduction of the steamship.

links (Grabher, 1993). The take-home message is that too many strong links are not good for your success in business.

Hansen (1999) examined the development and dynamism of strong links in firms and provided more examples of the adverse effects of a strong-link surplus. Strong links soon bring redundant contacts, since close colleagues of two strongly linked people start to make friends with each other independently of the first two contacts, and a number of redundant contacts will therefore be made. This is a consequence of the clustered and assortative nature of social networks. In life, redundancy helps self-assurance and makes contacts accountable. In life, redundancy is helpful. In firms, a certain amount of redundancy is needed for the backup functions, e.g., for the bridging people I mentioned above. However, if a problem is discussed by five different pairs of colleagues in an almost identical fashion, it is rather a waste of time. Furthermore, if inter-unit links are strong, members spend their time helping other units instead of doing their own tasks (Hansen, 1999). Too large a number of strong links may make the firm inefficient.

Once again let me stress that strong links are mandatory for any organization. When the question is not highly focused, or the information required is complex, a weak link will not help, since it would cost a lot of time which can only be provided by a strong contact. Still, weak links are important for solving diffuse problems and instrumental tasks. On the other hand, strong links are preferred for normative expectations (Hansen, 1999).[12] This ambivalence of link strength requirements may have helped the relativization of link strengths in modern industrial districts, making them more competitive (Sabel, 2002). I will return to the relativization of link strength in modern societies in Sect. 8.6.

8.5 Dark Networks and Terror Nets

Dark networks are antisocial networks organized for terrorism and/or smuggling arms or drugs, and counterfeiting currency. It is essential to learn about their specificities if we hope to fight them effectively. What are their special features? Dark networks are not nested or hierarchical in the usual sense. Almost all these networks can be divided into a highly integrated core and one or more peripheral circles. The core consists of deeply trusted ties, while the periphery has very weak, or

[12] A good example of a diffuse problem is the task: design a car with an alternative energy source! A good example of a normative expectation is: make more cars! To make more cars, we need strong links. However, if we want to make not only more, but better cars, we will need weak links, too (MacDuffe, 1997).

Fig. 8.4. If benchmarks set goals without specifying elementary steps or a method for achieving them, innovative solutions may develop better

even no links to core members. Members of the periphery are always expendable and the core always remains isolated. In dark networks, the deeply trusted core contacts are massively redundant. However, the core contacts become invisible when an attack is imminent or executed (Krebs, 2002).

Dark networks are not small worlds and they do not follow scale-free statistics. The 19 members of the Al Qaeda network directly responsible for the September 11 terrorist attack had an average path length of 4.75, which is very high in such a small group. In these networks long-range links do not exist. If such contacts are ever made, they will be merely transitory shortcuts. These shortcuts are rare events, after which the contact goes dormant and perhaps never becomes activated again. These networks are extremely flexible. In the event of any threat, they become instantly disorganized. Netsistance is not a question here. If the few people in the core remain untouched, the whole network can be rebuilt again. Redundancy is high to help the reorganization process. This is not a firm. Efficiency and money are not an issue here either (Milward and Raab, 2003; Williams, 1998). These peculiar networks call for intensive studies to understand their extreme dynamism and to find their weak points.

Terrorist attacks are scale-free. A recent report by Clauset and Young (2005) showed that the frequency and severity of terrorist attacks between 1980 and 2002 followed a scale-free pattern. This pattern allows the authors to make a number of thrilling comparisons and predictions. While in 1980 there was an average of 4 days between consecutive terrorist events worldwide, in 1998 the average terror-free interval was already down to 17.3 hours. Probabilistically, an event of at least the total severity of the September 11 attacks can be expected by 2012, if the scale-free pattern is not broken in the meantime. The reason for the scale-free distribution is not exactly known. The notion I quoted in Sect. 2.2 whereby successful completion of many subtasks results in scale-free probability and clustering for the overall success of the complete task (Montroll and Shlesinger, 1982; Shockley, 1957) may explain the surprising trend.

8.6 Pseudo-Grooming

In Sect. 8.1, we learned from the baboons that grooming is good for our health, both physically and mentally. Larger groups cannot maintain their social well-being by this action. Behaving in the traditional way, an average human being would spend almost half of the day grooming, which is clearly impractical. We have thus invented pseudo-grooming instead. The small-talk I mentioned in Sect. 8.2 as a key asset of grandmothers in helping the stabilization of society is a typical example of this.

Gossiping is another important type of pseudo-grooming, providing occasions for a joint experience of empathy-strengthening group cohesion (Dunbar, 1998; Szvetelszky, 2003). A direct proof of the stabilizing role of gossiping has been found by Gabriel Weimann. Studying a kibbutz, he noted that (Granovetter, 1983): "Gossip becomes one of the social forces, suppressing deviants and holding obedience to the common norm. [...] By the transmission of gossip items, mainly via weak links, as shown in this research, the kibbutz social system can keep solidarity, sanctions and obedience in a heterogeneous, segmented social group." Indeed, gossip is thought to be developed to control the 'free-riders', those who take advantage of cooperative and altruistic human society (Dunbar, 2005).

Gossip and Slander. I should make a clear difference between gossip and slander (Dunbar 1998; Szvetelszky, 2002). As mentioned above, gossip occurs in a spirit of understanding and therefore increases group cohesion. In contrast, slander seeks to isolate and exclude. Gossiping does not require the receiver to reveal his or her attitude towards the content. Slander stigmatizes and seeks to make the target an outcast. In summary, gossip produces weak links, increases group cohesion, and stabilizes the group, whereas slander produces strong links, if any, and may strongly destabilize a group, increasing the risk of group fragmentation. In other words, slander may overcome the netsistance of a group. It is quite remarkable that the above differences in the effects of gossip and slander are independent of moral judgments, i.e., that slander deliberately seeks to hurt and is often an outright lie.

In modern societies, not only weak, but even strong personal links have been deteriorating. When did you last visit your mother? Or your brother? *"What a silly question! I live with them. We just celebrated my 18th birthday yesterday. I would love to be just visiting them both, but not all dreams come true in life."* Spite, welcome back and congratulations! Now, as you are now an officially declared adult in our country, Hungary, many of my previous criticisms can be withdrawn. I missed you. As an inhabitant of a megalopolis, Budapest, you may find yourself in the middle of a contact desert. An inhabitant of a housing project in one of the suburbs will often have this experience. The growing personal isolation and lack of safety demands a much greater compensation by weak links. In the absence of traditional, 'human' links like personal contacts, real grooming is out of the question. Pseudo-grooming like small-talk and gossip cannot be performed directly either. But this presents no problem because modern technology has invented (pseudo)^2-grooming. (Pseudo)^2-grooming, as its name suggests, is a double substitution. Here not only is the physical act of grooming substituted by a chat, but even the personal presence aspect of simple pseudo-grooming is no longer required. (Pseudo)^2-grooming can occur via a wide variety of means, such as the telephone, Internet, etc. These elements of modern

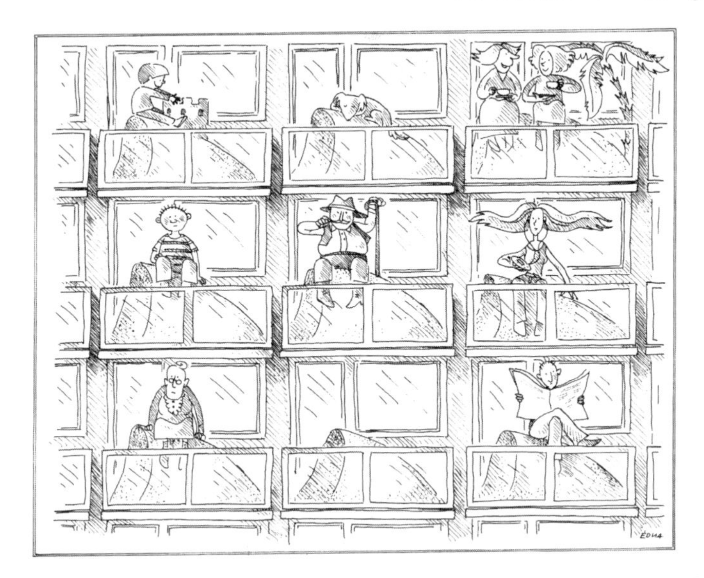

Fig. 8.5. An inhabitant of a megalopolis may find herself or himself in the middle of a contact desert

life have become the main stabilizers of psyches and societies in the 21st century.[13]

Cellular world. We have many examples of the importance of (pseudo)^2-grooming in our close neighborhood. Our kid starts to panic if the mobile phone accidentally remains at home. A rather serious depression may develop if the mobile phone gets stolen and the saved numbers are lost. The same depression may strike if our kid does not receive at least as many daily SMS messages as a best friend at school. We are no exceptions either. We carry not just one, but three mobile phones with us. What happens if the first suddenly loses power? What happens if even the second breaks down? Now a serious question again. Do you remember where the button is on your mobile phone which switches it off? Or are you also one of those who receive calls in public lavatories, public meetings, etc., and the list is long. I will never forget the Italian film where a mobile phone started to ring during a funeral. A real synchronization followed. Everyone checked to see whether it

[13]Pseudo-grooming is a double-edged sword. On the one hand, it helps us to survive the isolation of modernity and provides a whole arsenal of techniques for building and maintaining social links. On the other hand, pseudo-grooming reduces the complexity of our personal, face-to-face contacts to the simplicity of an SMS message. Moreover, pseudo-grooming gives an easy escape from the painful intensity of complex personal contacts, and tends to produce emotionally handicapped adults.

was hers or his that was ringing. Even the priest stopped the ceremony for a while to examine his mobile phone for a quick check. But the mobile phone kept ringing. Finally, the mourning relatives discovered (with great joy) that the mobile phone they thought to be lost had remained in grandpa's suit in the closed coffin, and by this accident it was serendipitously recovered. A happy ending, indeed.

Mobiles, SMS messages, and the Internet all serve as weak links, besides the strong links of personal contacts between people (see, e.g., Wellman, 2001), and bring an unprecedented flexibility to adjust link strength. It is now quite usual to break up a relationship with a farewell SMS. A string of 160 characters has become quite enough for this purpose. Andy Warhol's '15 minutes of fame' keeps shrinking. In the next part of the section, I will discuss several examples of modern habits and inventions for quick link strength adjustments.

(Pseudo)^3-grooming: radio and TV. Originally, radio and TV were a source of updated, sometimes live information and entertainment. Recently, these functions have increasingly been substituted by the newly acquired link functions of both. Call-ins and SMS votes all give a feeling of involvement and have led to the spread of (pseudo)^3-grooming. In this form of pseudo-grooming, not only grooming itself and actual physical presence have been replaced, as in the (pseudo)^2-grooming mentioned above, but the participants do not even need to know each other, and even the link between them is in fact a pseudo-link, since it is unilateral and at least partially virtual. Apart from the SMS surcharges, (pseudo)^3-grooming is a very inexpensive way to feel that one is part of the grooming community of the elderly female baboons in Sect. 8.1. Emotional involvement is low and the contact can be cut at any time without danger of retaliation. The involvement is self-centered, since the other partner is not even present, and it manifests a typical market behavior: when I need it, I take it; when I no longer need it, I dispose of it. You may actually call (pseudo)^3-grooming disposable grooming.

Mall kiss. Sex serves at least three purposes in primate communities. Besides giving a pleasurable means to repopulate the community, it is also a way of giving intimate and intense mutual joy to each other. Finally, sex is a source of stress relief. In its last function, sex actually serves in the same way as grooming (watch a bonobo, chimpanzee or real time show for a check). If you visit a plaza during the summer, you can witness the birth of a new type of schoolbreak sex: the mall kiss. During a perfectly executed mall kiss, the partners are keen to avoid even the faintest suggestion that they care about each other. Both of them are deliberately scanning the hemisphere behind the partner's shoulder to gather public opinion about their perfectionism, and to discover new subjects for the revised and enlarged editions of the mall kiss. If the partners are real professionals, they will from time to time make a 180 degree turn, allowing each other to scan the space all around. The mall kiss has nothing to do with the first two functions of sex mentioned above. Most of the time, the mall kiss does not even provide stress relief. Besides being an activity to kill time, it shows the signs of $(\text{pseudo})^3$-grooming: low, self-centered involvement, no endurance, and market behavior. You may actually call the mall kiss a form of disposable love.

Disposable love. Disposable love is a much broader phenomenon than the mall kiss mentioned above. Supermarket socialization has made disposability a key element of mate selection. The first step is to take it. I do not really care whether he or she wants it or not. The second step is to examine superficially what I have got. The third step is the decision. If I still like it, I use it. If I am dissatisfied, I dispose of it and go for the next round, hoping to get something better. I do not suffer, I do not try to develop myself for the partner or for the contact, I do not wait and I do not trust. If the relationship is not optimal from the start, if it does not immediately look like a movie romance, I quit, dispose of the love and look for a new one. Low, self-centered involvement, no endurance, and market behavior again. *"You are extremely biased. When I am thinking about my pretty girlfriend, Pity, whom I love with all my heart .... How dare you!"* I understand your feelings, Spite, and fully believe that you and Pity make a very balanced couple. Moreover, I did not mean to suggest that the behavior I describe is general. Fortunately, it is not. However, this behavior is spreading, which is interesting from the standpoint of social networks. This is the only reason I have included it here, and devoted such a lengthy space to it. Please insert the following standard sentence after what has been written and will be written in these boxes: This is not typical but gives us interesting lessons on social networks. Now, it is time to continue.

Signs for our link hunt. With such a fast merry-go-round of disposable contacts, new methods have to be developed to get the required number of new links. Fashion, smell, forehead bumper stickers about favorite singers, bands, teams, car makes, sports, and about a thousand other things make our teenagers a kind of open inventory for easy contacts as they move around in town.

Net links. In the last two decades entirely new forms of social organization have appeared: net groups, forums, chatboards, web logs (blogs), the list is long. These forms are all highly flexible. They easily establish social networks with each of their characteristics, such as modularity, hierarchy, etc. (Hallinan, 2003; Liben-Nowell, 2005). However, this network changes from day to day, from minute to minute. These groups do not restrict themselves to the virtual world. From time to time, some members want to meet in person. A chat party is then organised. A remark on a blog may be the first step towards a marriage. Everything is possible. Everything is flexible.

A postmodern sync: flash mob. Flash mob is a form of organization using stickers, Internet, or cellular phones to meet and do something unusual for a few minutes. As an example, imagine thirty people coming down on a subway escalator, one after the other, each of them demonstratively reading the same daily newspaper – upside-down. Flash mob collects together a pseudo-group. The participants seldom talk to each other, and very few know each other either before or after. Flash mob very quickly became global (`www.flashmob.com`), organizing worldwide events. But what is this phenomenon? In fact, it is a new form of synchrony which gives the participants the feeling that they belong somewhere – with the added security that their individuality and freedom will not be damaged. However, it is definitely not forbidden for two flash-mob participants to invite each other for a drink after the show. With joint participation, alienation has already been cast out.

Age relativism. If everything becomes mobile and flexible, the participants should also be mobile and flexible. Youth as a state is mobile and flexible. If you do not want to be petrified, if you do not want to be left out, you have to stay young. We are extremely close to the time when octagenarians will also be engaged in chat parties and flash mobs. Age-related hierarchy is about to be erased.

Gender relativism. Our bets on the gender of a teenager are growing increasingly risky. Same hairstyle, same cloths, same perfume. The situation is no better with older generations. Women should excel in our modern competition for life, while men have to socialize day and night to survive. Gender-based links are becoming less and less well defined.

Information relativism. A century ago, great-grandpas and great-grandmas had the newspaper, or from time to time a letter arrived, and if something really important happened, the village drummer came and announced the news. Information was unique and demanded strong links. Now, information has conquered our kitchen, bedroom and bathroom. The act of accessing information has become relative. Moreover, image flows have erased the differences between dogs and dinosaurs. For a modern kid both are toy figures. The content of information has become relative. Women and men love and die every minute in front of our very eyes. Fake love, fake death. The virtual is set up as real (Pléh, 1998).[14] The validity of information has become relative. We have developed immunity. All our links have been softened.

Athough all the above examples extend the number of ways of satisfying our increasing need for new contacts, most of these novel contacts are in fact weak links. Notwithstanding, in almost all these possibilities, there is a chance of making these links stronger. In fact, the illusion may arise that modern network members are masters of link strength adjustment in the sense that they may make the desired link as strong as they want. Let me note that this illusion shows the same low, self-centered involvement, without endurance, and typical of market behavior, as many of the symptoms I mentioned above.

In spite of the extremely flexible range of link strengths, it is rather obvious that the plethora of weak links established by the above inventions cannot be stable without strong links. The desire for strong links and the lack of strong links in most modern, broken families as well as the largely missing strong links of spirituality, induce a number of responses. One such is the appearance of postmodern tribes. These may be disguised as a religious sect, a political movement, or a gang. Alternatively, they may come in the form of an excessive habit. I go bowling every other day with my friends. Not because I like them so much, nor because I love bowling so much, but because my family life is a disaster and I do not want to witness it each night. In these modern tribes, totems and taboos are rather prevalent. A feature of our sect, movement or gang is the totem, and part or whole of the other sect, movement or gang is the taboo. This segregation of modern tribes can seriously undermine the stability of society. However, most of these strong links are in fact only pseudo-strong links. Why are they only

[14] The website of a popular chocolate quotes that school children believe that cows are lilac-colored. For them the difference between the real and the chocolate cow has become relative.

pseudo-strong? The reason is as above. Participants often have a low and self-centered involvement here as well. *"Do you really mean to say that a suicide bomber has a low and self-centered involvement?"* Well, Spite, we might actually start to argue over this, since I definitely see here a strongly self-centered behavior, and even the involvement may sometimes be low – I agree, seemingly paradoxically. However, my statement was not primarily about suicide bombers. I was only talking about *some* of the strong links.

Link strength has undergone a rather large-scale relativization in social networks. Link strength relativization in modern industrial districts makes them more competitive (Sabel, 2002). But is link strength relativization good in general? Park and Burgess noted in 1925 that transportation and communication "have multiplied the opportunities of the individual man for contact and for association with his fellows, but they have made these contacts and associations more transitory and less stable." We all feel the empty space that the erosion of strong links in the past has left behind.

Link relativism. Working conditions underwent a radical change in the 21st century. Many of us (who happen to be reading such a book) work at home or in some other kind of hermitage, and our only connection with the world comes in the form of broadband Internet, a mobile, or a combination of the two. Our food is stuffed into the fridge for weeks. Heating and hot water are taken entirely for granted. In some ultramodern homes, the microchip in the fridge orders the missing items automatically, and they are brought to our 'incoming fridge' without further ado. Modern life does not require personal contacts. Requirements and instructions arrive by email, and the result of our work is sent by another email a little later. Neither weak nor strong links are a matter of any relevance today. I am not forced to have a chat with the shopkeeper, because I can afford not go to the shop at all. We may plan both strong and weak links, and we do plan both in our modern life. The enormous liberty allowed by modern technology has significantly contributed to the relativization of link strength in society. Just as a final note, let me add that these thoughts are not restricted to the top (?) 20% of our world. Look around you. In China and India, there is an ongoing air-conditioning revolution. More and more people leave the unbearably hot streets in the summer and run to the enclosed safety of air-conditioned offices or homes. In parallel, they leave the occasional chats and links of the past, and start to live in our modern, air-conditioned isolation.

It is time to collect our luggage. We have just finished our fourth trip into Netland. Opening our luggage (Well, Spite, yours is full of photos of your girlfriend, Pity, I see. Oh! Sorry for my indiscretion!), let me summarize what we have collected here. We have got rich this time. We have found a large number of weak links in ant, bee, baboon, and dolphin communities. Some of these weak links were highly beneficial to stabilize the respective community. Human societies are definitely stabilized by weak links, starting with the role of women, through the intermodular weak contacts first discovered by Mark Granovetter (1973), to some more exotic forms such as the recently developed generations of pseudo-weak links. Diversity of the kind that arises from division of labor facilitates the formation of weak links. However, diversity is helpful only if it is tolerated by the majority. We have learned that both our firms and our kids can only be successful if they make enough weak links, since weak links make our world small and help in the spread of innovations. There were many take-home messages as well. It is our own duty to keep a large number of long-range, weak links in our personal social network, not only because it gives us a chance of being successful and stabilizes our strong links (e.g., our family life), but also because this is one way we can enhance the stability of our country. We also learned that tolerance and support for women's rights movements are not merely altruistic acts, but comprise further important elements to preserve and develop our own stability.

Now take a deep breath, drink a glass of crystal clear water, relax, and most importantly: think. When we started this chapter with bees, ants, and lion grandmothers, would you have thought that they would teach us such an important lesson? Next time, when you go to the zoo, pay them a visit and shake hands with them. (Do not worry, Spite! I only meant a virtual handshake with the lion grandmas.)

9 Networks of Human Culture

Now that you are back from the zoo, where you tried in vain to shake the six hands of all the ants for several hours,[1] it is time to start our fifth trip into Netland. Let us go and see what those macroscopic ants known as human beings can achieve. I will show you what type of networks we have figured out to support the last variety of social networks from the previous chapter.

9.1 The Language Net

Ants build nests, bees have hives, and both dance if they want to pass on information. We talk. "Words in human language interact within sentences in non-random ways, and allow us to construct an astronomic variety of sentences from a limited number of discrete units" (Ferrer Cancho and Sole, 2001). Indeed, the human language is a complex network, where stability can be defined as the stability of meaning.

The words of our texts have both a high level of clustering and a small path length, which makes the language net a small world (Ferrer Cancho and Sole, 2001; Steyvers and Tenenbaum, 2005). Moreover, words display a scale-free distribution, the so-called Zipf law (Skinner, 1937; Zipf, 1949). In the Zipf analysis, all words are arranged in a rank order from the most frequent to the least frequent. Interestingly, the most frequent words have the least semantic content (e.g., 'the', 'of', to name but two). The scale-free distribution of words can be explained as a balance between the minimal effort of the speaker who aims to use as few words as possible, and the minimal effort of the listener who aims to understand the message in as unique and precise a way as possible (see Fig. 9.1) (Ferrer Cancho and Sole, 2003). This explanation shows a kind of self-organized criticality phenomenon in

[1] If you have just opened the book here, you might think that this is an autobiography from a madhouse – if you feel uncomfortable with this notion, please turn back a page and read the end of the last chapter.

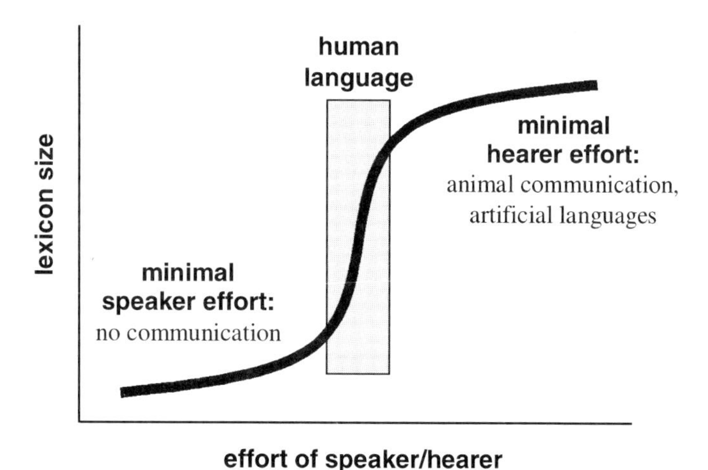

Fig. 9.1. The scale-free distribution of words: a balance between minimal efforts. The scale-free distribution arises as a balance between the intention of the speaker to use as few words as possible, and the minimal effort of the listener to understand the message in as unique and precise a way as possible (Ferrer Cancho and Sole, 2003)

language development. However, there may be other contributors to the surprisingly universal basic structure of our languages. Music has scale-free loudness and pitch fluctuations. Similarly, our speech also has a scale-free loudness pattern (Voss and Clarke, 1975). Whether the tonality of pre-languages also had a scale-free pattern, and if so, whether it influenced the development of the scale-free textual pattern of modern languages, is not actually known. Moreover, the systematic analysis of the possible influence of scale-free neural and cognitive patterns on the development of their textual representation is also in its early stages (Steyvers and Tenenbaum, 2005).

Besides the minimum number of words contra their maximal uniqueness, language has the same inherent ambivalence at a higher level. Indeed, it has to have a high degree of structure and at the same time it has to be highly informative. The high degree of structure gives us a minimum number of structural variations, while the high information content assures the maximal uniqueness of the meaning. These ambivalent features increase the complexity of language (Crutchfield, 1994).

Weak links are necessary for complex communication. If we imagine a hypothetical language where all words have a unique and exclusive meaning, with no ambiguity, no explanatory digressions, and no redundancy, then we end up with a rather simple language. As an example, you may think about the 'language' you use with your dog. "Sit!", "Stop!", "No!" are pretty clear messages, provided the dog is in the right mood. The transmitted information is focused, but not really complex. As another example, I remember when staying in London for a language course with my father and brother, we swore that we would talk to each other only in English. Keeping strictly to our promise as we walked into the Tate Gallery, we could only make the very same remark to one miracle after another: "How nice!" We soon felt like a bunch of morons and switched to Hungarian. Indeed, the meaning was focused and correct, the Turners and Blakes were indeed all nice there, but the complexity of their beauty required a larger vocabulary, with more degeneracy and ambiguity, and hence with more weak links between the words.

The language net can be divided into modules. Language modules have been defined via at least three highly different approaches. One of these stems from studies on cognitive deficits and language development (Levy, 1996). Another approach uses the relationships between the order of words conveying a meaning (the syntax) and the concepts covered by the words (Chomsky, 1957; 1968; 1975; Maynard-Smith and Szathmáry, 1995). The third type of language module definition arose from language usage and social dimensions. I will elaborate on this third modular description in some detail.

In language modules defined by social dimensions, we have a core module with a starting set of 300 to 400 words and all its variable extensions right up to Shakespeare and Ray Bradbury. There are various parallel modules, like slang, subspecifications of the scientific language, etc., which are all used by smaller modules of the society. These are extended, modified or replaced by regional modules which sometimes remain quite isolated, like versions of pidgin English.

Members of a society simultaneously use different language modules. The number of language modules depends on the number of social dimensions of their owner. The higher and the more diverse the social dimensionality of a society member, the richer the language she speaks. In this context it is not surprising that, from early childhood, rich and complex expression is an important factor of social success. Groups tend to select their leaders from those who speak more language modules, who have a higher number of social dimensions, and whose cognitive flexibility is greater. As I will describe in the synthesis of Chap. 12, cognitive flexibility leads to a better resolution of conflicts. Besides vision and willpower which set and maintain the goals of the group, skill in conflict resolution is a highly valuable leadership quality in a group which hopes to solve a complex task and to do it in an optimal way. This skill is usually first revealed by the number of language modules the owner is able to master.

Weak links help to renew a language. The contacts between the modules of the language net are usually provided by weak links of the social net. [As an example of this, a rare visitor to the Papua New Guinea tribe brings the word 'nettouck' (i.e., network) to the pidgin English vocabulary.] This role of weak links in language renewal (Milroy and Milroy, 1985) is very similar to the role of weak social links in the spread of innovations, as described earlier.

Ambiguity in our expressions arises from the same degeneracy that was detailed in Sect. 4.5 (Edelman and Gally, 2001) and leads to the emergence of weak links in the language network. Plural meanings of the same word provide an opportunity for more delicate expression, where the environment of the word profoundly influences its final meaning. Ambiguity is highly characteristic of the most frequent words with the least semantic content (Ferrer Cancho and Sole, 2003). An optimal amount of ambiguity, paradoxically, but in agreement with the role of the weak links it generates, stabilizes the meaning of the text as a whole. In the following points, I will list some examples relative to this statement:

- If we would like to get a proof of understanding, we do not ask for word-for-word repetition, but require a reformulation of our thoughts using different words (Pleh, 1988).
- If we want to know the meaning of a simple word in a foreign language, it is not enough to check 'the' corresponding word in a vocabulary list. First of all, there will be many possible meanings. To train our minds to grasp the extensions, modalities, and frequencies of the 'meaning field' of a particular word, we need to do quite a bit of 'networking'. When doing this, we check synonyms, expressions, etc. Moreover, to figure out the exact meaning of the word in its original place, we need to know its context in detail (Deacon, 1997). Many of the steps described here invoke a large amount of ambiguity, which helps to shape and stabilize both the complex and the actual meaning of the word. The more unusual the meaning of the text, the more stabilization it needs.

Ambiguity is all the more important in the stabilization of the language, since our language changes extremely rapidly due to the fast changes of social structure and the links it helps to represent (Dunbar, 2003).

Not surprisingly, ambiguity is an important element of any creative behavior, be it poetry, science or any other. As one of the related examples, scientific excellence has been shown to parallel tolerance of ambiguity (Stoycheva, 2003; Tegano 1990).

As an indirect example of ambiguity-induced stabilization, ambiguity helps arbitration. When two parties find a difference in their interests which cannot be bridged by any reasonable compromise, "ambiguity can be employed as a tool in achieving an adequate, if imperfect, settlement of the dispute" (Honeyman, 1987). Here ambiguity gives stabilization at another level, not by building weak links within the text, but by building weak links of both parties to the text. Although these links point to different segments or meanings of the same text, the fact that a commonly accepted text could be achieved without any negative ramifications tends to stabilize the situation.

A minimal ambiguity may act as a destabilizer. Ambiguity has to reach a threshold to act as a stabilizer. If ambiguity is below this threshold, the amount of weak links is not enough to make the meaning stable, but the remaining ambiguity is enough to cast doubt on the meaning itself. A rather absurd situation comes to mind to illustrate this. Our neighbor, Tante Sissi, was unbelievably old (at least for my seven-year-old mind) when she died. She had quite a number of friends living in Austria. One of them, a very respectable old lady, arrived a few weeks after Tante Sissi's death. I was the only person at home. My mother had taught me not to open the door to strangers. I observed the respectable old lady through a small window as she inquired about Tante Sissi, who had not written to her for a long time. My German was not that great, and I was also rather embarrassed by the situation, so I could tell her only the basic fact: "Tante Sissi hat gestorben." (Tante Sissi died.) The brutality of the statement made it unbelievable. Noticing the misbelief, I began to doubt whether my German was correct. Here the situation grew into an absurd modern drama: I disappeared into the room to look for a Hungarian–German dictionary, found 'died' in Hungarian, and pointed to the German version: 'sterben' in the dictionary. The lady began to cry. So did I. We kept crying for a while, each seeing the other's crying face framed in the small window. Here ambiguity was not enough to stabilize the message by putting it into a context. There were no polite preliminary sentences to prepare the lady for the bad news, and the short sentence had no emotional background to show her my participation, sympathy and consolation. However, lacking any skill in the language, there was still a remaining element of ambiguity (bad pronunciation, inappropriate intonation, etc.), and the message had therefore to be reduced to the level of absurdity to get it through.

Degeneracy is not only present in the form of ambiguity in our language. Small-talk also brings degeneracy to the semantic context of the discussion. The core of the information transmitted by small-talk is redundant due to previously known facts in most cases. However, the way the core information is presented in this form of communication builds up several layers of extended context and enriches the meaning. The degeneracy that arises develops weak links, which help to stabilize both the meaning and the community circulating the message (Dunbar, 1998).

9.2 Novels, Plays, and Films as Networks

When we construct a meaningful text, we build up each sentence from words which support each other and make the meaning of the sentence into a coherent whole. "Similarly, when we construct a paragraph of several sentences we make them support one another to form a meaningful paragraph. And the process continues, throughout a meaningful document, so that a meaningful text is evidently constructed by a recursively defined process: a meaningful text is an interdependent sequence of words and/or meaningful texts" (Scarrott, 1998). *"To be honest, when I am trying to follow your ideas, Peter, I have to say that quite often I do not see this 'recursively defined process' resulting in an 'interdependent sequence of words'."* Spite, you will be surprised. In spite of your continuous efforts to interrupt the interdependent sequence of my words, the text of this book is still a network. Let me give you just two significant examples to support this. If you carry out a statistical analysis of the references from this book (as a member of the LINK group, István Kovács, did), you will find that their frequency distribution follows a scale-free pattern. Moreover, the cross-references make the book into a small world. While writing these sentences, which seem to be fragmented to your young mind, Spite, I have made a well-organized network.

Stiller and Hudson (2005) showed that scenes from Shakespeare's dramas all describe a small world. Small-worldness is probably a general feature of all texts and serves the cognitive needs and capabilities of the audience. Moreover, the small world of the Shakespearean drama is constructed by ..., but perhaps I should ask you to guess? Yes, you figured it out correctly. The small-world of the Shakespearean drama is constructed by weak links. Scenes are connected by 'keystone' characters transmitting a minimum of necessary information to understand the connections between actors of the previous and the present scene. Actors of various scenes are involved in an intensive interaction with each other (developing strong links), but are linked only weakly to actors of other scenes by the keystone characters. Thus the keystone characters and the weak links they provide stabilize the network and the meaning of the whole play. In fact, this feature significantly contributes to our impression of complexity in Shakespearean drama, where the structure provides an optimal mix of segmentation and continuity (Dunbar 2005; Stiller and Hudson, 2005).

Superman as a WEAKLINKER. Modern cartoons are not about Hamlet, Gloucester, Romeo or Buckingham. Superman, Spiderman, or whateverman is the hero of modern times. He is omnipotent and generous. He helps everyone. He does not usually have a particularly stable family. He has lost his parents and had platonic love affairs, and he is a kind of angel figure. From the network point of view, supermen are typical WEAKLINKERS. They not only stabilize the story network and the diegetic landscape of the cartoon or film. They stabilize the diegetic world as well. Wherever they go, things get back to normal.[2] Everything seems to be quite clear, except one element. If social networks traditionally have women in the role of WEAKLINKERS, when shall we have a spiderwoman? As a possible answer, Hollywood may be promoting gender relativism with this message. Men should take over a part of the female tasks that go to forging weak links in order to allow women to participate better in male tasks. This also accelerates the link relativization mentioned in Sect. 8.6.

Let us return to Shakespeare. The comprehension of drama (novels, films, etc.) is governed by a match between the drama and the audience (Gallese and Goldman, 1998; Stiller and Hudson, 2005). In the optimal case, a very interesting and novel type of synchrony may develop between the dynamics of the drama network, and the neural state of the audience, both as a number of isolated individuals and as a synchronized network in itself. Obviously, the synchronized network

[2] I am grateful to István Kovács for this idea.

Fig. 9.2. In the optimal case, a very interesting, novel type of synchrony may develop

of the whole audience may only develop if several people watch the drama or movie at once. The probability and extent of this synchrony is determined by the pre-set cognitive and emotional schemes of the target person or persons. A synchronized synchrony of the whole audience requires a resonance between the drama and the joint cognitive and emotional schemes of most of the audience. If such a synchrony is achieved, a synchronized psychic relaxation may occur, which is called catharsis.[3]

At first glance, it was somewhat easier to reach catharsis in the ancient Greek and English times than it is now. People ate the same, had more or less the same troubles, and many of them knew each other personally and formed a rather closed community. Whom do you know in the darkness of a modern movie or traditional theatre? Still, we had masterpieces more than two thousand years ago, and we have them today. The question arises: what makes a masterpiece a masterpiece? *"My literature teacher drives me mad with a similar question: 'What might Shakespeare*

[3]Catharsis is similar to the synchronized laughter quake mentioned in Sect. 3.5.

have been thinking about when he wrote these lines?' Do you want to start the same game here? I am just asking you this because Pity and I are then leaving for a while." The answer to this question will be rather limited, adding only two new aspects, namely, networks and their synchronicity, to the wealth of previous attempts to solve the mystery.

What makes a masterpiece a masterpiece? Synchronization and resonance can obviously be achieved at the level of the story, values, etc. However, most of them are not network properties and are not even general enough to grant masterpiece status to the work for any extended period. Dunbar (2005) lists properties such as the correct reflection of evolutionary principles (like the protection of kin, rules of mate selection) as an important element of success. Using my limited network-oriented view to explain a long-lasting impression, sync can be achieved by mastering general network properties, i.e., small-worldness, scale-freeness, etc. We have already seen the small-worldness of the Shakespearean dramas (Stiller and Hudson, 2005). Does this property reach the masterpiece level? Dunbar and his colleagues' response is affirmative. They note that cast size usually matches the size of human networks, 25 to 35 individuals, which is the number of your average colleagues and pals, and the number of characters active in a scene at any one time is identical to the size of natural human conversations, involving a maximal number of 4 actors. Moreover, the internalization of the intentions, motivations, words and deeds of the 4 actors requires the audience to think to the fifth order, so to speak. A typical sentence to reflect this complexity is the following: I believe that A supposes that B intends to guess how C understands what D thinks. According to Dunbar's measures, fifth-order thinking is the usual cognitive limit. Thus a masterpiece works right at the cognitive limits of the audience. Why is a master a master? She (or in Shakespeare's case, he) has to work one order higher! The typical sentence is changed here to the following: the master supposes that the audience will believe that A supposes that B intends to guess how C understands what D thinks. Why do we not have more Shakespeares? There are only a small number of people who are able to work to the sixth order (Dunbar, 2005; Dunbar et al., 1994; Stiller et al., 2003).

More secrets of masterpieces. *"Preparing my final exam in high school, I have to read a lot. Let me assure you, Peter, that not all masterpieces are dramas. We have quite good novels with a few actors or with no actors* *at all. How do these novels achieve their synchrony?"* The language net provides a lot more options for obtaining a sync feeling from the reader. Now here comes my other hero: scale-freeness. In Sect. 2.2, I mentioned a number of evolutionary and other reasons explaining why we like scale-freeness. I also mentioned that music contains a lot of scale-free distributions. Novels may not be exceptions. I already mentioned that the occurrence of the references of this book also has a scale-free distribution, which probably reflects the rank of importance of the works quoted. There seems to be nothing special in this, since many scientific papers examined so far by LINK member István Kovács have the same distribution statistics. However, real masters may add several layers of scale-free complexity. They may employ a scale-free distribution of link strengths between various actors.[4] The master may incorporate scale-freeness by the use of structural elements (conversations, dreams, changes in the line of the plot, etc.), motifs (metaphors, unusual words, symbols, colors, smells, etc.) or conceptual elements (ideas, values, etc.). Scale-freeness may be present in all these, simultaneously giving a rich pink-noise network to the novel, the masterpiece. *"Peter, do you really think that poor old Tolstoy sat with a 'schoty' (just for you Peter, this is the Russian name for an abacus) calculating whether he had already reached a simultaneous scale-free distribution of the thousand things you listed in* **War and Peace***? Peter, this is absurd."* No, Spite, have you ever seen me checking whether my references have reached a scale-free pattern? I have never done that. But the references still have a scale-free distribution. In Tolstoy's head, the whole statistics was just there in its whole marvellous complexity. Moreover, great masters will certainly introduce novel quakes by building up a tension in the novel and than allowing a delayed but sudden relaxation. They will also use topological phase transitions in the textual network of the novel, synchronicity of all the elements listed above, and many more.

☺ **Network secrets of Greek and Roman mythology.** A recent report by Choi and Kim (2005) analyzed the virtual network of 1 647 gods, heroes, monsters, mortals and fairies in the Greek and Roman mythology. They found a scale-free degree distribution of the actors in the mythology network, which was a hierarchical small world.

---

[4]We usually have a main actor, a few key actors, some important roles and several minor roles; in inverse proportion, we learn a lot about the character of the main actor, somewhat less about the key actors, usually a single feature concerning the important actors, and practically nothing about the minor roles.

When the master goes too far. What happens, if the master is too unique? How will she be understood if she has inherited an exceptional brain which works not only to the sixth, but also to the seventh, eighth or even higher order (Dunbar, 2005)? We have two options here: the sync of her work (1) either contains a sub-network, which produces sync at the fifth order, and it will be a masterpiece, or (2) it contains a masterpiece at the seventh order which can be understood and enjoyed by approximately a dozen people on Earth (all fellow masters themselves, who are obviously biased). In the latter case, the master can only hope that evolution will make our brain capable of working to higher orders in the future and that she will be rediscovered in the 33rd century.[5]

Masters of the future. Oral communication proceeds using multiple layers of metacommunication. On the other hand, its transcribed versions have to cope with a one-dimensional chain of information (Nyíri, 2003). Over the millennia, masters have become able to construct a multidimensional rhythm and excitement cross-connecting the one-dimensional flow of words. Digital multimedia documents have the advantage of starting from the network-type organization of hypertext and from the multiple dimensions of sounds and image. Mastering and understanding the single dimension took us several thousand years. Hopefully a little respite in today's technical developments might allow the human brain to discover the further cognitive abilities required to create and enjoy the synchrony of multidimensional possibilities.

Are the great masters healthier? The complex interaction of actors and their possible reflection in the audience's and the master's brains makes me think about its similarity with the idiotype network introduced by Jerne (1974) and mentioned in Sect. 7.1. This network can be described by: a lymphocyte I acts on antibodies of lymphocyte A, acting on antibodies of lymphocyte B, acting on antibodies of lymphocyte C, acting on antibodies of lymphocyte D. Is this not rather similar to the network: I believe that A supposes that B intends to guess how C understands what D thinks? Immune cells are similar to neurons in a number of ways. Could this be one of them? In most cases, the cross-reaction of lymphocytes decreases rather abruptly after the third–fourth interaction (Ab3/Ab4),

[5] Actually, it would be extremely interesting to make an 'order analysis' of some masterpieces recently rediscovered just to see if we are already a bit better off than our ancestors. Reading Shakespeare, one doubts this.

and the sixth–seventh interaction (Ab6/Ab7) can be regarded as exceptional (Bona et al., 1981; Weisbuch et al., 1990). Complex idiotype repertoires are characteristic of early childhood and they decay later. Do some people keep them much more than average? Do great masters have a more complex anti-idiotype network? Are great masters healthier? *"Peter, I am beginning to doubt whether it was worth staying. First of all, why should there be relationship between a more complex anti-idiotype network and better health? Secondly, you yourself have quite often mentioned Mozart. He was a great master, as I hope we may agree. But was Mozart healthier? The poor guy died at the age of 35 probably from eating undercooked pork and getting trichinosis, which is a bunch of ugly worms enjoying themselves in your muscles (Hirschmann, 2001). Do you call this exceptional health?"* Well done, Spite! Thanks for the reference. However, even if this was the case (which is debatable, I suggest you read Dupouy-Camet, 2002), I do not think that a better immune system can necessarily fend off a massive parasite contagion. However, in answer to your first question, I must admit that I have no evidence for saying that a more complex immune network leads to any advantage in health.

The network approach has taught us an important lesson: a masterpiece is a masterpiece because it provokes a synchronized psychic relaxation of the audience,[6] ensuring both group cohesion and the safety of its members (Dunbar, 2005). Most masterpieces work at the limits of our cognition, and provide a complexity of network structure and dynamism which by itself provokes a relaxation, i.e., catharsis, in the reader or spectator.

The nested sync of masterpieces. Spite, this is the moment to take your leave with Pity. In Sect. 7.5, I was talking about synchronization events which may occur between various networks, proposing the hypothesis that, if the synchronization within a network is exceptionally well organized and strong, it may provoke the synchronization of a network one or more levels higher. A real masterpiece certainly reflects the sync of the master's mind and in turn induces the synchronization of the reader's neurons.[7] However, a masterpiece also induces the synchronization

[6] To understand the generality of this, you may remember here the cathartic effect of any 'Aha!' phenomenon or extensive laughter you have experienced in the past.

[7] It would be a nice experiment to compare the neuronal synchronization level when reading Tolstoy and a bad contemporary Russian writer. Moreover, the comparison of the sync of Tolstoy's mind and the bad contemporary writer's mind would certainly have been a great piece of study.

of a whole audience, and maybe of mankind as a whole! (Before you protest I remind you again that we are in a three-smiley box. Watch my hands! I am cheating!)

Now take a deep breath, drink a glass of crystal clear water, relax, and most importantly: think. *If you want to be a master*,[8] you have to practice. *"What should I practice?" Practice empathy*. Try to understand how others think. What are their beliefs? What are their motivations? Why do they do what they actually happen to do? Play around with it. Move the center of your thoughts from your own head to theirs. Look around with their eyes and try to get an idea of what the world looks like from this new vantage point. And make a careful note here: this is *not* only an altruistic act, which will help to stabilize the world around you. This is also the key to your own success. However, there is one more important note here. If you succeed in understanding how a person thinks but that person is similar to you, then you will have achieved practically nothing. If you succeed in understanding someone who is separated from you by a world of cultural features, customs, and beliefs, *that* will be the point when you may begin to aspire to being a real master.

9.3 Our Engineered Space

When I first inserted the title of this section in the book, I wanted to find the answer to a question which has haunted me for a long time: Why do people like Budapest? As I organized a world congress in my home town during the preparation of this book, I often heard the remark: "Oh, Budapest! I have been there, and I return whenever I can." But why do they want to return? Whenever I ask the question, I do not really get a convincing answer. Yes, life is reasonably cheap, the food is good, the wine is fine and Hungarian girls are pretty (Spite, you may stop your glorious smile and vigorous nodding). But you can find all this in a great many places in the world. So I kept on asking and in the end my poor respondents could not say anything else but: "You know, Peter, Budapest is a cosy place. We feel well when walking in the streets and looking around."

They might be right. I feel extremely well myself when walking around in Budapest, Paris or Venice, just to name a few of the great

[8] The term 'master' does not refer only to a master of drama or creative writing here. As you will see in Sect. 10.1, a market guru uses the very same cognitive abilities.

cities of our world. In contrast to this, I remember my discomfort when walking around in Kuala Lumpur. The buildings were beautiful and life was ultramodern, but still, I could not enjoy it. It took me days until I realized what the main reason was. I could not walk straight without feeling perturbed. After 3 to 4 meters I either had to change direction, or step down from the enormously high sidewalk and then step up again. I discovered in Kuala Lumpur that space distribution can be crucially important for one's personal comfort. Let me mention two more personal examples, and then I will revert to the scientific approach. Space distribution is not only an issue for town planning. It is also a key point of architectural design. When a new plaza was opened in Budapest, I felt awful when I visited it first. It was alienating and empty. I went there half a year later and entered a different world. I felt quite at home. Wow! Was I in the same place? What had happened there? Practically nothing. Only a few kiosks, benches, and flower boxes had been added. But could it make such a vast difference to break up an empty space in such a subtle way? My last examples are Epidauros and the Summer Palace in Beijing. It may be surprising to find them in the same paragraph. However, it was not the building which made this everlasting impression on me in these places, but the landscape. The harmony and the rhythm of the hills one behind the other was the real treasure of these two magnificent places. I learned that space distribution is extremely important for a good life.

Let me start as I did two sections ago: Ants build nests, bees have hives, and both dance if they want to pass on information. In that section, we examined our human copy of dancing: language. Now we shall examine the human version of nest-building, i.e., our engineered space. Staying for a second with animals, I shall use mole-rats as my first example. Mole-rats dig a rather extensive place to live in. The burrow of the naked mole-rat may exceed 3 kilometers in length and has a complex structure with different parts serving as foraging galleries, nest chambers, food stores, and toilet chambers. Mole-rat burrows have scale-free distributions. They are fractal-type objects (see Fig. 9.3) (Le Comber et al., 2002). This structure is due to the very same Levy-flight search statistics (Levy, 1937) I discussed in Sect. 2.2. If a mole-rat wants to explore its environment for more food in the most efficient way, it has to make the same area-restricted search with a scale-free distribution as albatrosses, ants, bees, deer, fruit flies, jackals or monkeys (Atkinson et al., 2002; Cole, 1995; Ramos-Fernandez et al., 2004; Viswanathan et al., 1998; 1999).

Fig. 9.3. The fractal properties of mole-rat burrows. The figure shows a schematic representation of a burrow of the mole-rat, *Cryptomys hottentous*, in a mesic environment. The burrow has a scale-free distribution of route lengths, which infers a fractal property. (Reproduced with kind permission from Le Comber et al., 2002). Natural or architectural objects are not pure mathematical fractals and their self-similarity is rather limited

As already mentioned, Levy flight was developed as an optimal search strategy, and seems to be conserved throughout evolution. We are no exception. If we scan an image, our gaze shifts from one place to another, producing a Levy-flight distribution (Boccignone and Ferraro, 2004). If we browse the World Wide Web, we make Levy flights (Huberman et al., 1998). Humans display a consistent aesthetic preference for fractal images (Hagerhall et al., 2004; Spehar et al., 2003). The pre-set space is scale-free in our mind. When we build one, whether we like it or not, we have to follow this pattern if we want to make ourselves feel at home. Le Corbusier and other great architects may have wished to educate mankind differently, but no one should forget that this is a million-year project. Genes are slow. *"Peter, may I inquire what are you talking about? You seem to be fighting against something, but you have forgotten to define the goal!"* Yes, Spite, you are right again. Seeing our beautiful towns filled with the horror of block-shaped houses, and having dreamt in my childhood of becoming a sculptor or architect, this is a very important issue for me. I lost my temper and I am sorry about that. So let us start from the beginning.

In traditional architecture we have a lot of buildings which have a scale-free space distribution. This is usually called the fractal property after Benoit Mandelbrot, who discovered the generality of the scale-free space distribution of various natural (cloud, mountain, tree,

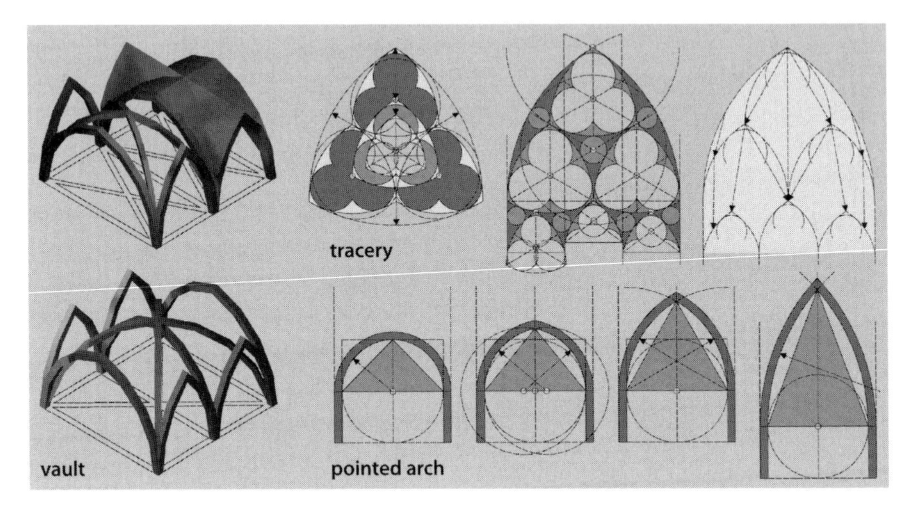

Fig. 9.4. Fractals in the Gothic style. The figure shows the fractal properties of the pointed arch, a major motif in the Gothic style. The variable, self-similar scales are highlighted. (Reprinted with kind permission from Lorenz, 2003.) Here again, self-similarity is not as pronounced as in a pure, mathematical fractal

snowflake) and man-made objects (Mandelbrot, 1977). Reims Cathedral in France, the Ca' d'Oro and the Palace of the Doges in Venice all have a scale-free pattern. They are self-similar. However, self-similarity is not always complete here, and does not extend to several scales but remains restricted to just a few. A great part of the exceptional beauty of these buildings comes from the fact that they have a fractal property (see Fig. 9.4) (Lorenz, 2003). Before Spite asks the same question as in the last section, when he mentioned Tolstoy's abacus, let me note that this part of fractal design was not planned. It came from the harmony between the old masters and nature. However, the architecture of the 20th century got rid of ornaments, porticos, gazebos and all the beautiful little adornments of past masterpieces. Several scales have simply been erased, striking a damaging blow to the scale-freeness of our evolutionary past.

A year after the English version of Mandelbrot's book (1977) was published, Peter Eisenman exhibited the House 11a project. Eisenman thought of fractals (scale-FREE-ness) as a mix of three destabilizing concepts: discontinuity, recursivity and self-similarity contradicting our presence, origin and aesthetic object, respectively (Ostwald, 2001). Fractal architecture became akin to chaos theory, turning the original idea upside down. Scale-freeness was not a result of organic development, but a negation of the order forced on us by the 20th century masters. Not surprisingly, the House 11a versions did not have much to do with comfort. The smallest house had human height but was obviously not a house, and the largest object was too large to be a house. Furthermore, the right-sized object was filled with an infinite series of scaled versions rendering it useless as a house. Still, the fractal design brought a great resonance, and within a decade several hundred works had been designed or published in the field (Ostwald, 2001).

Fractal architecture soon lost its appeal. In the 1990s more and more people started to criticize the initial zeal (Ostwald, 2001). However, the basic idea refused to lie down. When Carl Bovill (1995) published his well-balanced book on fractals in architecture, self-organizing, natural forms were slowly rediscovered and 'true' scale-freeness was reinstated: the scales that are really important to us in the centimeter to meter range regained their rightful place. Scale-freeness seems to be a key point of human-friendly architecture. However, the nodes of this self-similarity are also important. The non-metric system has some very good examples of this. If we miss, or under-represent the inch–foot–yard range in space, the building we get may conform to the purest 'fractalism', but it will lose any hope of being enjoyable. Space design has to achieve a synchronicity with our own human measures.

Cities are fractal (Batty and Longley, 1994; Portugali, 1999). One of the oldest scale-free distribution laws, Zipf's law of town-size distribution (Zipf, 1949) already shows the level of self-organization in cities. There have been many approaches to explain scale-free town development, such as the diffusion-limited aggregation model (Vicsek, 1989), the correlated percolation model (Makse et al., 1995), or the intermittency and cluster analysis models (Zanette and Manrubia, 1997). Many of these models reflect the fact that towns emerge from local actions, display a hierarchical modular structure, and grow in a way that obeys the Matthew effect mentioned in Sect. 2.2. Here the Matthew effect reflects the fact that the probability of the development of an area depends on the occupancy of its neighborhood.

What are the space representations of scale-freeness in our towns? We have the same situation here as in the great novels of the last section. In the case of organic development, a lot of measures become scale-free in a parallel way. We have a scale-free distribution of object sizes from the smallest element of shop windows up to the largest squares.[9] There is scale-freeness in the distance what we can cover

[9]This shows the utmost importance of keeping our shop windows full of exciting little treasures and not allowing them to be covered by a giant poster. It is not

undisturbed. Suburbs grow according to the scale-free rule (Makse et al., 1995). We have a scale-free distribution vertically as well, in the sense that most of the houses are rather small, but some are bigger, and a very few are really big. If we have beautiful hills around (but not all around!), that makes our situation much better. Towns are indeed fractal (Batty and Longley, 1994; Portugali, 1999; Vicsek, 1991).

"Let me ask a question here: are you suggesting that towns should be allowed to develop as they wish? Should we still swim in the organic sewage of the organically developed medieval towns?" No, from time to time we have to correct the organic development of our towns. When Haussmann designed the avenues and boulevards of Paris, cutting them into the flesh of the city, he just inserted the missing scale, which could not have been developed in the medieval town. Pope Sixtus V did the same with Rome. The same reason underlies the development of large parks. There can be no doubt that we have to help organic development. But we have to do it wisely. Nikos Salingaros (2004) listed a series of guidelines in his keynote speech to the 5th Biannual Congress of Town Planners in 2003:

- Re-establish and protect the small scales: pedestrian network, ornaments, kiosks, low walls for sitting, etc., to provide links with the town at the level of physical intimacy; increase contrasts to give visual excitement and make all this harmonically multifunctional to provide an emotionally nourishing physical environment.
- Slow down the big scales: highway construction, skyscrapers, suburban sprawl.
- Provide a scale-free green space distribution.
- Establish an organic integration of the 'electronic city'.

Other network features also operate in our engineered space. A good town is not only scale-free; it is also a small world, with a number of long-range connections, either in the traffic or at the level of the 'electronic city' (Salingaros, 2004). Moreover, a good town is nested. Its functions are mixed and intertwined (Alexander, 1965).

Architectural design reduced the diversity of 'modern' towns in the 20th century. Alternative routes were cut and *the* highway was continually widened and extended to serve increasing traffic levels (Alexander, 1965; Salingaros, 2004). The re-establishment of alternative paths would make a lot more weak links, which would help to stabilize town

---

surprising that properly arranged shop windows are another hallmark of the best cities around the world.

traffic. Similarly, the suggestion of mixing up the functional areas of the town, destroying the Le Corbusier-type home–office segregation (Alexander, 1965; Salingaros, 2004), also gives rise to stabilizing weak links, as we have seen in many previous examples.

Extra care should also be taken here with regard to fringe areas. Like the brain or business organizations, the crossover areas of urban traffic have to be developed with extreme care (Salingaros, 2004). As an example, the area where you leave your car and walk to the subway is especially important. Here we should put a number of stabilizing weak links, like places where people can stop, sit for a while, or perhaps wait for each other and have a chat. Here we should pay special attention to scale-free distributions by introducing a lot of natural fractal objects, like trees and other plants of various sizes. What do we have today instead of these green community places? We have dreary concrete multistorey car parks and asphalt wastelands.

The last few sentences have touched upon another issue here. With proper planning, we are not only promoting weak links inside the urban traffic network; we are also promoting the formation of weak links between the inhabitants of the town. Therefore, in the same act, we achieve stabilization of the traffic and help to stabilize town life as well.

A properly engineered space is a masterpiece. It requires the same multitude of scale-freeness and synchrony as the masterpieces of literature I mentioned in the last section. This will bring a resonance and synchrony to visitors and inhabitants (Mikiten et al., 2000). This is the same magnificent, serene feeling that all of us may experience upon entering St. Peter's Cathedral in Rome or walking around a zen garden in Kyoto. The design of a masterpiece requires a master. This master does not have to understand human relationships to 6th order, like Shakespeare. This master has to understand a similarly high level of spatial complexity. This master has to have a high tolerance to the ambiguity of overlapping elements and designs (Alexander, 1965). This master has to think at least to the 6th order with regard to space. Modern engineering has become an art, as I will discuss in Sect. 9.5.

I think I have now found the answer to my question. We like Paris, Venice, Budapest and all the real towns of the world because we feel the harmony and wisdom of generations of great masters there. These towns give us the same joy of collective relaxation as laughter, clapping, Mexican waves in a stadium, or a great novel.

9.4 Software Nets

We have arrived at the last section on human-made networks. *"I am missing something here. Actually, not just something, but a LOT of things. Do you have nothing to say about power nets, communication networks, transportation networks, the Internet, the World Wide Web, electronic circuits and the like? Are human achievements limited to language, novels, space and software in your mind? Although I am now trying to empty my brain after my successful final exam at high school, my mind is still definitely broader than this."* Congratulations, Spite, on your successful exam! I do not doubt that your mind is broader than mine, since the IQ declines with age. Therefore I should be making every effort to catch up. However, let me assure you that it is not my proverbial absentmindedness that makes me leave out all the networks you so wisely listed. Some of these networks, like the power net, have already been covered, while my literature search in other areas did not lead to non-redundant information, besides the small-worldness, scale-freeness and nestedness of these other networks.

Software segments manifest network behavior with scale-free and small-world connectivity (Valverde et al., 2002; Myers, 2003; Potanin et al., 2005). However, software design differs from the other networks mentioned above. Evolvability, which means the possibility of highly flexible further development, is a primary concern here. Modern software is designed through a parallel process by many teams and individual developers. Consequently, modularity and hierarchy are exceptionally highly evolved in the various program packages and have become a key issue for software quality. The multidimensional and multifaceted links between modules have to be well designed (Gamma et al., 1994; Pressman, 1992). Modules are usually 'pre-fitted' to a large number of possible cooperating modules by implementing 'design patterns', constructing the intermodular fringe areas mentioned in the context of brain organization (Agnati et al., 2004), and in the last section. Design patterns are a form of pidgin formalization, facilitating industrial cooperation (Sabel, 2002).

The need for modular building extends to smaller segments, inducing the emergence of several well-known network motifs, such as feed-forward loops. In fact, software reverse engineering is a process for extracting these simple motifs from existing software and applying them for general use (Myers, 2003).

Software modules form two major classes with a large asymmetry between their incoming and outgoing contacts. Certain modules have a small number of incoming links and a large number of outgoing links.

These are usually large modules, performing a complex function that requires the help of several smaller, more general modules. The latter modules form the other class, with a large number of incoming links and a small number of outgoing links. These are small modules performing a simple function which is reused in various contexts by many larger modules (Myers, 2003; Potanin et al., 2005). According to the idea of Valverde et al. (2002), this distribution might have developed as an effort to minimize development costs by introducing an optimal trade-off between a small number of large, expensive components and a large number of small, cheap components. The diversity of uneven contacts certainly increases the chances for the emergence of weak links. This increases the stability of software systems, which I will describe in the rest of this section.

Several pieces of software have a bad smell. As one example, excessive use of hubs, which means highly connected modules here, is considered to be bad design practice and called an antipattern (Brown et al., 1998). This agrees with the scale-free link distribution mentioned above, which reflects the fact that, in complex software, modules with only a few links predominate. The software deodorant is called refactoring. Refactoring is "the process of changing a software system in such a way that it does not alter the external behavior of the code yet improves its internal structure" (Fowler et al., 1999). This definition of refactoring satisfies the functional definition of weak-link manipulation I gave in Sect. 4.2, following Berlow (1999). Several refactoring[10] steps introduce more links into the system (Brown et al., 1998; Fowler et al., 1999), and these are weak links according to the above functional definition. As a consequence, the program becomes more stable. Refactoring is a typical reconfiguration of a larger system, utilizing weak-link-induced stabilization. Another form of weak-link-related program stabilization is degeneracy. Degeneracy appears in the form of polymorphism of object-oriented systems (Myers, 2003).

As the extensive efforts for refactoring show, software systems are notoriously unstable and fragile. As an example of this, a simple typographical error usually halts a whole program behemoth. (For other amusing examples, just switch on your computer, and wait a little.) According to Lehman's second law, software systems tend to develop a high level of complexity "unless work is done to reduce it" (Lehman et al., 1998). Complexity is fueled by finite computational resources (Crutchfield, 1994). However, complexity here is not the complexity

[10]Examples of such refactoring steps are the introduction of smaller, more concise, single-purpose fragments, split into two subfunctions, etc.

which would automatically invoke stability and error tolerance. The complexity of software systems serves their extreme evolvability. As I mentioned above, the need for autonomous parts and extreme modularization are key points of software complexity. There is a rather recent trend which recognizes the parallel need for degenerate functions in larger programs. Degeneracy will serve the costs of its implementation and running by allowing the design of a safer and more stable computer program (Myers, 2003). We may observe that software design has smuggled tinkering into engineering in the form of refactoring and extreme programming (Beck, 1999; Brown et al., 1998; Fowler, 1999). However, this leads to the next section, where I return to engineering practices and summarize what we have learned so far about achieving stability in designed systems.

9.5 Engineers and Tinkerers: An Emerging Synthesis

In Sect. 3.6, I summarized Jacob's (1977) powerful view of evolution as a tinkerer who, in contrast to the engineer, does not optimize the system in advance, making a blueprint, but assembles interactions until they are able to work. Here I will show that the complexity of technical development has led to the convergence of engineering and tinkering. In other words (Sole et al., 2003a): "The fact that even engineers become tinkerers in large systems illustrates how complicated is the achievement of optimal structures once they reach some complexity level."

Indeed, engineered systems have long reached the point where our cognitive limits make the logical analysis of the system as a whole quite impossible in one mind alone. Since the extension of our biological 'cache memory', e.g., in the form of computer–human cyborgs, is not yet a reality, we have three options:

- We may employ exceptional people, who have extended cognitive abilities. This attributes a novel meaning to the previous remark that modern engineering is becoming an art. In this case the result may be exceptionally creative, even optimal, but its further development may be risky, requiring an original talent.
- We may break the design process into comprehensible parts and get used to the extensive application of motifs and modules (Alon, 2003). We may extensively develop fringe areas, pidgin formalization or design patterns, as described in the last section.

- Finally, we may also sacrifice some of the economy of the design and build in degenerate parts (Edelman and Gally, 2001), diversity and weak links to stabilize the system.

Highly optimized tolerance (HOT) (Carlson and Doyle, 2002; Csete and Doyle, 2002) should tend to become the self-organized optimal tolerance of the system itself including degeneracy and numerous weak links, rather than a carefully orchestrated human effort to achieve all this.[11] In nanotechnology and electronic chips, the construction of such complex systems is within reach (Edelman and Gally, 2001; Ottino, 2004). Indeed, recent advances in the self-assembly of complex micro-semiconductor designs (Zheng et al., 2004) signal a new era of network-driven engineering practice.

There is one more important consequence here. The innovation landscape of traditionally engineered systems is rough, in the sense that their 'innovation equilibrium' is highly punctuated. The different designs sit in the deep energy minimum of their purpose-optimized structure and further development becomes difficult. On the other hand, self-organizing, designable, degenerate systems "would serve unpredictable environmental conditions, where recognition of novelty is important and programmed planning is not possible" (Edelman and Gally, 2001). Similarly to the smoothing of the energy landscape of proteins (Sect. 5.3) or the diegetic landscape of dramas and novels (Sect. 9.2), weak links of engineered designs bridge the saddles of the innovation landscape (Tyre and Orlikowski, 1994) and allow a faster (and cheaper) development of novel design. Achieving both an improved safety and evolvability of our engineered systems in one act is worth the extra costs which the degenerate weak links may involve.

Here we end our fifth trip into Netland, in which we have surveyed some human-made networks. We have learned that ambiguity and degeneracy stabilize the meaning of our texts, the language net. The interconnected synchrony of multiple scale-free distributions may be an important feature of textual masterpieces, allowing them to induce better psychic relaxation and catharsis. The space around us is an extremely important element of our well-being, which has to be well designed with the fractal property and numerous weak links to stabi-

[11] An additional form of engineered diversity has emerged from studies of protein evolution. Highly evolved protein structures have higher 'designability', meaning that the same structure (design pattern) may accommodate more individual protein sequences (elementary solutions; Li et al., 1996; Tiana et al., 2004). The increased diversity of more designable systems also leads to the development of more weak links.

lize both ourselves and our community in towns. Finally, modularity, degeneracy and weak links are important to improve both the stability and evolvability of engineered systems. One of the most important treasures we have picked up are further examples to show that weak links help both stability *and* change in complex systems. This statement may seem rather contradictory as it stands, but it will become quite clear in the synthesis of Sect. 12.3.

10 The Global Web

On our sixth trip into Netland, we will go a level higher in complexity. In Chap. 8, we examined social networks where the elements were usually individual people. In Chap. 9, we saw some of the networks these communities produced. In this chapter, we investigate 'supersocial' nets, in which the elements are the social networks of the previous chapter, in the form of firms, investor groups or whole nations. Chapters 8 and 10 together show another nice example of nestedness.

10.1 The World Trade Web

Markets behave like networks. During the last century, our world became really small, not only in the sense that a former adventurous expedition by Marco Polo is now the daily routine of many business executives specialized in Eastern markets, but also from the network point of view. The world market has small-worldness (Serrano and Boguna, 2003). The first scale-free market behavior was noted by Benoit Mandelbrot (1963) when looking for regularities in price fluctuations. His observations were confirmed later on a larger scale (Mantegna and Stanley, 1995). The market of our global village is also scale-free in the distribution of both trade degrees and trade link strengths (Li et al., 2003). Influence links also show a scale-free strength distribution, which means that a small proportion of actors have an exceptionally high influence on the behavior of others (Janssen and Jager, 2003). Markets are disassortative, showing an avoidance of linkages between hubs (Serrano and Boguna, 2003).

Markets are not in equilibrium. Devastating market crashes, like the tulip speculation in the Netherlands in 1637, the South Sea Company scandal in England in 1720, or the more recent world trade crashes of 1929, 1987, 1997, 1998 or 2000 (Sornette, 2003) may lead to a quasi-instantaneous evaporation of trillions of dollars. Market crashes may swallow years of pensions and savings in a minute. It is our common interest to learn about their development and to recognize the factors what might stabilize the markets.

Market dynamics shows the properties of self-organized criticality. Self-organized criticality may evolve from reorders triggered when a final product is completely sold out. This chain of reorders behaves very similarly to the avalanches typical of the self-organized critical state (Scheinkman and Woodford, 1994). In another form of the same phenomenon, the continuous inflow of energy is the never-ending penetration of new and reintroduced products. Tension may develop from various sources. Accumulating trade surpluses and deficits as well as chronic under- or over-evaluation of prices or stocks may all cause increasing tension in the market. A 'rational bubble' may develop if the actual market price depends positively on its own expected rate of change and the positive feedback leads to an increasing discrepancy. A similar scenario is the 'speculative bubble', where the hope of a large profit and the generated buying spree enforce each other for a while. Relaxation may occur smoothly or abruptly. The latter may lead to a market crash, a market quake. Price fluctuations and the extent of market quakes both display scale-free distribution. This scale-free distribution is not due to fluctuations in the incoming information, but is in fact caused by the collective behavior of the actors on the market. Here I have to recall what we learned about panic quakes in Sect. 3.2. When we panic, herding behavior occurs. A tension is given and relaxation is delayed. A typical self-organized criticality develops (Bak et al., 1997; Helbing et al., 2000; Saloma et al., 2003). *"I am preoccupied by my saving efforts to cover the costs of my canoe tour with Pity, and I am getting more and more anxious as I listen to you. Please, do not try to tell me that people are in a continuous state of panic on the world markets."* Do not panic, Spite! In most cases, there is indeed no panic on the markets. However, herding does occur – and in fact this may lead to a well-justified panic.

How does this work? If we do have 'business as usual', actors on the market behave in the 'usual' way. Most actors follow a fundamentalist strategy, i.e., they analyze the fundamental value of the market item, and buy (sell) if the market value is below (above) this value. Fundamentalists know what they want and they have a wide variety of incoming information, all of which influences their final bullish or bearish behavior. On the other hand, 'noise traders' attempt to identify price trends and patterns and consider the expectations and behavior of other investors as an important source of information. Uninformed investors also rely heavily on general market opinion, exemplified by the opinion of a few key players or by rumors. The less information they have, the more they follow the crowd. If changes become abrupt, even the fundamentalists, the key players, will not have enough time for analysis, since they have to act promptly. The action space becomes rather limited.

We may analyze the situation within the framework of weak and strong links. Spite, if you still have your aversions for weak links, feel free to go with Pity to raise some cash for your tour. In the 'business as usual' situation, a fundamentalist actor on the market has a number of weak links. She or he is weakly linked to several sources of information which all modulate her or his final decision (to sell or buy). Other actors are linked to different information sources and will act differently. There is a diversity of opinions and actions which leads to another set of weak links. All these links help market stability. If collective behavior (herding) develops, the market actor will rely on a rather limited amount of information which already makes a strong link. Moreover, this information tends to become just one piece of information, namely, what the majority does. As herding develops, each of the actors will rely increasingly on the same information. If the link strength distribution departs from the usual scale-free pattern where weak links dominate and becomes an almost all-strong network, the market will be significantly destabilized and its behavior will become critical. If (1) the market is overvalued, (2) actors start to believe that the bullish trend is not sustainable, and (3) they also believe that the others believe the same, then a market quake may develop (Bak et al., 1997; Buchanan, 2000; Krugman, 1989; Lux and Marchesi, 1999; Ponzi and Aizawa, 2000; Sornette, 2003).

Shakespeare on Wall Street. The last few sentences concerning the expectations of market actors call for a comparison with the great masters of Sect. 9.2. Great masters are people who have the exceptional cognitive abilities required to think to the 6th order: "The master supposes that the audience will believe that A supposes that B intends to guess how C understands what D thinks ..." (Dunbar, 2005; Dunbar et al., 1994; Stiller et al., 2003). What are market gurus supposed to do? It has already been noted by Keynes (1936) in his famous beauty contest parable that the optimal strategy is not to pick those faces that you find beautiful, but those the other players find beautiful. However, the other players also know this rule and will not think about their own choice, but about the choice of the others, and so on. *"I honor Keynes' genius but I will never change my strong belief that Pity is the* *prettiest girl in the whole world."* You deserved the kiss, Spite. This was nice of you. But let me give you some advice: when you invest your fortune to raise cash for your canoe tour, please avoid the stock markets. Returning to the market gurus, we may state that "the guru supposes that the market agents will believe that A supposes that B intends to guess how C understands what D thinks ...". The iterative loop developed here is rather similar to that of the great masters. *"Do you really believe that Shakespeare would go to Wall Street if he were alive today?"* I believe Shakespeare's and the market guru's motivations are different. Shakespeare was mobilizing his 6th order thinking to discover and show our world to us. A market guru mobilizes her or his nth order thinking to discover and show the market to us. I have to say that my world is not really restricted to the market and I do not think Shakespeare would have wanted to restrict his to it either. However, there is another possible answer here. Do we have a Shakespeare today? If you cannot see him, could it not be because he is already on Wall Street?

How much can you earn with 1 dollar in 70 years if you are a market guru? If in 1926 you had invested your dollar in one-month U.S. Treasury Bills which is one of the safest securities in the world, by 1996 your original investment would have grown to 14 dollars. If you had taken a greater risk and invested your dollar in the S&P 500, you would now have 1 370 dollars. What would have happened if you had known which of these two investments would give you a higher return in the coming month, and you had changed your investment accordingly for 70 consecutive years? Your single dollar would have made you a startling 2 296 183 456 dollars by 1996 (Farmer and Lo, 1999).

It is time to return to market quakes. Big market crashes show distinct characteristics. A crash is preceded by a slow buildup of long-range correlations leading to the global cooperative behavior of the market actors described above. This is marked by a series of unusual oscillations superimposed on scale-free behavior. When a market quake approaches, the oscillations become faster, showing the increasing aggregation of a multitude of agents into a kind of superagent. After the crash, oscillations start again and the superagent becomes fragmented as business gets back to usual. Our losses are not due to the markets. They are due to our own collective behavior (Buchanan, 2000; Krugman, 1989; Lux and Marchesi, 1999; Sornette, 2003).

Embargos as destabilizers of the global market. A diversity of actors in the market helps to avoid crashes. Diversity is beneficial to global markets as well, due to the development of weak links. Consequently, extensive embargo policies are detrimental to the main actors themselves, since they decrease market diversity and, therefore, stability.

Collective behavior is not restricted to people on the market. Most countries (with the notable exception of Germany, Austria and Japan) on the world trade web showed a rather significant synchronization of their economic cycles with those of the USA between 1974 and 2000 (Li et al., 2003). In our global village, whether we want it or not, both joy and sorrow are joint experiences. We have to care about each other ... *"and develop weak links to preserve our stability."* Yes, Spite, but the help here is more complex than weak links. The main lesson of this chapter is that we should avoid herding and try to behave like human beings in times of danger.

Behaving like humans is not always as easy as it seems. There are also ways to prepare for damage. Wealth distribution and bet-hedging are commonly used tactics to make our future safer in uncertain conditions (Bernoulli, 1738). As examples of this, both ownership diversification (Stark and Vedres, 2002) and portfolio diversification (Stark 1996) are widely used measures in uncertain market conditions. From the network approach, this bet-hedging strategy improves the stability of the economy by introducing a larger number of weak links instead of a few strong ones.

In summary, the maintenance of weak links between market actors and with their various sources of information seems to be highly beneficial for market stability. As an important form of this, market diversity has to be promoted if we hope to avoid the devastating crashes of the past. Moreover, we should gather together all our wisdom and strength and behave as human beings and not as part of a herd when danger threatens.

10.2 Turning Points in History

Not only markets, but also human history is full of crashes. Actually, we have almost nothing else in our written history but crashes. These were the events considered to be worth recording. In this section, I will use our knowledge of networks to show that many properties of the turning points in our history can actually be quite well explained with network dynamics. I will start with self-organized criticality, the herding behavior similar to what happens in market crashes, then continue with topological phase transitions, and conclude with an analysis of link strengths in these processes.

On 30 September 1920, King Alexander I of Greece wanted to save his gardener's monkey from a quarrel with his dog. The monkey accidentally bit the king on his calf. The wound got infected and the handsome king died in painful agony three weeks later. Alexander I was succeeded by his father Constantine, who returned to the throne after his abdication in 1917. King Constantine started and lost a war with Turkey where thousands of civilians were massacred and 1.5 million Greeks and Armenians had to be evacuated. *"This is a very sad story. I will make a note never to buy even a toy monkey for Pity at Christmas to save us from accidentally starting the Third World War. Can you tell me how monkey bites affect your networks? Do you wish to imply that networks get blood-poisoning?"* The key word is not the monkey here, Spite, but self-organized criticality. Wars follow scale-free statistics, meaning that we have had many small-scale wars in human history and, fortunately, only a few megawars, like World Wars I and II (Levy 1983; Richardson, 1948). Since the situation leading to a war is extremely complex, we still do not have a full explanation for this law, known as Richardson's law.[1] However, self-organized criticality may be a key point here, too. A continuous and uneven technological change plays the role of inflowing energy, allowing conquering desires to develop as tensions. A geopolitical avalanche is started, where more and more metastable countries join the bandwagon. The war (a society quake) propagates like a forest fire, another self-organized critical phenomena (Buchanan, 2000; Kennedy, 1987; Roberts and Turcotte, 1998). If the tension is big enough to be critical, the initial trigger may be a very tiny event, like the shot by Gavrilo Princip in Sarajevo or the bite of the gardener's monkey in Athens.

Self-organized criticality also invokes collective wartime behavior (Brunk, 2003), like the herding effects leading to the highly overestimated stock prices and the subsequent market crashes of the previous section. Collective behavior helps tiny events (shots, bites and the like) to explode into a widespread vox populis, forcing politicians to follow the 'nation's will'.

---

[1]Lewis Fry Richardson is also known for the famous paradox of the British coastline: if he measured it with smaller and smaller scales, the length of the coastline became longer and longer. This is a typical fractal property which helped Benoit Mandelbrot (1967; 1977) to formulate his famous concept of fractals.

Fig. 10.1. Collective behavior causes tiny events to explode into a widespread vox populis

Collective behavior and the good consumer. A rather significant amount of high-profit production builds on the collective behavior of consumers. If consumers avoided collective behavior, TV advertisements could win only a tiny fraction of people over to the idea that they should change their recently bought mobiles to the latest model. These would be those five people who dropped, lost, or otherwise incapacitated their mobile over the last few weeks. In reality, millions will buy the new model, feeling sure that the other million will do it and becoming afraid that by abstaining they will not be keeping up with the Jones's. Manufacturing just five new mobiles would not make a profit, but manufacturing a million certainly will. A great many interests tend to push average members of society into a state suitable for collective behavior. We should bear in mind that whoever becomes accustomed to engaging in mindless collective behavior in one aspect of life will be highly susceptible to herding in all other aspects of life, including market crashes, wars, and all other events which would certainly ruin the large profits of any business on Earth (Kopp, 2000).

"If wars are scale-free, should we fear for an even bigger event than World War II?"
Unfortunately, this cannot be excluded, Spite. Statistics may pacify us into believing that a gigawar is even less likely than World War II was, but the scale-free distribution says nothing about actual incidence. Indeed, it may happen tomorrow. The take-home message is that, when you start a war, you never know how big it will be in the end. What is the solution? We should build up trust and weak links between nations, as I will explain in Sect. 10.3. Another piece of advice also comes from our previous notes. When you see herding behavior start to propagate in your country, watch out and try to remain human – or otherwise, be prepared for a war.

Now take a deep breath, drink a glass of crystal clear water, relax, and most importantly: think. We have already had three examples of herding: panic, market crashes, and war. *Is it always useful to become immersed in synchronization with the crowd?* Should we not develop a better ability to discriminate between the synchronized relaxation of joyful laughter and the 'over-synchronized' relaxation of devastating crashes and wars? Should we not learn from networks and *confine* relaxation by allowing only information to go for a global ride? The advice at the end of the book might help us to accomplish these complex and demanding tasks.

Fortunately, human history is not only a history of wars. We have had many other turning points, too. The generalization of network properties is a very powerful tool for understanding the rugged landscape of the punctuated history equilibrium. In Sect. 3.4, I gave two examples of topological phase transitions[2] developed by Tamas Vicsek and coworkers (Derenyi et al., 2004; Palla et al., 2004), applying the idea to cells and animal communities. Here I extend this analogy to societies and show the possible involvement of weak links in stabilization.

- **Random graph phase.** A random graph develops if resources are plentiful. Surprisingly, the hunter–gatherer societies of prehistoric times were rather rich. Sixteen thousand years ago the average height of our Paleolithic male ancestors was 170 to 177 cm (Cohen and Armelagos, 1984), and this significantly declined later, only to be surpassed again in the last century. Association was rather loose, there were no classes, and no continuous discrimination occurred. This society was in the random graph, primitive community phase. But the population grew and, consequently, resources became scarce, whereupon a random to scale-free transition occurred, which may have grown into a scale-free to star transition. This may have corresponded to the development of the first classes in the society.[3,4]

[2]For the definition, see the glossary in Appendix B.

[3]In fact, a similar, but reversed change can be noted in modern, Western societies, giving a partial scale-free to random transition as a result of ample resources. This is continuously challenged, however, by population growth and by the large increase in population density.

[4]Successful empires may have lost their strength in part due to this scale-free to random transition, which disorganized and softened the original, strong links.

- **Scale-free phase.** The scale-free system is a 'borderline' case between the random graph and the star phase. It is very fragile and transient, but in spite of this, it is very robust. We call it democracy. This democracy net always keeps a delicate balance between anarchy (random net) and dictatorship (star net). Fortunately, in democratic systems, the society is not segmented and weak links flourish. Consequently, the fragile system becomes robust. Democratic systems show the greatest complexity of all. However, weak links and their buffering may grow too great. A democratic society remains flexible for smaller challenges, but occasionally may become overcomplicated and unable to make a fast response to a life-threatening danger.[5]
- **Star phase.** If the situation gets really bad and resources become exhausted or unavailable, the scale-free phase changes into a star phase. Democracy collapses and a dictatorship develops. In dictatorships, theocracies and meritocracies, the society is segmented, secondary contacts and weak links are sparse and buffering is small. The system is centralized and efficiently fights against big, well-known challenges. However, the system is unstable and cannot respond to complex challenges. In the background new social formations arise. They are neither buffered nor channeled, but oppressed. Tension often develops and self-organized criticality may arise, leading to a society quake. Depending on its size and form, the society quake may be a reform, a system change, a revolution, a civil war, and so on, and the list is long. The society quake may facilitate a disintegration of the society net into fully connected subgraphs.
- **Disintegration to fully connected subgraphs.** When the situation becomes tragic, resources become grossly unavailable, the star phase breaks and the network disintegrates into a number of small, but fully connected subgraphs. This is the end of the network, since its integrity (its netsistance) is lost. Dictatorships often end with a sudden, disintegrating event. In wars, revolutions and similar social cataclysms, weak links are all broken, and only fully connected subgraphs like core families, small military units, etc., try to survive or escape together. However, the situation here is vastly different from that in our body or a cell. Disintegration is called

[5] You may recall here the analogy with muscle unit synchronization in Sect. 7.3. There I mentioned that weak links are necessary for changing, unknown tasks of high complexity, e.g., playing music, while strong links are required for the fast execution of well-defined tasks, e.g., weight lifting. A life-threatening danger is a rather well-defined task where survival often depends on the speed of the response.

death in that case. Disintegration here inevitably leads to terrible suffering. However, it is also a new opportunity. Here the network elements (people) remain functional, even if they are temporarily alone. Unlike our cells or proteins, they still have their ability to survive, even under these conditions. They can regress to a more primitive life form. In social cataclysms, buffering and order cease. Members of isolated subgraphs have a new opportunity to reform a big net, their society. Hidden social formations come to light, competing violently with one another, and a new strategy is selected. After this the system may return to a dictatorship, or become a democratic formation, if the anticipated challenges are drastic but simple, or small but complex, respectively.[6]

French absolutism and revolution: An example of a star → subgraph → scale-free transition. Degenne and Forsé (1989) gave an interesting network analysis of the French monarchy, where the initial segmentation of the society became increasingly dysfunctional as noble titles were sold and resold in massive quantities during the 17th and 18th centuries. Although only strong links were allowed, weak links developed as well. Institutional gridlock made the French Revolution a credible option. This example illustrates the benefits of the subgraph phase. As Montesquieu (1734) noted: "[in war] each person places himself in a suitable situation, whereas in times of peace, he is positioned by others, and generally very injudiciously." Indeed, social cataclysms reduce the links of the atomized society and facilitate upcoming change. Is the situation really so 'suitable' in the middle of a cataclysm? For Montesquieu, who lived 66 years of peace in an outgrown society anticipating a big change (the French Revolution), a cataclysm was obviously more desirable than it would be to us, for we regularly witness the past and present horrors of the world on the small screen in our living room.

Social classes and Karl Marx revisited. A social class can be regarded as a network module. When a large number of network members find themselves in the same situation, with strong links to each other and rather weak links, if any, to the other modules of society, social classes will develop. Social classes presuppose uniformity within segments of the society. This is only temporary, since technological development requires

[6] A subgraph to star phase transition might have occurred when the Greek towns became more dependent on a mother-town, called a metropolis, when threatened by an outside danger. As an additional example, challenged empires may have partially disintegrated due to the star to subgraph transition.

greater specialization. Diversity and weak links then develop. Consequently, uniformity, exclusive strong links and social classes slowly dissolve. The diversification of internal links may also be triggered by the diversification of external links. In the long run, globalization certainly helps democratic change from the network point of view. In addition to these ideas, Karl Marx might have been right in stating that all classes will be erased and communism will appear if resources become plentiful again. Indeed, in this case the scale-free or star network might shift to a random graph pattern and its structures might dissolve. Thinking about the increase in human population and the situation in Africa, this will not be achieved on a global level in the near future. Moreover, we have grown quite used to the state of 'organized criticality', the scale-free, democratic and highly complex social net. The low complexity of a random graph might seem extremely dull to us. From the network point of view, communism is not an exciting alternative.

The end of history. On the basis of the above interpretation, democracy is certainly not the ultimate equilibrium constituting "the end of history ..." (Fukuyama, 1992). Democracy is a highly robust, but at the same time very fragile, complex system, which requires constant change to maintain its self-organized state. Democracy is by definition not in equilibrium. The random graph pattern may pose a system of boring and very low-complexity equilibrium. If all of us ever find ourselves with plenty of resources, that will be THE END of our history.[7]

The extremely hypothetical picture of the topological phase transitions exemplified above is based on the important work by Tamás Vicsek and colleagues (Derenyi et al., 2004; Palla et al., 2004). Many of the examples lack sufficient detail to be fully convincing. Nevertheless, I hope they demonstrate that topological phase transitions of networks provide a useful tool for analyzing and understanding the key trends and events in our past.

Before I shift to another useful tool, link strength analysis, let me analyze the effects of globalization on the interactions of social nets in a few examples. My first example is the exportation of democracy. Weak links are important in society, but they are not useful all the time. Weak links will not be much help for societies which are not in their scale-free phase. Weak links and democracy may not fit a resource-poor society which requires a star phase to survive and develop. Moreover,

[7] The almost certain boost in human population under these conditions makes this highly unlikely.

the complexity of the scale-free system needs a certain level of practice to maintain, and development of this practice requires time and self-organization.

The development of democracy is a slow process requiring several generations of resource expansion. We may have two rather widely separated phenotypes of BIGS and SMALLS adapted to low and high resources, respectively (Bateson et al., 2004). The reversal of these phenotypes is fairly slow, requiring a change in resources for several generations.[8] In Sect. 7.5, I put forward the idea that BIGS tend to build weak links, while SMALLS keep mainly strong contacts. Following this thought further, I may conclude that the growth of resources in the Western part of the world will inevitably lead to a gradual shift towards the BIG phenotype and the development of multiple weak links. This leads to an increased speed of innovation and to a better stabilization of society. It thus serves as an important reason for making democracy viable. If we accept this view, in countries where resources are not abundant, or were not abundant up until two generations before, democracy is not viable in its well-developed, pure form. Let me add an important remark here. Obviously, we have a multitude of reasons which lead to one social formation or another. Link strength is just one of them. However, I am quite sure it is worth thinking about this in the context of exporting democracy. We may have voluntaristic demands to build up stable democracies in countries where poverty was or is prevalent. In a society where the majority still belong to the STRONGLINKER personality trait, this will not be stable. Now take a deep breath, drink a glass of crystal clear water, relax, and most importantly: think. *Has the time not come to slow down a bit?* If certain changes require three generations, why do we want to accomplish them in just one year?

Are civilization diseases our penalty for achieving democracy? In Sect. 7.5, where I introduced the concept of SMALL and BIG phenotypes (Bateson et al., 2004), I noted that a similar concept, the concept of thrifty genotypes and phenotypes (Hales and Barker, 1992; Neel, 1962), found a correlation between our change of lifestyle and many of the civilization diseases such as diabetes, obesity, high blood pressure, atherosclerosis, etc. The question arises as to whether civilization diseases might be unavoidable consequences of our change of lifestyle, leading to a modern society with weak links, better stability and an improved potential

[8] You may live in luxurious conditions but, if your grandparents lived in poverty, you may still harbor the SMALL phenotype.

for achieving innovations.[9] Our 21st century Hamlet in the third world countries may have a rather difficult dilemma:

> To eat, or not to eat: that is the question:
> Whether it is nobler in the mind to eat less,
> Stay healthy and preserve the old society,
> Or to take arms against a sea of troubles,
> Start to eat, get diabetes and build democracy?

Fortunately, it is not yet clear whether these two changes are related in any way. A rather intensive study would be required to see whether there is a range of slow increases in welfare and nutrition for which civilization diseases can be avoided, and the SMALL phenotype (and STRONGLINKER personality trait) can still be changed. *"This is an excellent and very important research program! Shall we start it tomorrow?"* Spite, I am happy to see your enthusiasm, but I have to discourage you. Obviously, we can start animal experiments of this sort even today. However, animal experiments have a very low predictive value (if any) for the human situation when it comes to questions of nutrition. Carefully planned nutritional experiments with fully informed and consenting participants can be envisaged with human beings. However, by the very essence of the problem, the outcome will be revealed only after two to three generations. Think about the situation if the results show that the original setup was badly designed and the experiment has to be repeated in a modified form! No third world countries will be left (I am always super-optimistic as you know) by the time our great-grandchildren eventually finish this project.

Globalization brings freedom of travel, and this in turn brings foreigners. If we integrate them, they will increase our stability. If we try to assimilate them, we do not gain very much. If we segregate them, our future is in danger. Integration of foreigners brings the development of several stabilizing weak links into the society. Assimilation consumes diversity and destroys the chances of making weak links between society members. Segregation is even worse, since it does not give the smallest chance for any links to develop and gives rise to relaxation barriers. Tensions develop and the danger of unpredictable society quakes increases. I quoted the Admonitions of the Hungarian King, St. Stephen in Sect. 8.3: "A country with one language and one custom is weak and perishable. Therefore, I order you, my son, to show goodwill to our visitors, and protect and cherish them, so that they

[9] I am thankful to Gabor Szegvari for this question.

will prefer to stay with you rather than to live elsewhere." What was a major source of modernity a thousand years ago has become a major source of stability today.

Pagel and Mace (2004) gave an interesting analysis of human cultural diversity. They note that group homogeneity, group cohesion and the development of common norms help altruistic, cooperative behavior and effectively filter and deter cheats. However, this requires the minimization of intergroup migration. Foreigners break cohesiveness by introducing the danger of escaping cheats. A period of experience (i.e., that cheating is not highly prevalent among foreigners) as well as greater resources (occasional cheats do not pose a great threat) are both required to ease our historical wariness of strangers. You may take your glass of crystal clear water and note once again the need for patience. If we wish to achieve stability, we should slow down, not only in our deeds, but also in our judgements.

But diversity is a double-edged sword. Cultural diversity also means customs like slavery, female genital mutilation, child exploitation, etc., which cannot be tolerated in our globalized, modern world. General norms should be observed by all of us. However, with these limitations, diversity requires enhanced protection today. Both experimental model systems (Kerr et al., 2002) and game theory models (Axelrod, 1997) suggest that the majority culture 'naturally' endangers minority cultures, even without any bias in social influence. This danger is further increased if the population becomes well-mixed, as humankind tends to do today. Due to our increased freedom to make links only with those we wish to, the extinction of the minority cultures is growing even closer. Moreover, the free choice of links greatly increases the chances of segregating the remaining cultural groups. Same-interest groups and same-opinion islands are increasingly formed, and intersegmental communication may easily deteriorate. There is an urgent need to make special and directed efforts to educate ourselves and our societies to build intercultural, bridging, weak links.

Analyzing social formations from the standpoint of link strength, the stabilizing force of weak links can be found mainly in the scale-free (democratic) periods of our history. Dictatorships (star phase) traditionally hate weak links and try to destroy them. Weak links cannot be monitored, cannot be controlled, and pose a continuous threat to any dictator (Arendt, 1973). Disintegrated or random graph societies neither need nor display weak links. Thus in both pure star-phase and random graph formations the strengths of all links are more or less alike. Links are close to 'all-strong' in the star phase, and 'all-weak' in random graphs, if 'strong' and 'weak' make any sense in the absence of another link as a reference strength. Fully connected subgraphs do not form a society, so the strength of their long-range links is zero.

Weak links and modernity. Medieval society was very hierarchical and had very few weak links. The spread of weak links started with the widespread use of money, developed through the economic necessities of the 16th century, and was generalized by the growing social life and the Enlightment of the 17th and 18th centuries. The first step in these changes was the widespread use of money at the end of the Middle Ages, which allowed greater flexibility in society and, bringing with it a wide variety of roles and interactions, turned many formerly strong links into weak links (Simmel, 1990).[10] The development of overseas commerce in 16th century England depended on the creation of joint stock companies. Joint investment allowed investors to raise the high costs and spread the risks of pirate attacks or bad weather. As an additional effect, the joint stock company developed weak links between its investors. By the 18th century, life flourished in English and French society. In London, 20 000 inhabitants met every night in public places, while in France the multiplication of social roles became extremely well developed. As soon as a person could not be reduced to a single role, the idea of personality was born. Determined, normative and publicly sanctioned expectations gave rise to the ambiguity of the individual. Weak links were both signs and instigators of modernity (Degenne and Forsé, 1999; Fukuyama, 1995; Seligman, 1997).

Weak links and transcendency. Spite, here you may turn your critical attention elsewhere for a while! A rather provocative idea will follow. If I put together the idea of Dunbar (1998; 2005) that the registry of our social contacts requires a tremendously high brain capacity and the notion that extensive emotions as well as meditation require a high level of neural synchrony (Aftanas and Golocheikine, 2001; Damasio, 1994; Orme-Johnson and Waynes, 1981; Rolls, 1999), I may provoke your mind with the statement that these may balance one another. The increase of weak links and the decrease of religious belief in the 16th, 17th and 18th centuries was not just a mere coincidence, and was not only a struggle between the rising bourgeois class and the power of the church. It also reflected a decrease in the synchronization intensity as our brain became occupied by the increasing registry of social links and developed the 5th order thinking of

[10] I am grateful to Eszter Babarczy for this idea.

Shakespearean theater. Our great-grandparents might sacrifice the wholeness of their transcendent link for the incoming flood of social links. This may be an important reason behind the benefits of hermitage. *"In spite of your warning, I stayed and listened patiently to your idea. You describe this transition as a tragedy. Let me play with the idea that you are right. In this case, our great-grandparents lost their superstitious nightmares in parallel with the decrease in their transcendent experiences. The same emotional super-synchrony you mentioned may be needed for both. Do you really want witch-burning back?"* Witch-burning is not my favorite Sunday brunch special. However, the wholeness of transcendence is a great loss for many around us. A great loss. The loss of social links would also be a great loss (as an important example of this, I could not write this book!) and would ruin the whole society. What can we do? You know, Spite, I am forever an optimist. I hope we will develop a good enough brain to have them both.

10.3 Weak Links: A Part of Social Capital

In the last section we saw that our society net may undergo quite profound changes from a random graph to the scale-free, star and disintegrated phase as resources become sparse or stress develops. In this section, I will make a few remarks on the competitiveness of societies using their network structure as a guide.

The formation of a top network demands the stability of its elements, the participating bottom networks. We are currently witnessing the globalization of the world. The world economy as a top network requires the stability of the bottom networks, the participating countries. Our globalizing world is highly competitive. Evolvability of societies, and their readiness to discover and implement technological developments is a key element in their competitiveness. Here I will continue with the examination of both requirements.

Social stability demands the stability of the participants. Our psychological stability is a sensitive balance influenced by a large number of effects, such as hormonal status, social links, etc. Studies by Kunovich and Hodson (1999) demonstrated that weak links in the form of informal social nets, such as sports clubs and social activities, improve mental health after a traumatic event, such as the civil war in Croatia. However, participation in formal groups, providing strong links, such as church, party organizations, and unions was shown to be detrimental to mental health. Similarly, stress due to evacuation of an Israeli community was efficiently counteracted with the *perception* of social embeddedness (weak links), and did not require actual reassuring contacts (strong links) with the stressed people (Steinglass et al., 1988). While our common sense agrees with the help of social links to cope with stress, the detrimental effect of strong links may seem paradoxical. As an explanation, strongly linked communities may revive and relive stressful events, thereby hindering the healing process (Kunovich and Hodson, 1999). Moreover, a strong link expects a future reciprocation and this obligation may pose an unbearable burden on the stressed individual (Kawachi and Berkman, 2001).

In contrast to the stabilizing role of weak links, social segregation and tensions are detrimental to our well-being. Negative social experiences destroy our well-being more than an equal amount of positive experiences can help us (Rook, 1984). Segregation invokes loneliness, helplessness and the feeling that the individual has lost control over her own life. Stress provokes an attitude of rivalry in men which is destabilizing for their health (Kopp and Réthelyi, 2004). A self-perpetuating vicious circle may develop. Male participation in civic organizations may break this vicious circle by decreasing both rivalry itself and the fear of insufficient rivalry. Fortunately, women respond to stress differently. They do not compete – they socialize (Skrabski et al., 2004).

The above examples may be generalized. Not only psychological stability, but health and longevity are also improved in a society which is not segregated, where the backbone of strong links is helped by a large number of changing, informal weak links. Social links decrease both hypertension and the probability of heart attacks. Building a social net roughly equals the health benefits of non-smoking (Kopp and Réthelyi, 2004; Marmot and Smith, 1989; Putnam, 2000; Skrabski et al., 2003). Self-efficacy is an important element of health, allowing us to cope with the difficult situations in our life (Bandura, 1997). This is extended by resourcefulness, which is our hope that we will be able to get help when it is needed (Antonovsky, 1985).

Innovativeness is a key element of competitiveness. The fast spread of innovation depends on long-range society contacts, which are usually weak links. Fukuyama (1995) described the difference between traditional French and German companies. Due to the long history of French absolutism, governmental control and title-based hierarchy (Degenne and Forsé, 1999), in a traditional French company the links between the workers and supervisors were mostly formal and strong. Detailed regulations often made sudden, ad hoc changes impossible, thereby hindering competitiveness. As an example of this, the regulations governing the dyeing of cloth run to 317 articles. In contrast, in a traditional German company the manager knew the workers and their contacts were less formal. The same parallelism can be drawn between the traditional Chinese family business and traditional Japanese collective behavior (Fukuyama, 1995). *"Are you sure about these judgments? Germany is also well-known for tight regulations, and the love of French people for style and art is common knowledge."* Many thanks to you Spite, for this warning. It is not the actual nation that is important here, since generalizations never catch the complexity of real life. You should only pay attention to the two extremes of possible behavior. The take-home message is quite clear: in all cases, weak links help flexibility and the development of innovation.

Competition and innovation both increase and continuously reshape specialization and division of labor (Durkheim 1933; Fukuyama, 1995). Concomitantly, social roles become more complex and differentiated. Proliferation and transience of roles erode their stable definitions. Role expectations become ambiguous and negotiable (Seligman, 1997). All these challenge social stability and require a constant rebuilding of the social net. To achieve this, we need weak links as a sign of modernity, since strong links can tie only stable and established roles.

Innovation needs good communication between the segments of society. To respond to complex challenges, novel and unusual links have to be built. At first these links are by definition weak. However, the transfer of complex, tacit knowledge requires more than a weak link. It also needs benevolence- and competence-based trust (Hansen, 1999; Levin and Cross, 2004).

In summary, both social stability and competitiveness are helped by a complex halo of links in society. Importantly, it is not only the links, but also their perception, our trust, which stabilizes our mental and physical health and makes complex information accessible. "Trust enters into social interaction in the interstices of the system or at system limits, when for one reason or another systemically defined role expectations are no longer viable." Trust starts, where confidence ends (Seligman, 1997). Trust makes it easier for us to invest energy in building up long-range, weak links. Trust is the probability of weak links, the probability of the development of the modern social net.

Trust is a part of social capital. Social capital refers to the institutions, relationships, and norms that shape the quality of a society. It is important to note that social capital is not just the sum of the institutions which underpin a society – it is the glue that holds them together. Social capital has several key ingredients (Fukuyama, 1995; Putnam, 2000; Seligman, 1997):

- a well-developed social net,
- participation in civil organizations,
- cohesion of local communities,
- solidarity,
- trust, reciprocity and community support.

I hope that the present book convinces the reader that weak links are also a part of our social capital. Weak links are the same "lubricants of the social system" (Arrow, 1974), just as the weak-linker water was for proteins (Sect. 5.3).

Modern times have multiple effects on social capital. Our increased opportunities for building weak links certainly help the development of the social net, trust, support, solidarity and cohesion. In contrast, market behavior and relativization (Sect. 8.6) cause the erosion of social capital. We are increasingly shareholders and not stakeholders in our societies. Our commitments and responsibilities are short-term, market-oriented and private, rather than long-term, generalized and public (Dahrendorf, 1968). The erosion of social capital paves the way to the development of collective behavior and vice versa, herding helps to destroy social capital.

Robert Putnam's powerful analysis *Bowling Alone* described the erosion of social capital in the United States (Putnam, 1995; 2000). Television, suburban sprawl and all the changes of modernity curtailed the once so common and rich American social life with its picnics, bridge clubs, churches and the like. September 11 and the subsequent response made another attack against social capital: it eroded trust. Trust is a treasure of a modern society. Trust is actually more than a treasure. Trust is a necessity for stability and competitiveness. *Whoever destroys a single bit of social trust after three beers, whether a terrorist, a politician, or any one of us, harms the stability and competitiveness of the whole nation.* Trust is like our well-being, discussed at the beginning of this section. It has to be built up over generations, but it can be destroyed in a second. Trust-killers push their modern society back into the Middle Ages. Now take a deep breath, drink a glass of crystal clear water, relax, and most importantly: think. Think about your responsibility to save social trust, and especially before the next time you start to speak.

This is the end of our sixth trip into Netland. Spite, it is your turn to make the summary now. *"OK, Peter, I will try. I was fascinated how the sync between people which you called herding determined market crashes, peace and war. It was also interesting that big events at the stock exchange or in history followed the same self-organized criticality as sand piles or our favorite pastime with Pity: sex*

Fig. 10.2. September 11 and the subsequent response made another attack against social capital: it eroded trust

quakes. Your analogy of topological phase transitions with major changes in history was rather striking. Pity and I have agreed to read a few more books to see whether we believe it or not. I liked your thoughts on the role of patience in the exportation of democracy, and on the network backgrounds of conservatives and liberals. Finally, I was truly touched by the idea of trust as a part of social capital and as a basis for our mental and physical health. Pity and I have promised each other that we will trust each other forever." Spite, I am happy to hear the decisions you have made regarding your own life as a result of this chapter. However, do you not think you have forgotten something? *"Like what?"* Weak links! We had quite a lot of evidence to suggest that diversity and weak links stabilize both markets and societies. I also discussed the development of weak links and argued that they are part of our social capital. The trust that you and Pity feel for one another is called confidence, because your link is strong (at least I hope so). Trust is our opportunity to make stabilizing weak links.

11 The Ecoweb

"There is something I do not understand here. Why are you writing about ecosystems after the world market and the whole of human history? So far I thought we were going from small to big." And you were right, Spite! We are still going from small to big. Humankind with all its global glory and a few thousand years of written history only occupies a part of Gaia's ecosystem, which has been developing for almost 4 billion years (Schidlowski, 1988; Holland, 1997). On our seventh and last trip into Netland, we shall learn about the properties of the oldest and biggest network of all: the ecoweb.

11.1 Weak Links and the Stability of Ecosystems

The ecoweb is not a birthday party. Here links are built between predators and prey, consumers and food. You either eat – or you are eaten.[1] Ecowebs display the usual network characteristics, such as small-worldness and scale-freeness, but differently. When we eat, we get really close to each other: the now proverbial six degrees of separation of human societies go down to two degrees of separation in ecowebs (Williams et al., 2002). However, path length and other small-world characteristics, like clustering of ecowebs, do not typically differ from those of random graphs (Dunne et al., 2002a). A bona fide scale-free behavior can only be demonstrated for smaller ecowebs with low connectivity or for modules of larger ecological networks. In fact, many ecowebs display a degree distribution with a single characteristic degree, which makes them similar to random graphs (Dunne et al., 2002a; Jordan and Scheuring, 2002; Jordano et al., 2003; Montoya and Sole, 2003).

[1] We usually consider ourselves to be at the top of the food chain. However, if we really think about it, the King of Nature is the pit bull. (Pit bulls occasionally consume humans. On the other hand, oriental cuisine has other dog delicacies than pit bulls, as far as I know. The same cannot be said for sharks, which consume us more systematically, but may also end up on our table on an unlucky day in the history of sharks.)

The large variability in these network properties may be explained by the differences in size, connectivity and complexity of the ecowebs studied. However, the topological phase transitions of networks we saw earlier (Sect. 3.4) (Derenyi et al., 2004; Palla et al., 2004) may also work here. The characterization of resource-rich or resource-poor states has not yet been systematically addressed in the ecoweb analysis. The random-like, scale-free and high-connectance, few-node networks of Dunne et al. (2002a) may actually correspond to the random, scale-free and fully connected subgraph phases of Derenyi et al. (2004), showing a response to increasing stress. Even if this assumption is not valid, it would be interesting to see whether a random-like ecoweb structure can be shifted to a scale-free-type network if resources are curtailed.

Another type of ecoweb scaling emerges from the transportation network analogy due to Garlaschelli et al. (2003). By decomposing the loops of ecowebs in a renormalization-type operation, they constructed a tree-like energy flow connecting the various foods and consumers, starting from the external environment. After this transformation, the ecowebs become similar to the fractal-type transport systems we saw in Sect. 7.2, and show the general scale-free behavior of the allometric scaling laws (Sect. 2.2) (Kleiber, 1932). Although the exact value of the scaling exponent has recently been debated, all authors agree that food webs are very efficient resource transportation systems (Barbosa et al., 2005; Camacho and Arenas, 2005; Garlaschelli et al., 2005b).

Ecowebs show a rather complex dynamism. I have already mentioned the synchronization of hare and lynx populations fluctuating in sync over millions of square kilometers in Canada (Blasius et al., 1999). The variation in population size as well as in the lifespans of bird species both display a scale-free distribution, which also invokes cooperative behavior (Keitt and Stanley, 1998).

As is generally true for non-random, scale-free systems (Albert et al., 2000), the removal or addition of an element to an ecoweb may lead to a wide variety of consequences. The removal of a keystone species which may bridge separate eco-modules or may have many connections, can be catastrophic. On the the other hand, a random failure does not cause secondary extinctions in most cases. Sequential damage often shows a threshold, beyond which the system displays extreme sensitivity to removal of any further species. A rather simple rule of nature conservation may follow from this: We should find the very few keystone species and protect them, and then we may kill the rest. Fortunately, this is not so easy. A seemingly redundant species playing only a minor role in a particular ecosystem may suddenly assume the role of the keystone species as environmental conditions change. The unpredictability of keystone species status resembles the yeast metabolic network, where 40% of the genes become essential under special conditions only (Papp et al., 2004). Facultative essentiality is a general feature of complex networks and reflects the same stabilizing power as weak links. As an additional warning sign, it is extremely difficult to predict the sensitivity threshold of ecowebs. In other words, we never know whether the ecosystem will become completely unbalanced after the extinction of the second or fifth species (Allesina and Bodini, 2005; Dunne et al., 2002b; Holling, 1973; Ives and Cardinale, 2004; Jordan and Scheuring, 2002; Jordan et al., 2002; Montoya and Sole, 2003; Paine, 1969). Since we cannot put the Earth into an incubator, we need to preserve its diversity to protect the ecosystem under a variety of possible conditions.

Ecological systems are by definition diverse. Diversity may further increase due to either immigration or specitation. In bacterial communities, a large part of diversity is microdiversity, i.e., an abundance of genetically almost identical species (Acinas et al., 2004). Bacterial diversity is fine-tuned by the vivid lateral transfer of DNA (Ochman et al., 2000) between various bacteria. Diversity and especially microdiversity invoke differential contacts between system elements, which give rise to strong and weak links. Symbioses, like that of the mycorrhiza fungi with terrestrial plants may – weakly – link whole forests (Wiemken and Boller, 2002). Weak links are also perceived in ecowebs as indirect, higher order interactions (Abrams, 1983). Indeed, data analysis of food webs suggests that most interactions in complex ecowebs may be weak (Berlow, 1999; McCann et al., 1998; Montoya and Sole, 2003; Paine, 1992).[2]

Molecular sources of ecological diversity in situations of need. Stress induces an increased variability of ecosystems (Warwick and Clarke, 1993). As one of the possible reasons for this, when an ecosystem becomes unbalanced, participants begin to consume unusual foods. This is a perfect scenario for acquiring unusual viruses and prions, which often cross the species barrier (Scott et al., 1999). These and perhaps other mechanisms change the genetic or epigenetic status of the reorganized ecoweb and open new ways to unleash a surge in diversity (see Sect. 6.3). As a conse-

[2]I am grateful to Márton Tóth for many of these ideas.

quence, diversity helps to stabilize the ecoweb and promotes its survival, by increasing its netsistance.

The dangers of partial extinctions: The distribution of genetic diversity is highly uneven. Diversity is not uniform. We may feel this from the uneven distribution of diversity at the phenotype level. As an example of this, think of the human-made monocultures of wheat or corn. However, the really important diversity of diversity lies at the genetic level, since this forms the basis for future stability and changes, the evolvability of the species, as described in Sect. 6.3. Rauch and Bar-Yam (2004) showed that the distribution of genetic diversity is scale-free. Thus, most of the potential future development of a species may be concentrated in a tiny environment. Moreover, the borders of diversity sanctuaries are often not visible. Therefore, a 'small', 'negligible' partial extinction may wipe out most of the reserves in genetic diversity and may lead to a dangerously low chance of survival for the whole species in the long run.

Top predators stabilize diversity. Sergio et al. (2005) found that the presence of five different predators (the goshawk and 4 owl species) was consistently associated with a higher diversity of birds, trees and butterflies in the Italian Alps. Their results justify conservation efforts concentrating on top predator species. In agreement with this, Bascompte et al. (2005) suggest that overfishing of the top predator sharks may have induced trophic cascades leading to the degradation of Caribbean reefs.

Recent data indicate that diversity enhances the stability of an ecosystem (Hughes and Stachowicz, 2004; McCann, 2000). However, this statement has had a rather eventful past, called the diversity–stability debate (McCann, 2000). Before the 1970s, ecologists believed that diversity enhanced the stability of ecological networks. The famous ecologist, Robert May (1973) and other colleagues challenged this view by showing that the more species the system has, the higher are the fluctuations that occur if we remove or add a constituent. May and others were partially right. The instability of overconnected systems is a well-documented effect, as mentioned in Sect. 3.3 (Fink, 1991; Siljak 1978; Watts, 2002). The establishment of new links is costly, since new territories must be explored for the new food, new hunting techniques have to be employed, etc. Therefore, it remains an open question whether real ecosystems ever reach the unbalanced, overconnected state.

May (1973) built his networks as random graphs and calculated their equilibrium. But links in the ecoweb are not random and ecosystems never reach equilibrium (McCann, 2000). If assessed by dynamic methods, weak links clearly dampen oscillations between consumers and resources and decrease the likelihood of extinction (Berlow, 1999; McCann et al., 1998).

On the basis of a large amount of data, diversity won the diversity–stability debate (McCann, 2000). Diversity does not only act over intervals of weeks or decades. The diversity of 3 300 biogenic reefs stabilized them over a 542 million year timescale (Kiessling, 2005). Diversity not only stabilizes ecosystems, but also gives them a greater potential for evolution (Earl and Deem, 2004). Functional and spatial diversity all help system stability – by using weak links. Strong and weak links may have complementary roles in ecowebs: while strong links build up the frame of the network, weak links provide its robustness (see Table 11.1) (Garlaschelli et al., 2003; McCann, 2000).

Our forests are superorganisms connected by invisible weak links. Trees in the forest are nursed by the wood-wide web of symbiotic fungi, called mycorrhiza. This concept was put forward more than a hundred years ago (Frank, 1885) and has been proven by recent studies. As an example of the multiple benefits, trees which are well-exposed to light feed trees growing in the dark by supplying assimilated carbon through ectomycorrhizal bridges between them (Wiemken and Boller, 2002). Mycorrhiza makes our forests into a superorganism which is stabilized by weak links.[3]

11.2 Omnivory

What is best for the stability of our ecosystem? Are we helping system survival more if we restrict our diet to three liters of fresh orange juice per day, or should we eat everything? As a first approximation to the answer, humans are by nature omnivorous animals (like pigs). In this section, I will use omnivory as a special case to enhance and extend our knowledge of the stabilizing role of weak links in the ecoweb.

Omnivorous animals are typically inefficient in consuming their prey (Sole et al., 2003a), which means a low-affinity interaction. Adaptive foraging generally leads to the development of a few strong and many weak links (Kondoh, 2003). The omnivore feeds largely on the

[3] I am grateful to Márton Tóth for this idea.

Table 11.1. Stabilizing effects of weak links on ecowebs

Weak links	Effect on stability	References
Weak predator effects in field experiment	Removal of weak links induces higher noise of network parameters	Berlow, 1999
Weakly coupled oscillation of hare and lynx populations	Resistance against perturbations	Blasius et al., 1999
Strong and weak link patterns together	An appropriate pattern of strong and weak links is needed for network stability	de Ruiter et al., 1995; Yodzis, 1981
Strong and weak link patterns together	Reduced trophic cascades	Bascompte et al., 2005
Loop-forming weak food fluxes	Network stability increases with the number of weak links	Garlaschelli et al., 2003; Neutel et al., 2002
Weak and intermediate strength links	Dampened oscillations of model web densities	McCann et al., 1998
Indirect effects	Approximately 40% of system stability is derived from these weak links	Menge, 1995

lowest trophic level. As a result, it has strong links to this trophic level and makes weaker and weaker links at higher and higher levels (Neutel et al., 2002). Thus omnivory helps the development of a wide range of weak links in ecological networks.

In parallel with the diversity–stability debate, the contribution of omnivory to ecoweb stability was also questioned over a long period. Pimm and Lawton (1978) concluded to a destabilizing role for omnivory, while Dunne et al. (2002b) found that omnivory and robustness are independent in ecosystems. On the other hand and in agreement with the stabilizing role of weak links, numerous studies reported that the presence of omnivory exerts a stabilizing force in the dynamics of ecosystems (Borrvall et al., 2000; Fagan, 1997; Holyoak and Sachdev, 1998; McCann et al., 1997; 1998). Omnivory reduces trophic cascades (Bascompte et al., 2005) and extends the range of topologies in which the food web remains stable (Kondoh, 2003). Omnivores, like typi-

Fig. 11.1. Omnivory-induced system stability probably helped us to conquer this planet

cal weak-linkers, are thus reminiscent of water in Sect. 5.3 and may also smooth the saddles between local stability regions of the eco-landscape.

What can be the reason for the differences in the proposed or observed consequences of omnivory? Some of the discrepancies arises from the same sources already mentioned in the diversity–stability debate. Earlier models designed to monitor secondary extinctions lacked a detailed description of interaction strength and system dynamics. Moreover, overconnectedness (Sect. 3.3) may have overturned the stabilizing effect of omnivory in some systems. In reality, omnivory is often facultative, which means that an omnivorous animal may spend a long time being a vegetarian, for example, if there is plenty of fruit around. Additionally, an increased variability of data due to the increased system imbalance also disturbed the final conclusions.

Omnivory as our increased chance for survival. We eat everything possible and even more (watch TV or order fast food if you do not believe this). Omnivory-induced system stability probably helped us to conquer this planet. Had we been exclusively herbivorous or carnivorous species, system imbalance would have wiped us out a long time ago, or it would have forced the diversification of humans to herbivorous and carnivorous sub-types, generating a coexistence with our Neanderthal or other cousins and making human history an unending war between the two. Omnivory also provides us with a greater variety of plant toxins and other xenobiotics than a carnivorous or herbivorous diet, and this creates more weak links between these compounds and our cellular proteins. Omnivory seems to be a good strategy for staying well-balanced.

To answer the opening question of this section, it is better not to keep to strict promises in hard times. Eating our first fruit after years of all-burger days will both serve our own survival and help the ecosystem to get stabilized. On the other hand, the above scenario gives a good example of the nestedness of the ecoweb. In the long run, my choices from the eco-menu on the local level determine the survival of my species at the planetary level. The brave extension of existing knowledge and the discovery of large-scale, trans-network interactions led to the concept of Gaia described in the next section.

11.3 The Weak Links of Gaia

James Lovelock formulated the Gaia hypothesis at the end of 1960s. Gaia is the whole ecosystem of the Earth. Everything around and inside us belongs to it: the biosphere, atmosphere, oceans and soil, all making up a highly regulated complex network. At first, the self-regulating nature of the whole Earth ecosystem was received with skepticism. However, the increasing number of demonstrations had led to wide support by the end of the 1980s (Lovelock, 1979; 2003; Lenton, 1998).

"As it grows older, the Earth system weakens." Our Gaia network is "elderly and we should treat it with respect and care" (Lovelock, 2003). How can our aging Gaia be stabilized? Most of the existing examples of the self-regulation of Gaia involve negative feedback mechanisms (Lenton, 1998). However, it is quite clear that these are only the very first proofs of system stabilization at the global level. Gaia is the most complex network we have ever met, and has most probably developed a lot more stabilizing effects than we have ever found or even thought of. The extreme robustness and resilience of ecosystems (as discussed in Sect. 11.1) may encourage a general neglect with regard to preserving the stability of this largest ecosystem, Gaia herself. One often hears: "Without much care for the consequences, we should do what we need to do. If some species become extinct, we will find them elsewhere. If they become endangered worldwide, we can put them into sanctuaries or save videos and DNA samples until we can. The ecosystem will survive anyway. It has survived worse scenarios in the last 3 to 4 billion years." These assumptions are true. Gaia's resilience is indeed amazingly strong. However, there is an important point here. Does it make us happy to think that Gaia will survive – without us?

The stabilizing effect of weak links in ecosystems gives an additional, powerful argument for maintaining diversity. Diversity serves not only to preserve a larger genetic pool to combat future challenges, and it serves not only to prevent a general system crash at some time in the remote future. It serves also to avoid extreme fluctuations *today*. Pfefferkorn (2004) warns in his recent article on the Permian catastrophe: "Today we are living through another mass extinction." According to a recent estimate, climate warming scenarios may commit 15 to 37% of species to extinction (Thomas et al., 2003). If we pauperize ecowebs, they will be enriched in strong links. The depletion of weak links leads to larger oscillations and further extinctions, which may propel a downward spiral (Bascompte et al., 2005; McCann, 2000; McCann et al., 1998; Scheffer et al., 2001). Moreover, decreased diversity lessens the chances for degenerate functions which may substitute for one another during system fluctuations (Edelman and Gally, 2001; McCann, 2000). As a conclusion, we cannot risk the extinction of even a single species. Not only because of some altruistic, God-substituting pattern such as: "If we have become the rulers of the Earth, we should take care of her beings", but also because the more species vanish, the closer we get to unbearable fluctuations.

"I do share your thoughts, but let me ask a question here. You have already noted most of this in previous chapters, so why do you repeat it here? More pages, more money?" No Spite, fortunately this is not a string of sausages, but a book. Here the basic rule is almost the reverse: the more words it has, the less it may mean. Why do I stress the importance of diversity yet again? First, it is important. *If I am allowed to sum up only a single lesson from the stabilizing strength of weak links described in this book, then it should be the commitment to diversity.* However, the commitment to diversity is not only to preserve the ecosystem from large fluctuations. It has an even more serious reason. The concept of Gaia means that *all* these networks are connected. The anticipated extinction cascade due to climate warming will not only induce fluctuations in ecowebs. Eco-fluctuations will likely contribute to the destabilization of all other physicochemical networks around us. Among other things, it will make weather fluctuations wider. Larger weather fluctuations may induce an even larger imbalance in the ecosystem. We have another grim possibility here for a downward spiral.

I would make one more remark. Diversity here is not only the diversity of the species around us. Cultural diversity (Pagel and Mace, 2004) is also included. Cultural diversity is important, not only because it stabilizes societies and the social mega-net of our globalized world. Cultural diversity also makes weak links with the ecosystem. Our fast-conquering, omnipresent Western culture is not only pauperizing the cultural heritage that all previous generations have so far collected (Axelrod, 1997). It is also destabilizing the whole ecosystem, Gaia, around us. Population and economic growth may lead to a destabilized period around 2060–2080, according to some estimates (Johansen and Sornette, 2001). However, the situation might be worse. If we are truly unlucky, we may face a critical phase in the inherent growth of human networks, in the destabilized ecosystems, and in the destabilized complexity of Gaia all at the same time.

"Peter, you frighten me. Thinking about Pity, who is expecting, I am deeply concerned. We do want to take our responsibility for our future, for the largest known network, Gaia." Wow, a child! That was quick work, Spite. Congratulations to both of you! In Sect. 12.4, I will make some remarks as to whether Gaia is the largest known network or not. Apart from that, I am very happy to see your commitment. To end this chapter, let me list some advice.

Ecosystem management (I could say Gaia management) usually behaves as a human substitute for the Le Chatelier principle in thermodynamics. Has the equilibrium been changed? Too bad, we should add an extra measure to push it back to the original. Prevention of perturbations is a primary concern. This is wrong, and not only wrong, but fundamentally wrong. First of all, the system was not in equilibrium. It is not at all obvious that the previous state is more stable or more desirable under the present conditions. Secondly, as mentioned in Sect. 3.2, life is a relaxation phenomenon. If we reverse the change, we may prevent relaxation. Thirdly, even if we assume that the system was in a quasi-equilibrium, this equilibrium is not the simple equilibrium of the chemistry textbooks. The rules governing this equilibrium are much more complex than the Le Chatelier principle, and will be mentioned in the synthesis of Sect. 12.2. If we use simple logic and fight against the changes observed today, we may induce cascading changes leading to an even greater disturbance. Moreover, even if we are successful in our fight, this may not prevent the next perturbation occurring tomorrow at an unexpected point and level. *"Peter, by now I am not only frightened, I feel helpless. Was this your advice?"* Do not worry, Spite, there is more advice to come:

- **Learn system logic.** You are built from networks, you make networks, and you live in networks. You should understand how they work and how you can influence them.
- **Let it burn.** Once you have learnt how the system works, you should not block relaxation processes. If you do not take care over this, a netquake will follow with all the unpredictability I mentioned in Sect. 3.2. *"What is the 'burn' here?"* If you do not let small forest fires burn, much larger, devastating fires will follow (Malamud et al., 1998; Sornette, 2003). Similarly, you should leave mild illnesses untreated. Let your immune system relax. If you have a problem with someone, say so! Chronic resentment may cause a chronic disease. You may continue the list and help the easy relaxation of tensions wherever you go.
- **Redirect growth.** Since the beginning of our history all human networks have been expanding. Material growth may have approached the limits where it exceeds the resilience of the nesting network, Gaia. You should slow down, act less and think more. It is brainless logic to jump up and do everything you can. Patience and the strength of abstinence are sadly missing from our current world. However, stasis is not an alternative. You cannot stop growth. But you may redirect it. The Western world,[4] which consumes most of the world's resources, should make a shift from material growth to intellectual, artistic and spiritual growth. Knowledge is nonrivalling (Johansen and Sornette, 2001): you can share and develop it without exploiting the outer networks. You have to redirect your links, too. It is not your treasured goodies, but your fellow human beings who deserve your links. Caress your spouse, not the buttons of your computer. (*"Pity, don't you think he should stop writing here?"*) How can we accomplish all this? The West may learn from the traditions of the East and from some primitive societies. At least, while both still exist.
- **Protect diversity.** Diversity is the key to system stability and development. You should make every effort to protect it. If a complex system experiences trouble, it may begin to shift towards a star phase, and it may move towards collective and critical behavior. None of this helps diversity. If you get into trouble, stay calm and think. Remember the wisdom of hundreds of generations before, and remain independent. Do not go with the herd, and think twice before accepting the rules of a strict hierarchy. Like the foreign-

[4] And here I mean not only the USA, the EU and other G7 member countries, but more and more the fast-developing China and India, too.

Fig. 11.2. The West may learn from the traditions of the East and from some primitive societies. At least, while both still exist

ers, those who are different, those who are strange. Your tolerance will not only stabilize the many, but will also isolate the intolerable fanaticism of the few. If you can link future fanatics into your current networks, you have achieved real success. Stabilization of your present and future both require proper links: the last piece of advice and the main point of this book.

- **Balance your links.** Weak links help you to avoid unbearable fluctuations. However, you should not blindly start to make all the weak links possible. This will lead to an overconnected network, and will not decrease, but probably increase fluctuations. You should keep the delicate balance between a few strong and a lot of weak links. Moreover, you should make a few long-range links connecting social groups which were poorly connected before. The promotion of their understanding is your only ticket for a safe present and future. How can you do all these? Turn back to the first piece of advice and start again: learn system logic.

With this set of advice, we have reached the end of our last trip into Netland. I do not think I can give a better summary here. I can just ask you to take a deep breath, drink a glass of crystal clear water, relax, and most importantly: think about this advice. The next chapter will be the concluding chapter of the book. The coda I promised earlier.

12 Conclusions and Perspectives

"As Pity and I have been attending some religious education before our marriage, I have found you an excellent quotation: 'And further, by these, my son, be admonished of making many books there is no end; and much study is a weariness of the flesh' (Ecclesiastes 12:14). Don't you think it is time to finish your book?" I have good news for you, Spite! We are indeed approaching the end. In this, the last chapter I will first summarize everything we have learnt about the weak-link-induced stabilization of complex systems. Then I will spend some time redefining weak links and stability. Finally, another set of advice will follow: how to build proper links in your own life. Keep this advice, Spite. Your child may need it even more than you do.

12.1 The Unity of the Weakly-Linked World: A Summary

Tables 12.1–12.3 summarize the most important effects of weak links discussed in all the preceding chapters. It is clear from the tables that we have more than a dozen examples in which the stabilizing effect of weak links has been established for a wide variety of networks. Additionally, in another dozen examples, weak-link-induced network stabilization seems to be quite plausible. Although a general proof for weak-link-induced network stability is still lacking, I hope you will agree that this wide array of demonstrations cannot be a matter of pure chance. Weak links do indeed stabilize a lot of complex systems.

Weak links came as a great surprise to me. It was not so much their existence, which is obvious, but their importance that was the new message from my reading and which constitutes the main message of this book. These links form most of the links inside and around us and stabilize most of the networks we have ever had or accompanied.

I owe a great deal to weak links. Besides stabilizing me during the process of writing this book, weak links have given me a good opportunity to demonstrate the strength of the network approach.

Table 12.1. General effects of weak links in networks

Weakly-linked elements	Stabilized network	Stabilized function	Chapter and key reference
Network modules and networks	Various types of network	Help to link modules and to build upper networks	Sect.2.4; Csermely, 2001a; Degenne and Forsé, 1999; Granovetter, 1973; Maslow and Sneppen, 2002; Rives and Galitski, 2003; Spirin and Mirny, 2003
Society members	Social nets	Establishment of small-worldness; better network navigation; better innovation	Sects. 2.1 and 2.4; Dodds et al., 2003a; Granovetter, 1973; 1983; Skvoretz and Fararo, 1989
'Distant' network elements	Various types of network	Help relaxation by distributing perturbations	Sect. 3.2; Ghim et al., 2004
Network modules	Various types of network	Help netsistance (network integrity) by disjoining modules	Sect. 3.2; Sethna et al., 2001; Sornette, 2002
Network elements	Various types of network	Serves as a 'reserve' which can be broken during stress and rebuilt after	Sects. 3.4 and 6.3; Granovetter, 1983
Oscillators	Various types of network	Helps to achieve stable synchronicity	Sect. 3.5; Blasius et al., 1999; Bressloff and Coombes, 1998; Gao et al., 2001; Lindner et al., 1995; 1996; Strogatz, 2003

I hope I have convinced you that networks and their general rules provide an extremely useful conceptual framework to understand the world inside and around us.

The different levels of the nested networks behave as a multidimensional puzzle and offer us a great tool to identify their missing elements or discover the missing links between them. Recent advances in genetics give excellent examples of cross-network annotation of existing, but in its present form rather useless information (Bergmann et al., 2004; Stuart et al., 2003). This cross-fitting works best at the level of analogies, rather as this book often builds on analogies. I have to note here that I agree with Rose (1997), who warned of the danger of analogies as a source of wishful thinking.

☺☺ **The most important and most favored hypotheses of the book.** Let me list some of the analogies that highlighted novel ideas in the preceding chapters. Most of them are only guesses at the moment, so I have marked this box with two smiley faces. I apologize for the fact that one or two items are actually only three-smiley dreams. The small smiling faces in front of each statement indicate its potential importance according to my current subjective judgment.

Table 12.2. Demonstrated examples of weak-link-induced network stabilization

Weakly-linked elements	Stabilized network	Stabilized function	Chapter and key reference
Water and atoms of proteins	Proteins	Protein structure and conformational transitions	Sect. 5.3; Barron et al., 1997; Brinker et al., 2001; Csermely, 1999; Klibanov, 1995; Papoian et al., 2004
Proteins (e.g., chaperones)	Cellular protein net	Cellular phenotype	Sect. 6.2; Rutherford and Lindquist, 1998
Motor unit oscillators	Muscle	Movement precision	Sect. 7.3; Semmler and Nordstrom, 1998
Social actors	Psyche	Psychological well-being	Sects. 7.5 and 10.3; Degenne and Forsé, 1999; Freud, 1915; Kawachi and Berkman, 2001; Kunovich and Hodson, 1999; Veiel, 1993
Animals	Animal community	Group survival, group cohesion	Sect. 8.1; Noe, 1994; Silk et al., 2003
Females after menopause	Social net	Infant survival, group cohesion	Sect. 8.2; Connor et al., 1999; Lusseau, 2003; Silk et al., 2003
Society members	Society	Social cohesion, innovativeness, efficiency	Chap. 1, Sects. 8.3 and 10.3; Degenne and Forsé, 1999; Granovetter 1973; 1983; Putnam 1985; 2000
Firm employees	Firm social net	Firm efficiency, resistance, innovativeness	Sects. 8.4 and 10.3; Cross and Parker, 2004; Dodds et al., 2003b; Fukuyama, 1995
Firm owners and firm	Ownership network	Firm survival	Sect. 8.4; Stark and Vedres, 2002
Owners and firms	Portfolio	Profit	Sect. 8.4; Stark, 1996
Participants of (pseudo)-grooming, small-talk, gossip, etc.	Social net	Psychological well-being, social cohesion	Sect. 8.6; Dunbar; 1998; Szvetelszky, 2003
Scenes	Drama	Plot	Sect. 9.2; Stiller and Hudson, 2005
Items of town architecture (districts, houses, etc.)	Town	Traffic, town life	Sect. 9.3; Salingaros, 2004
Software elements	Software (especially after refactoring)	Program	Sect. 9.4; Brown et al., 1998; Fowler et al., 1999
Consumers and prey	Ecosystem	Ecosystem stability	Sect. 11.1; Berlow, 1999; Blasius et al., 1999; Garlaschelli et al., 2003; McCann et al., 1998; Neutel et al., 2002

Table 12.3. Conjectured examples of weak-link-induced network stabilization

Weakly-linked elements	Stabilized network	Stabilized function	Chapter and key reference
Atoms	Crystal	Cohesion	Sect. 5.2; Ball et al., 1996
Water and atoms of RNA	RNA	RNA structure and conformational transitions	Sect. 5.3; Csermely, 1997
Enzyme functions	Metabolic net	Cellular metabolism	Sect. 6.2
Transcriptional modulators, DNA	Gene expression network	Transcription	Sect. 6.2
Cytoskeletal proteins	Cytoskeleton	Cell rheology	Sect. 6.2
Hormonal effects	Hormone network	Hormone response	Sect. 6.4
Immune cells	Immune system	Immune response	Sect. 7.1; Brede and Behn, 2002
Blood vessels	Blood circulation	Blood supply	Sect. 7.2
Neurons, neurotransmitters, astrocytes	Neural net	Neural function, learning and consciousness	Sect. 7.4
Words	Language	Meaning	Sect. 9.1
Electronic circuit elements	Electronic circuit	Circuit function	Sect. 9.5
Market actors	Market	Market	Sect. 10.1
Elements of all ecosystems and the environment	Gaia	Our past, present and future on this Earth	Sect. 11.3

- ☺ ☺ ☺ The basic statement of the book: weak links stabilize all complex systems.
- ☺ Levy flights represent our evolutionarily conserved sense of scale-freeness, showing a disturbed relaxation: the scale-freeness of games, music, and all types of art are reflections of this evolutionary heritage (Sects. 2.2 and 9.2).
- Music helps learning due to a better synchronization of our neurons. This is achieved by the stochastic resonance of neuronal oscillations with the pink noise of music (Sects. 3.1 and 3.5).
- I suggested a large variety of potential self-organized criticality phenomena (adolescent psycho quake, coughing, crying, firm reorganization, gossip, laughter, lightning, sex, wooing, etc.; many sections starting from Sect. 3.2).
- ☺ I gave three novel examples of topological phase transitions of networks: cell death, animal communities and turning points in history (Sects. 3.4 and 10.2).
- ☺ The rather controversial, but intriguing idea was raised that synchronization can be achieved between networks at different levels. This

trans-network sync is especially induced if the sync is very strong at one of these network levels (Sects. 3.5 and 7.4).

- ☺ ☺ Weak links make the stability landscapes smoother, e.g., energy landscape, fitness landscape, innovation landscape, diegetic landscape, etc., and facilitate transitions of the respective networks between their local equilibrium points (many sections from Sect. 5.2).
- ☺ Chaperones and other proteins buffer cellular noise and diversity due to their weak links (Sect. 6.2).
- Organelle diversity in cells acts as a stabilizer of higher eukaryotes (Sect. 6.2).
- ☺ Planned and efficient disintegration of the cellular net occurs during apoptosis (Sect. 6.3).
- ☺ Pink-noise-regulated treatment protocols are helpful against cancer (Sect. 6.4).
- ☺ Low-affinity, multitarget drugs are more helpful in several cases than single-target, high-affinity drugs (Sect. 6.4).
- ☺ The aging process displays a preferential deterioration of weak links, which leads to increased noise and destabilization of aging networks (Sect. 6.4).
- Intermediate motor unit synchronization is a weak-link-induced form of movement stability. Birth is characterized by an extensive sync and high fluctuation rate between motor units of the uterus (Sect. 7.3).
- ☺ ☺ The psyche can be assessed as a network with strong and weak links. Weak links are stabilizing here as well (Sect. 7.5).
- ☺ ☺ The SMALL and BIG phenotypes correspond to the STRONGLINKER and WEAKLINKER personality traits, respectively (Sect. 7.5).
- ☺ Grandmothers have more time and opportunity to establish weak social links. Weak-link-induced stabilization of the animal and human social groups may be a novel explanation for the existence of the menopause (Sect. 8.2).
- ☺ ☺ ☺ Diversity-tolerance is extremely important for the stabilization of society (Sect. 8.3).
- ☺ Women play a central role in the stabilization of society. (This may contribute to the development of many critical areas of the world, like Afghanistan, the Middle East and the Balkans; Sect. 8.3).
- ☺ ☺ Due to the slow change of the SMALL phenotype to the BIG phenotype, which is able to build stabilizing weak links, we should wait 2 to 3 generations of well-being before attempting to build democracies in developing countries (Sect. 8.3).
- Fringe areas have a general importance in the organization of complex networks. Brain intermodular fringe regions, engineering pidgin formalizations, architectural transition areas, and software design patterns are all very important examples of this same phenomenon (Sects. 7.4, 8.4, 9.3, 9.4 and 9.5).

- Gossip and slander build weak and strong links in a social group, respectively. This may be what underlies their stabilizing and destabilizing roles (Sect. 8.6).
- The 21st century developed a wide variety of pseudo-grooming and link strength relativization (Sect. 8.6).
- Superman exemplifies weak-linking and gender relativization (Sect. 9.2).
- Similarly to crying, laughter, clapping and stadium waves, synchronized relaxation acts as a source of catharsis (Sect. 9.2).
- Great masters, great architects and market gurus are people with exceptional cognitive abilities who think at least to 6th order (Sects. 9.2, 9.3 and 10.2).
- ☺ Scale-freeness in architecture is a source of improved relaxation and psychic well-being (Sect. 9.3).
- ☺ ☺ Weak links are key contributors to social capital (Sect. 10.3).
- ☺ ☺ ☺ This is the place for the synthesis of the next section, the coda I promised before.

I must admit that most of these analogies are rather hypothetical. Some of them are unbelievable, some may stretch the analogy beyond an acceptable measure, and some may eventually be proven to be wrong. However, the list is rather impressive and demonstrates the benefits of the network approach, going beyond the reductionism that pervaded most areas of science in the second half of the 20th century.

Scientific reductionism helps to conceal the fact that we are unable to grasp the whole, and need to analyze the details. Gould and Levontin (1979) warned against the dangers of extreme reductionism in their famous essay on the spandrels of San Marco. Rose (1997) describes numerous elements of reductionism:

- reification, which converts a dynamic process into a static phenomenon,
- arbitrary agglomeration, treating different reified features as arbitrary exemplars of a common character,
- improper quantification, adding an arbitrary number to agglomerated reified features,
- confounding metaphor with analogy or homology.

In contrast to reductionism, networks help us to regain a holistic view of our life. There is every reason to hope that 'networkism' will never develop into a new 'religion'. The network approach is an immensely beautiful general method, and not a monolithic belief. Moreover, it is not a holistic view without details. Networks offer you both the generality of rules *and* the delicate details. You may remember the nested networks. If you keep your distance, they behave like a point. If you go close to them, they will suddenly become a whole world.

12.2 Revisiting the Definitions: A Synthesis

In the last section I summarized our knowledge of the basic hypothesis of the book, i.e., that weak links stabilize all complex systems. I still owe you a definition of weak links and stabilization. In this section I will try to say more about both of them.

Warning! Danger zone! As you proceed through this section, you will find some rather far-fetched ideas on the possible use of strong and weak links as a new approach to describe the stability of complex systems. Perhaps I should have written the whole section in a smiley box. However, I think these ideas are rather novel and important enough to deserve a larger font than that here. I have thus kept to the regular typography for the text in what follows, and ask you to bear in mind that most of the statements are only hypothetical ... as yet!

Weak Links: The Definition Remains Unchanged

I must admit right at the outset that, even after so many chapters and examples, I have no better definition for weak links than the one I gave in Sect. 4.2. The examples of Tables 12.1–12.3 gave me the clear impression that the actual ratio of stabilizing weak links is highly dependent on the system properties and cannot therefore be generally defined. The 'weakness' of the weak links may also vary along these lines. Moreover, the exact position of weak links is also important. It is quite plausible that we may need different amounts of weak hubs (date hubs; Han et al., 2004; Luscombe et al., 2004), weak bridges or 'simple' weak nodes to stabilize a complex system. Much more experimental work, methodological precision and modeling effort is needed to see whether any generalization or classification can be made with regard to these numbers.

In the absence of adequate knowledge for a theoretical or numerical approach, I can only resort to the experimentalist definition of Sect. 4.2: A link is defined as weak when its addition or removal does not change the mean value of a target measure in a statistically discernible way (Berlow, 1999).

Stability of Complex Systems: A Reintroduction

Fortunately, much more can be said about the definition of the stability of complex systems after the numerous examples of the preceding chapters. Reiterating the remarks of Sect. 4.3, network stability can be defined either as parameter stability or as network persistence (netsistance). The criterion of netsistance is rather simple: the network has to keep its integrity, meaning that most of its elements should stay linked to each other. In other words, the so-called giant component of the network should be preserved. The studies I have read so far have not provided any clues as to how to connect weak links and netsistance. Perhaps further development of the topological phase transition theory of Tamas Vicsek and colleagues (Derenyi et al., 2004; Palla et al., 2004) will give more data to be able to discuss this field better. But in the absence of such data, I will restrict my thoughts to the parameter stability of networks.

The parameter stability of networks usually 'overshadows' netsistance. If a network can stabilize most of its parameters, it is rather unlikely that it will grossly change its topology and undergo a massive topological phase transition.[1] Assessing the role of weak links in the parameter stability of networks will give us some rather big surprises and novel thoughts by itself. Let me approach this enormously complex field by starting from the simple and moving towards the more difficult. First I will consider only the topology of weak links in complex networks, and I will propose a novel type of Le Chatelier principle showing how networks may counteract the effect of destabilizing perturbations. Next I will introduce the energy-type network parameters, and I will describe various stability landscapes, proposing a novel role for weak links in making the local minima of these stability landscapes more accessible. Finally, I will move one step closer to the real world and show what happens if a change in a network element provokes changes in other network elements. Under these conditions the stability landscape itself is also continuously changing. Several simplified subtypes of this incomprehensible complexity, such as the Nash equilibrium or supermodularity have become fundamental elements of game theory. I will end my discussion by showing that weak links help to simplify the multiple stability landscapes and also make the otherwise hopeless search for equilibrium easier under these complex conditions.

---

[1] I should note here that the connections between these two types of stability have not yet been worked out either.

Le Chatelier Principle for Networks

If a perturbation hits an element of the network, it becomes unstable and will be unable to make strong links with any of its neighbors. Sometimes strong links will turn into weak ones, but in most cases a large number of links will break and only a few, mostly weak new links will form instead. In the next step, more and more perturbations hit the network. In consequence, the ratio between strong and weak links shifts towards the latter. As a result of the increasing amount of weak links, the network becomes more stable. Consequently, if the perturbations have not overwhelmed the network, the network automatically stabilizes itself by converting strong links into weak ones (Fig. 12.1). This may be regarded as a Le Chatelier-type principle for complex networks.

"Granovetter (1983) showed that, in the case of high unemployment or any other type of increased everyday stress, weak links are broken and people tend to rely on strong links. Don't you think that there is a contradiction between this statement and your Le Chatelier principle?" Well done, Spite! Many thanks for this remark. The contradiction here is only apparent. The Granovetter statement referred to static network behavior which can be observed as an average

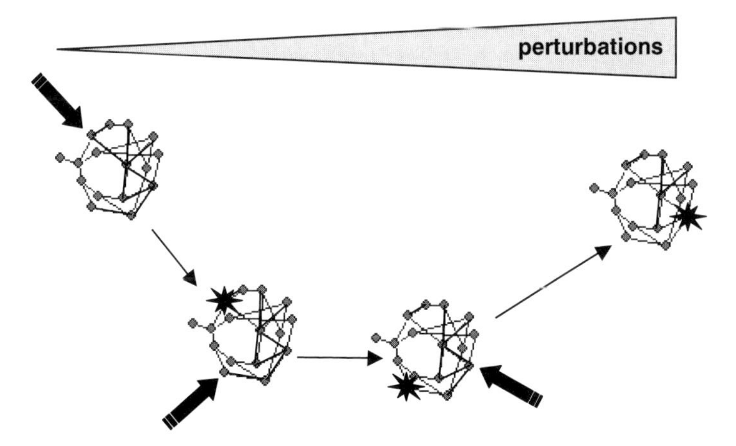

Fig. 12.1. The network Le Chatelier principle. Incoming perturbations (*thick arrows*) first destabilize the bottom networks of the perturbed network elements (*black stars*). The unstable element turns some of the strong links (*thick lines*) into weak ones (*thin lines*). As a result of the increasing amount of weak links, the network becomes more stable. Note that only two link strengths (strong and weak) are marked in the illustration. In reality, link strength has a continuous spectrum and the modification of links after a perturbation has a much more complex pattern than the one depicted here

in the long run. Indeed, it is true that in the absence of resources, the network does not have enough energy to build up weak links on top of the most essential strong ones. However, the network Le Chatelier principle refers to dynamic processes. Even these 'all-strong' networks turn some of their contacts into weak ones if perturbations arrive. In their case, this may be more transient than in other networks.

Stability Landscapes, Punctuated Equilibrium, and Netquakes

In spite of the initial statements above, network topology is clearly not enough to analyze the stability of networks. We may get a better picture by introducing the stability landscapes. The landscape approach usually picks two parameters of the network and plots the stability criterion (e.g., energy) as a function of these two parameters in the third dimension (see Sect. 6.2 and Fig. 12.2). Obviously, this three-dimensional approximation is an oversimplified version of real life, where hundreds of parameters may change, and we would need a hundred-and-one dimensions to describe them. The landscape approach was first introduced by Sewall Wright in 1932 and has been successfully applied to describe the stability of proteins (energy landscape; Bryngelson and Wolynes, 1987; Bryngelson et al., 1995; Dill, 1985; 1999), species evolution (fitness landscape; Kauffman and Levin, 1987) and the innovation process (innovation landscape; Kauffman and Levin, 1987; Tyre and Orlikowski, 1994).

The stability landscapes of complex systems are rugged. They have many local minima, separated by smaller or bigger saddles. Often the minima are hierarchical and show modularity (Ansari et al., 1985). The heights of the saddles (the extent of free activation energies) may often show a scale-free distribution (Yang et al., 2003).

Due to the rugged surface, the temporal dynamism describing the transfer of the network from one local minimum to another on the stability landscape does not show a continuous change. Shorter or longer segments of relative stasis are followed by a sudden shift to a novel stability landscape minimum which may often result in a drastic change of most network properties. This phenomenon is called punctuated equilibrium, following its powerful description for the majority of evolutionary processes (Gould and Eldredge, 1993). Punctuated equilibrium was proposed to characterize the development of many complex systems like protein folding (Ansari et al., 1985; Yang et al., 2003), economics (Schumpeter, 1947), innovation (Tyre and Orlikowski, 1994),

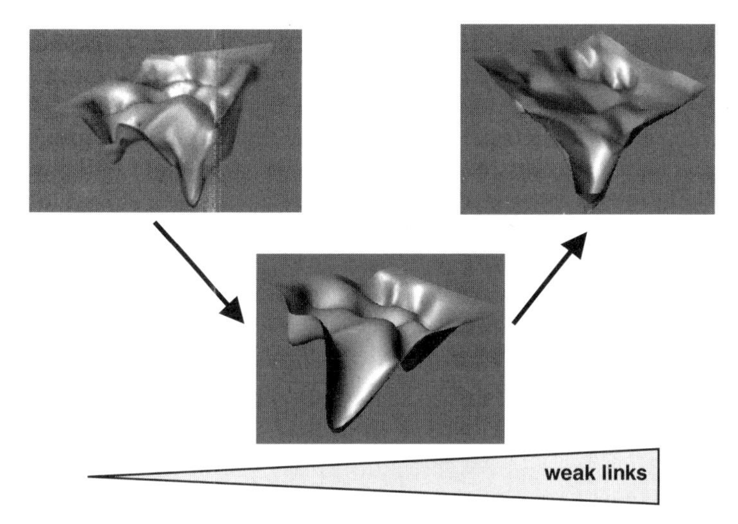

Fig. 12.2. The stability landscape of networks. Weak links smooth the stability landscape

institutions (Aoki, 1998), software design (Crutchfield, 1994), and scientific progress (Kuhn, 1962). In punctuated equilibrium, stasis, especially the stasis during intensive environmental changes, is not just a period when nothing happens, but an active phenomenon (Gould and Eldredge, 1993) where the network requires stabilization by weak links.

The dynamism of punctuated equilibrium resembles the various quakes we have met since Sect. 3.2. Netquakes are signature properties of self-organized critical systems. Indeed, self-organized critical networks have been shown to have punctuated equilibrium (Kauffman and Johnsen, 1991; Bak and Sneppen, 1993; Sneppen et al., 1995). *"This seems to be so general that we may say we are living in a life quake."* Thanks for the life quake, Spite. This is a beautiful expression. However, it may not be true. As I mentioned above, self-organized criticality is in a punctuated equilibrium. However, not all punctuated equilibrium is a result of self-organized criticality. If the interaction strength of network elements is very diverse, too weak or too strong, self-organized criticality may not develop (Sethna et al., 2001). However, the complex system might still manifest the features of a punctuated equilibrium under these conditions.

Weak Links Improve the Accessibility of Stability Landscapes

Following this introduction, I now come to the novel part of the discussion. How do weak links influence stability landscapes and punctuated equilibrium? Table 12.4, which extends Table 4.1 in Sect. 4.1, summarizes the features of network stability in the presence and absence of weak links. The characterization shows the clear differences between networks having many or only a few stabilizing weak links. The take-home message is as follows: in the presence of weak links, stability landscapes become smoother. In other words, weak links make the punctuated equilibrium less punctuated.

Too many weak links may destroy equilibrium. The involvement of weak links in lowering the saddles of a stability landscape gives a novel meaning to the overconnected and therefore unstable networks mentioned in Sect. 3.3 (Fink, 1991; May, 1973; Siljak 1978; Watts, 2002). If the network has many weak links, the stability landscape becomes very smooth. This means that most saddles become low in the stability landscape and practically all energy minima become accessible. If there is a well-defined, absolute energy minimum, the network will instantly go there and remain there, so that equilibrium is easily achieved. However, if such an energy minimum does not exist, the network may shuffle between various energy minima which are close enough to each other to offer an equally good place to stay, and equilibrium is never achieved. In the latter, multi-minimum landscape, the presence of higher saddles and increased roughness may make most of the competing local minima inaccessible, and may help one of them to be found as a possible equilibrium.[2]

Weak links help the convergence of bottom network stabilities to top network stability. When speaking about network stability, we should not forget about nestedness. All elements of the top network are themselves bottom networks. All of their links are in fact a part of a complex synchronization (or desynchronization) phenomenon. I have already mentioned cross-network stabilization from top to bottom networks and from bottom to top networks in Sect. 4.3, referring to the stabilizing mechanisms of the top networks exerted on bottom networks,[3] as well as to the stabilizing

[2] I am grateful to Cleopatra Ormos for this idea.

[3] These were the mechanisms to (a) stabilize, (b) segregate, or (c) disassemble the unstable bottom networks.

Table 12.4. Network stability in the presence and absence of weak links. The content of Table 4.1 in Sect. 4.1 is repeated at the top of the present table, while novel statements are placed at the bottom of the table. The term 'energy' may refer here to fitness, design efficiency, market value, story integrity, etc., depending on the type of stability landscape (energy, fitness, innovation, economic, or diegetic landscapes, respectively)

Network has many weak links	Network has few weak links
Long-range contacts give small-worldness, modules are well-connected	Average distance between elements is large, modules are sparsely connected
Behavior of bottom networks is optimally synchronized, giving small fluctuations	Bottom networks are either tightly coupled with large fluctuations or behave independently of each other
Communication is good in the network, relaxation goes smoothly	Communication is restricted in the network, relaxation is disturbed, relaxation avalanches may occur
Noise is easily dissipated or absorbed in the network	Network is noisy and noise stays in segments of the network
Network is integrated and behaves as a whole	Network is segregated and behaves as an assembly of its constituent modules and bottom networks
Changes are dissipated and occasional errors are isolated, so that the network is stable	Changes and noise persist, the network is error-prone and unstable
The saddles of 'activation energies' are transiently lowered. The transition between local energy minima is smooth	The saddles between local energy minima are high. The transition between energy minima is difficult
Relaxation proceeds undisturbed	Relaxation is blocked
The parameter space of the stability landscape is 'discovered' by the network. Chances of finding the minimal energy state are high	Large segments of the parameter space of the stability landscape are inaccessible to the network. Chances of finding the minimal energy state are low
The network is not noisy. 'Unusual' transitions between energy minima do not occur. The most probable status of the network is generally predictable	Due to the blocked relaxation, the network is noisy. On rare occasions the network noise (especially pink noise) allows highly unusual jumps between energy minima. The most probable status of the network is difficult to predict
The network is plastic. After a jump to another energy minimum the reconfiguration of the network is easy. Novel weak links are established, which stabilize the network in the novel state, deepening the newly found energy minimum. Once the network has reached a new state, it may stay there, unless a neighboring saddle gets lowered again	The network is rigid. After a jump it does not change. Consequently, it will not 'adjust' to or 'fit' the novel state. Since the energy minimum of the novel state has not been deepened, the energy can only be lowered by another jump

efforts of the bottom networks on the environment,[4] respectively. However, there is also a remark here from the standpoint of weak links. If the top network is not integrated, the individual bottom networks will remain in their own optimum state, and will not be forced to observe the optimum state of

[4]These were the stabilization efforts of the bottom networks on any environmental parameter or their symbiosis.

the top network. Integration of the top network is only achieved by the proper ratio of strong and weak links between the bottom networks. In other words, the cooperation of bottom networks needs weak links. This is related to the proper level of bottom network synchronization, which is also promoted by weak links, as I mentioned in Sect. 3.5.

Sorting and selection: a new difference. Vrba and Gould (1986) emphasized the difference between sorting and selection. They described selection as one of the causes of sorting. Using the effect of weak links described above, I may add that, in the presence of stabilizing weak links, the actual sorting will converge upon (obey) the underlying selection sooner than in their absence. Weak links facilitate the search for the optimum state on the fitness landscape (selection) and provide a better fit between the actual outcome (sorting) and the long-term goal (selection).[5] Obviously this does not change the difference between selection, which may operate at a certain level of nested networks, and sorting, which may be observed at the same level or any other levels above or below.

System Complexity Induces a Rugged Stability Landscape Which Makes Further Development Difficult

The complexity of the structure of the energy minima of the network may increase in parallel with the complexity of the network itself. Using the landscape representation, this means that complex systems have a more 'rugged' stability landscape. These stability landscapes may have more local energy minima and these minima may be separated from each other by larger saddles. Kauffman and Levin (1987) reached similar conclusions when they analyzed their fitness landscape model. They showed that, with increasing complexity, access to the best optimum becomes more difficult. Let me give a few examples to make this general statement clearer:

- Complex proteins have a more rugged energy landscape. A complex protein has more folding problems than a simple one, and the search for the energy minimum becomes more demanding as complexity develops (Bryngelson and Wolynes, 1987; Bryngelson et al., 1995; Csermely, 1999; Dill, 1985; 1999).
- Complex organisms may have a more rugged fitness landscape. A bacterium in our bowels may find a new phenotype much more easily than a squirrel.

[5] If this assumption is true, the evolution of omnivores (with lots of weak links) should be generally smoother than the evolution of herbivores or carnivores.

- Complex designs may have a more rugged innovation landscape. A washing machine can be redesigned more quickly than a Boeing 777.
- Complex novels may have a more rugged diegetic landscape. *Little Red Riding Hood* can be more easily made into a video clip than *War and Peace.*

"This sounds tragic. According to this, the more complex we become, the more we get trapped by the 'frozen accidents' of our great-grandparents." Do not despair, Spite. First of all, without the frozen accidents of our great-grandparents we could not develop and maintain our complexity (Gell-Mann, 1995). Secondly, there is an almost automatic solution to escape from this catch-22 situation. I will give a possible answer below.

Four Mechanisms to Facilitate Changes in Complex Systems Despite Their Rugged Stability Landscape

Complex networks learn to extend their local stability islands (escape route 1). Self-organizing complex systems have the luxury of degenerate subsystems, which give a better error tolerance to the whole. Better error tolerance will extend the local stability island. A larger error tolerance allows higher designability.[6] As an example, more complex protein structures have a higher designability, i.e., there are more individual structures fitting the stability criteria of the same native structure (Li et al., 1996; Tiana et al., 2004). Increased designability is a consequence of the increase in the local stability island in the energy landscape of the protein,[7] and it allows the development of diversity in compliance with local circumstances.

Complex systems learn to jump around the stability landscape (escape route 2). Besides higher designability, the buffering (stabilizing) effects of weak links, also allow the development of hidden diversity (see Sect. 6.3). Weak links prevent the detection of this hidden diversity by the environmental selection process. Stress decreases

[6]Designability is the maximal number of individual solutions accommodated by the given design pattern, i.e., the maximal number of individual solutions fulfilling the stability criteria of the network environment (see discussion in Sects. 3.6 and 9.5).

[7]The increased thermal stability of the most popular 'superfolds' occurring in the folding simulation of a 27-mer heteropolymer provides an excellent example (Zeldovich et al., 2005).

the total amount of weak links, leading to a decrease in buffering capacity and forcing most of the previously collected hidden diversity to appear. This is the point when selection begins to act on diversity, when it is suddenly exposed, erasing most of it in the long run.

Let me transcribe part of the take-home message of Sect. 6.3: by the stress-induced release of diversity, the system may bridge the distance between local minima on the rugged fitness landscape. So far we see no difference between simple and complex organisms. However, the frequency of local, observed stress may have changed during the development of complexity. If bacteria form no biofilm, their stress is unavoidable, since the protective microenvironment is missing. With symbiotic events and with the development of multicellular organisms, local stress at the level of individual cells becomes less common. For our own cells, stress is a tiny island in the middle of the ocean of a more or less undisturbed homeostasis. Consequently, cells of more complex systems have more opportunity to collect hidden diversity, since the 'discharging' stress comes more seldom. However, when it does come, it leads to the discharge of a much wider spectrum of previously hidden diversity. The conditions for the complexity-dependent extra 'jump' are now fulfilled. In spite of the fact that the next stable island of the fitness landscape is less accessible due to the complexity of the network, the same increase in complexity means that the network can make bigger jumps around the fitness landscape and a new optimum can thus be reached.

Modernity mandates a safety net around innovators. The development of modern designs requires much better conditions for innovators than before. Not only does complexity increase as we go from simple prebiotic bacteria to human cells, but a Boeing 777 is definitely more complex than a coffee grinder from the early twenties. We have every reason to assume that the innovation landscape of the Boeing 777 is much more rugged than that of the coffee grinder. It is rather plausible to expect that a major innovation on a rugged innovation landscape requires an extended stasis (in the form of undisturbed creative freedom), compared with an innovation on a smooth innovation landscape. CEOs and science policy makers should learn from this. Quite paradoxically, if they want a faster innovation process for designs with high complexity, they should allow a longer period of undisturbed creative freedom to the developers than before. The establishment should be alerted to the need for more long-term science grants to provide an undisturbed creative environment, and also to the need to separate think tanks in the high-tech industry if we hope to develop complex systems further. *"If I have understood correctly, undisturbed creative freedom is not enough here. From time to time, good management should also create an unexpected, high-stress situation to reveal the diversity of hidden ideas which have been accumulating during the long stasis of undisturbed freedom."* Great idea, Spite, I could not agree more. However, premature stress will uncover premature ideas, and they may not span the required distance to the next stability island on the innovation landscape. To sum up, Spite, establishing the right rhythm for the jumps in the innovation process needs a great deal of wisdom.

Weak links help the stability of both the bottom and the top network. Weak links in the bottom network thus generate increased diversity by various means (increased designability and hidden variation). Increased diversity generates more weak links in the top network. This provides another nice example of the cross-stabilizing effects of nested networks.

Complex systems may have learned to optimize their search for the absolute stability minimum (escape route 3).[8] In random networks relaxation is global. In complex networks relaxation becomes restricted to the local environment. As mentioned in Sect. 2.2, decreased network relaxation may increase the diversity of jump lengths[9] on the stability landscape. In this process the jump-length distribution approaches a scale-free distribution. This means that the network usually makes a rather small shift on the stability landscape. However, the network will occasionally make larger jumps on the landscape, and on very rare occasions, a very large jump will follow. Thus, optimally, the walk followed by the network may become a Levy flight, which is probably the most efficient way to explore the stability landscape (Vishwanathan et al., 1999). Thus a network may have three basic scenarios for exploring the stability landscape:

- If the network has a lot of weak links, it is close to a random network, relaxation is undisturbed, the stability landscape is smooth, transitions have a high probability, and the network may 'allow'

[8] This escape route is more hypothetical then the first two routes. To mark this difference I have inserted two smiley figures. However, I think the content is important, and I have therefore kept the usual font size.

[9] The jump length here is obviously not a physical length, but denotes the difference between the parameter sets of the two local minima on the stability landscape.

itself the luxury of choosing a search strategy closer to a random walk than the optimal Levy flight.[10]

- If the network has fewer links, these links will differentiate into strong and weak links, the relaxation becomes restricted, the stability landscape is rugged, transitions have a low probability, and the network has to develop a hidden variability and compensate itself for the reduced number of jumps. The network will thus select an optimized, scale-free search strategy, i.e., a Levy flight.
- if the network has only a minimal amount of weak links, it will remain in the vicinity of the original minimum on the stability landscape and become 'frozen'.

It should be noted here that we need much more modeling and experimental effort to determine whether:

- these assumptions are true,
- they are general to all types of stability landscapes (energy, fitness, innovation, diegetic, etc.),
- there is a continuous shift between the three extreme search strategies (random, optimized scale-free/Levy flight and 'frozen'),
- the Levy flight is the most efficient search on stability landscapes,
- there are other search strategies for the absolute minimum of the stability landscape which have not been mentioned here.

Complex networks have rather elaborate mechanisms to regulate the amount of weak links in them (escape route 4). This is something we discussed in Chap. 6. As we saw in escape routes 1–3, not only do weak links improve the chances of complex networks achieving an inherent diversity (designability, route 1), and not only do they give them the opportunity to make jumps on the landscape (hidden diversity, route 2), but they may also help them to find the optimal strategy for these jumps (Levy flights, route 3). In addition, complex networks are able to increase and decrease the amounts of weak links in them as circumstances may require (route 4).

Weak Links Increase the Evolvability of Complex Systems

The propensity to evolve is called evolvability (Kirschner and Gerhart, 1998). This gives me a chance to restate the above take-home message

[10] Obviously, this is not a conscious choice. Due to the fast relaxation processes, the network cannot do anything else but to shift the search strategy from an optimal scale-free search towards a random search.

in another form: weak links help to set the degree of evolvability of the network. This is very much along the lines argued by Earl and Deem (2004), showing that evolvability is a selectable trait.

Weak links regulate the smoothness of the stability (energy, fitness, innovation, diegetic, etc.) landscape. The absence of weak links makes networks frozen, while many weak links make networks plastic. If we have an intermediate quantity of weak links, we end up with complex networks, where the above four escape routes promote the evolvability of the system.

Le Chatelier-type principle for the evolvability of complex systems. Let me restate the take-home message of the above discussion in a shorter form. As the system grows more complex, it has more chance of stabilizing itself and its environment. As a consequence, it acquires a higher evolvability, which helps to preserve its chances for development in spite of the stricter requirements to accomplish this task.

The evolution of evolvability and our future. Alternating periods of stasis and sudden change are needed for the evolution of evolvability. In other words (Earl and Deem, 2004): "Populations that are subject to more severe environmental changes can produce lower-energy individuals [with greater fitness] than populations that are not subject to environmental changes." As a hypothetical consequence of the above, the extinction of dinosaurs might not only have happened because of the cataclysms after the asteroid impact in the Yucatán Peninsula (Alvarez et al., 1980). They may also have been preceded by a longer stress-free period beforehand. We humans started to enjoy a rather stress-free period some time ago. Although the 'stress-free' centuries behind us are nothing on the evolutionary scale, the long-term consequences speak for themselves. *"Now I am beginning to regain confidence in our future. For us personally, this might be a disaster, but for humankind as a whole it might supply just the tension needed to keep our evolvability fit."* Spite, I love your global optimism. However, we do not know very much about Gaia yet, so we would do better to avoid any critical events. If you keep diversity today, you help Gaia to stabilize *and* you increase the general evolvability, including your own (Earl and Deem, 2004).

Link management in scientific discoveries. Alternating periods of stasis and evolvability can be observed in a slightly different context if we survey scientific progress. When we try to describe a new observation, we often use a sequential approach of alternating 'focused' and 'fuzzy' segments. When we force our brain to work in the focused mode, we use Occam's razor to cut everything from the network of thoughts which is not absolutely necessary for the exact description of the subject. The focused mode makes the description concrete and scientific, but it cannot usually insert a brand new element into the solution. In contrast, the fuzzy mode is usually a brainstorming process, when we start to play and add a wide variety of possible ideas to the existing core of the solution to the problem. The fuzzy mode destroys the clarity of the description, but often advances our understanding in unexplored fields. The focused mode increases the roughness of the innovation landscape, while the fuzzy mode decreases it, most probably by decreasing and increasing the amount of weak links in our thought network, respectively. The fuzzy mode may be helped by sleep. Interestingly, birds needed a fuzzy mode of quality-deteriorating sleep besides their focused mode to achieve a high quality song (Derégnaucourt et al., 2005).

Stable Networks Can Participate at the Next Level of Self-Organization

In all the above discussions relating to stability landscapes, an energy-type function was used as the criterion for assessing how 'good' a network was. This function may be energy itself, or fitness, or design efficiency, or market value, or story integrity, etc. What is common to all these? Why do they have to be optimized?

I should note here that all these functions are our own fabrications to make the 'goodness' of our systems tractable and measurable. This raises the following question: what *happens* when a network is 'good' and it reaches the minimal energy of the respective stability landscape? Networks become stable. Is this something that actually *happens*? In fact it is not. It is just another parameter we use to talk about goodness. *"I do not understand the big hassle about the 'what happens' question. Can you not figure out for yourself what happens, Peter? Survival happens. That is what happens – or not, as the case may be. It's not that difficult!"* Wow, Spite, you got really emotional this time. But you are right, survival happens, at least in the long run. But for the moment, it is only a probability. By the time survival actually 'happens', the stability of the system will have changed a thousand times. Survival is an integrative measure of an average stability, and it is not what I am looking for.

The prompt, energy-dependent and visible action of stable networks is self-organization. If the bottom networks have found their energy minimum and are able to stay at least minimally stable, they can become a member of a top network. Designs are used, firms are traded, proteins build cells, cells build organisms, organisms build ecosystems, and so on. So how far can this chain be continued? In fact, the chain goes on until we reach Gaia. And what does Gaia build? Where is Gaia used or traded? What is the link, what is the reward here? What makes the discrimination between the 'stable' states of Gaia possible? What is the larger network in this case?

The links of Gaia. Here comes the three-smiley part. We may opt for three answers. The first answer is spiritual. Gaia builds God. If Gaia is stable, God accepts her. If we make her unstable, as we are doing right now, we had better switch to the second answer before we end up with an unintended scientific 'proof' of the upcoming Doomsday. The second answer poses Gaia ready for marriage. From time to time Gaia gets very close to the point when she should develop further by getting integrated into a larger network. Hopes are raised, Gaia looks around excitedly – and then sinks back into her unhappy spinsterhood again and again. In the absence of external stabilization, an inner reorganization follows. Another version of self-organized criticality may bring into play the recurring criticality of a lack of self-organization at the higher level. Are we approaching such a point? Is the next round of global frustration and reorganization close? Is it not high time for us to shift our resources from wars to exploring other worlds, other civilizations, to find the missing link that might help Gaia to stabilize and avoid the next criticality? We know practically nothing. These questions are either grossly premature or completely misdirected. But let me ask again here. Are these questions really premature? Are they really misdirected?

"Peter, this is thrilling." I agree with you, Spite. If I could, I would have attached seven smileys to the above box, like the stars we have on Metaxas. I have an excuse though. The third answer. Gaia does not build anything. More than that. Gaia cannot build anything. From the network point of view, we are a closed system here, being lonely, linkless, helpless and measureless. Gaia is slowly cooking in the greenhouse effect, with a logo on her T-shirt which says: "Watch out, you guys! Self-organization ends here." Does this sound sad? Looking at the good side of it, we have to be responsible. We have no outside evaluation, no influential link to take and no synchrony to grab and enjoy. Either we set and keep our own measures, or we collapse. Sorry, I have no better news for today.

When the Stability Landscape Changes: Network Games

We have moved very far away from weak links and network stability criteria. I am sorry about that. Networks are so general and so logical that they make you think the unthinkable. Nevertheless, it is high time to return to science again. In the last part we had quite a nice characterization of stability landscapes with or without weak links. However, the *whole* of the last part was wrong. Spite, do not look at me like that. I was not lying. I made an approximation. Most of the science we know is only an approximation. Approximations are our bread and butter here, since they make problems tractable.

Now the trouble with stability landscapes is that they change. Here comes nestedness again. When bottom networks develop, they are not in a sterile space. They interact with other bottom networks forming elements of the top network. As they interact, the responses of all the other bottom networks change the environment, and consequently change the stability landscape (Nowak and Sigmund, 2004; Ruthen, 1993). Imagine a hundred, no, a thousand bottom networks. As they start to interact and the stability landscape starts to change in ten thousand dimensions. What's the matter, Spite? Are you OK?

"Peter, if you continue, I am leaving. This is too much. It has become incomprehensible and unimaginable. How can our world be so complex?" I agree with you, Spite. The complexity of multiple stability landscapes (I will call them multi-landscapes in the rest of the book) is really beyond our cognitive abilities. I have some good news though. We are not the only dumb guys around: this is beyond everyone's cognitive abilities on Earth. Spite, do you remember Dunbar (2005) and his 6th order thinking? According to his assumptions, to understand the changes in a stability landscape with six interacting elements would require Shakespeare's cognitive level. I have very bad news here. A network with six elements is not a network. It is a mere network embryo. For typical networks, we would need to think to the 1 000th order, or maybe to the 10 000th. *This* is clearly impossible. We have to make things simpler. Some of the most brilliant human minds have been involved in this simplification for some time now. The result is called game theory. Obviously I have neither the knowledge nor the courage to make an overview of all aspects of game theory from the standpoint of networks and weak links here. Let me restrict myself to two examples which exemplify the logic of some of the available solutions to this exciting problem.

My first example is the Nash equilibrium. The concept of the Nash equilibrium was described in an extremely elegant, 28 line publication by John Nash (1950). If a game is in a Nash equilibrium, no players will gain any benefit from a change in the playing strategy, if all the other players keep their original strategy. In other words, in a Nash equilibrium there is a balance between the gains and costs related to aggressive behavior. From the point of view of multi-landscapes, the Nash equilibrium is a very special case in which every element of the network sits in a local minimum, so that there is no need to imagine the multiplicity of the stability landscape. If the game is able to reach a Nash equilibrium, the multi-landscape has been simplified to the (mono-) stability landscape we had in the last section.

The second example is supermodular games. These are non-cooperative games. In these games the actions of various players are completely independent of, but complementary to one another. This is assured by a set of criteria. The most important of them is that the game has multiple positive feedback: an increased activity by some players raises the benefits of a similar activity by others. Under these conditions an avalanche-like behavior will develop. Such a situation might be the adoption of a new standard in an area of technology or the collective withdrawal of money from an ailing bank. Under these conditions these games have and can reach a Nash equilibrium (Milgrom and Roberts, 1990; Topkis, 1979). If we imagine this concept on the multi-landscape, we may say that most of the multiplicity of the possible stability landscapes has been erased by the complementarity condition, and the remaining multi-surfaces may converge to a situation where the above condition that everyone sits in a local minimum is set. As the above examples show, the equilibrium may be a rather short-term equilibrium here since, once the standard is adopted or the bank goes bankrupt, the game comes to an end.

Generalizing the logic of supermodular games, we see that game theory often uses a principle of rational human behavior for a drastic reduction of the multi-landscape to a few dimensions, where the search for an equilibrium point (determined again by expecting rational behavior on behalf of all the actors) can be accomplished. A beautiful example of this approach is the book by Harsanyi and Selten (1988), which contains the core of their work establishing several basic concepts of game theory and resulted in their winning the Nobel Prize with John Nash in 1994.

Our efforts to reduce the multi-landscape towards a mono-landscape in human relationships are not just based on ad hoc decisions. We have a rather large set of jointly internalized behavior codes and strategies. The process in which we acquire these codes and strategies is called socialization. It happens in families, schools and all human groups we join throughout our lives. *"Wow! Are you saying that, when our despised geography teacher tried to discipline us, he was actually providing help to achieve a better and faster outcome in our future games?"* Yes, Spite, you have hit upon the point. Obviously, in many cases you may leave your original group and go to another, if you feel you cannot converge to the common mono-landscape model of the original group. However, if you decide to stay (or you cannot leave), you should converge to the mono-landscape, otherwise the group will not be efficient. I must note here that the geography teacher should have done the same. But it is not only geography teachers that have been invented in our culture to accomplish this task. Institutions, rules, norms, laws and roles were all developed to simplify the multi-landscape and to make the responses of our partners predictable in a given situation. All these elements of human culture reduce the dimensionality of multi-landscapes, and make the situations of our everyday life cognitionally tractable.

However, we still have to simplify the remaining multi-landscape after the application of pre-set rules. Segments of the multi-landscape may also be continuously excluded as mutually optimal strategies develop between players during a game. This then becomes an iterative learning process (Axelrod, 1997). As an example, in supermodular games, complementarity may develop as players increasingly restrict themselves to a subset of strategies which prove beneficial, and abandon those strategies which have performed consistently badly (Milgrom and Roberts, 1990).

As a last remark I do not want to give the impression that games or reduced multi-landscapes will always lead to a Nash equilibrium. In several cases, the network may just remain unstable, showing oscillations or completely irregular dynamics. Moreover, the complexity of multi-landscapes may hide a number of surprising equilibrium conditions.

Obviously the above lines were not an attempt to summarize game theory in a one-page manner. This extremely complex and exciting topic deserves a whole book by itself. I only wanted illustrate the type of logic game theory uses to solve the cognitive problem of finding equilibrium conditions from amongst the ever-changing multitude of dimensions and stability landscapes.

Weak Links Facilitate the Convergence of Multiple Stability Landscapes

To assess the role of weak links in the complex equilibrium between interacting network members, I will first make an artificial approximation in which I dissect the interwoven texture of weak and strong links of a network into a weak subnetwork and a strong one. Let me give a brief analysis of the role of game theory in the treatment of these two subnetworks:

- **Weak subnetwork.** Elements of the weak subnetwork do not interact appreciably. If one of the weak elements changes, the other network elements will neither be affected, nor change their playing strategy, whence the overall stability landscape of the network will not change. The stability conditions of the weak subnetwork are mostly predictable. Since elements do not interact appreciably, the use of game theory is not needed to find the stability conditions of the weak subnetwork.
- **Strong subnetwork.** Any change in an element of the strong subnetwork affects all its neighbors. As summarized in the previous part, even the mono-landscape of the strong subnetwork is largely unpredictable. Since any change in a given element may lead to a change in the game strategy of all its neighbors, the stability conditions of strong subnetworks are governed by the laws of game theory.[11]

The game theory of proteins and cells. So far I have not mentioned consciousness as a condition of game theory. The rules of the game are not only the rules which the players of the game accept, set or discover as conscious actors. Rules in this broader sense are all the simplifying conditions that can be applied to two or more networks when they engage upon a strong interaction with one another. Thus the application of the basic simplifying approach of game theory will help us to define the changes in the energy landscapes as proteins start to interact with each other (Kovacs et al., 2005), but also the stability landscape of individual cells, such as those of the immune or neural systems, as they form larger networks, and so on. Consciousness of the network elements 'only' gives the additional level of complexity that the network elements do not necessarily have to make physical contact in order to learn about each other's presence and incentives. From the assembly of the hundred thousand types of protein molecules which we usually call a cell, the only ones that will cause a significant change in each other's energy landscapes are those which make a protein complex and get into physical interaction. If these proteins were members of a conscious

[11] Importantly, if the elements of the strong subnetwork are not complex enough to show any appreciable change after a change in their strong neighbor, there is no need to use game theory even for the equilibrium conditions of the strong subnetwork.

society, the cries for help from their fellows adjacent to the breaking mitochondrial membrane would certainly alarm most of them before the actual damage arrived in the form of oxidation or proteolytic attack.

After restricting the applicability of game theory to elements of the strong subnetwork, I will now add the two subnetworks together again and, keeping the strong and weak identification of the elements, assess the effect of weak links on the games of the strong subnetwork members. As we have seen before, weak links lower the saddles of the stability landscape and may even eliminate several local energy minima. They thereby reduce the 'roughness' of the mono-landscape. What might correspond to this situation in real games? Elements of the hypothetical weak subnetwork should not participate in the games themselves. However, they help the players to find the optimal solution faster. What do we call them? In real human games, they are usually called mediators, moderators, negotiators, arbitrators or appraisers. In fact, all the people mentioned in Sect. 8.3: friends, gossips, hairdressers, madams, priests and psychologists (the list can be continued!) are elements of the weak subnetwork helping the various games of our life.

As discussed in Sect. 8.3, all these weak-linking actors help to stabilize society. Now we have an additional element to explain how they do it. They simplify the multi-landscape by helping a faster and more efficient exclusion of the stability landscape segments containing non-profitable plans for potential actions. Due to the weak links, this action is not unilateral, but is in most cases extended to all participating players of the game, which is not necessarily helped by the same weaklinkers. Thus weaklinkers allow a faster convergence of the 'real players' of the strong subnetwork towards equilibrium.

Do all weak linkers help this action? They most probably do. At least, they may provide an opportunity to simplify the multi-landscape *without* adding their own extra dimensions to it. In other words, weaklinkers do not increase multi-landscape complexity, since they do not have strong links and hence do not participate in the game.[12]

[12]In the real world, my initial approximation does not hold: weaklinkers do participate in games themselves. In other words, all mediators participate in the game, since a fully impartial mediator may not even be efficient (Kydd, 2003). If weaklinkers have a few strong links, they will obviously also increase and not only decrease multi-landscape complexity. Since there are far more weaklinkers than strong – this is not only due to the scale-free distribution of link strengths described in Sect. 2.4, but also due to human cognitive limits, which in most games

Two examples. Although I promised not to give extensive examples from various games, I would like to mention the fact that in the minority game and in the binary evolutionary game, the game manifested a much smaller fluctuation if the players were non-identical (Challet and Zhang, 1997; 1998). Non-identical players may develop weak links and enhance the convergence of multi-landscapes.

Weak Links May Explain the High Assortativity and Clustering of Social Networks

Are all weaklinkers equally efficient at inducing the convergence of multi-landscapes? Most probably they are not. Fast convergence of two or more multi-landscapes towards each other is best achieved if weaklinkers persuade the two or more players of the game to adopt the same type of strategy. This works best if the weaklinkers really 'link' the two players with a strategy preference that provides a bridge between the players' strategy preferences. Thus the most efficient weaklinkers should be rather similar to both potential players to ensure a fast and efficient simplification of the multi-landscape. How will this requirement be reflected in the network structure? Similar elements should group together, because otherwise their decisions will never converge, and they cannot help each other to achieve this. This grouping is called assortativity and clustering.

As already mentioned in Sect. 8.3, social networks have these two rather specific features (Newman, 2003b; Newman and Park, 2003). The reason for this specificity remains largely unexplained. Games between similar strong subnetwork members, as well as efficient moderation received from similar weak subnetwork members, may significantly accelerate multi-landscape convergence. Members of technological or biological networks[13] experience a much smaller range of change in their stability landscapes as a result of the action of another network member. Consequently, they are not such a great help for multi-landscape convergence. Multi-landscapes may grow to a level of complexity requiring the help of assortativity for landscape convergence only beyond the level of social nets, ecosystems, 'evolutionary nets', and perhaps the networks of cells. Amongst these, only social networks have been sufficiently studied to be able to characterize their

seriously restrict the number of strongly linked players – it is quite plausible that the net outcome of weak link contribution is a simplification of the multi-landscape.

[13]These networks all have a negative assortativity.

assortativity. Efficient simplification of stability landscapes may provide an additional and important explanation as to why assortative and clustered social nets were selected and kept by evolution.

Link relativization and game theory. *"I have an objection again. Clusters of social networks imply strong links. If I have two friends and they become friends with each other, then all three of us make strong links. Your theory that weak links simplify the multi-landscape cannot be applied here!"* Well done, Spite! This is a nice objection. However, you must bear in mind that you keep your strong links until the three of you are in good agreement. In this case the identical strategies themselves simplify the multi-landscape and you do not need any weak links to help the process. However, if a disagreement starts between two of the three, the third friend may want to help to rebuild the friendship between the two who disagree. This is now a typical weak-link situation, where the differing stability landscapes of the disagreeing friends are helped to converge by the mediation of their common friend. You may observe that the links were temporarily weakened between the mediator, who is not part of the disagreement, i.e., the game, and the other two friends.[14]

Open Questions and Summary

Many exciting questions remain open, such as:

- How do the above weaklinkers simplify fitness multi-landscapes?
- How do ecosystems tackle this problem?
- Can we make anything out of this at the level of our lovely old spinster, Gaia?
- Do protein complexes or cellular nets give us a new understanding of multi-landscapes?
- To what extent is this set of ideas valid for the diegetic landscapes of our best dramas, novels and films?

"And??? I am so excited! You cannot just leave these questions like this!" Sadly, I will have to leave them for now, Spite. We must leave something for our next book, you know.

Let me restate the take-home message here. In the first step, weak links make the definition of the equilibrium conditions possible with the development of mono-landscapes. In the second step, weak links make the search for the now-defined equilibrium possible by smoothing the mono-landscape. Without the multidimensional convergence

[14] I am grateful to Zoltán Borsodi for this idea.

provided by weak links, the network would never reach equilibrium in most cases.

Link strength, probability and thermodynamics. Probability and thermodynamics in their classical interpretations assume the absence of strong and conditional interactions between the participants of the observed network. Thus, using the above distinction between weak and strong subnetworks, both concepts describe only the behavior of the weak network. We have to be extremely careful if interpretations, consequences or direct use of classical probability and thermodynamics 'sneak in' to our thoughts when we speak about the properties and behavior of strong subnetworks, where the 'conditioned' interactions of network elements often make 'simple' thermodynamical or probabilistic predictions useless.

Spite, please help me to check how comprehensible this important section has been. May I ask you for a summary? *"You trust me too much, Peter. This has been a rather difficult section. However, I will give it a try."* Spite, I have to make something clear. It is not your fault if you do not understand this section or any of the others. It is my fault. My ideas were either badly formulated or simply not ripe enough to talk about them, if a person like you, after qualifying at a good high school, does not understand what is written here. Science is not science, or at least, not important science, if the essence of it cannot be formulated in common language for the layperson. *"It is nice of you to say that. I feel more comfortable about beginning my summary. First you described an interesting negative feedback effect in networks, hypothesizing that a perturbation may cause a shift from strong to weak links, thereby increasing the chances of the network getting stabilized. You also gave a detailed summary showing that weak links can smooth stability landscapes (you called them mono-landscapes) and make punctuated equilibrium less punctuated. As the last element of these equilibrium ideas, you extended the 'guiding effect' of weak links towards equilibrium to the multiple stability landscapes (multi-landscapes) of game theories. I was discussing this with Pity and we decided to talk more about our problems to each other and to our friends. It is rather silly not to use the available weak links to find the equilibrium of mutual respect and love in our everyday games. I liked your idea about the evolution of evolvability as complex systems develop and the links of Gaia were thrilling – as I noted there already."* Spite, your summary was better than I could have made it myself. Many thanks!

Completing this section, I feel like a curious and stubborn child. Walking around in the attic I found a colorful and intriguing thread: weak links. I felt that there must be something exciting and unknown at the other end of the thread, and I started to pull it. I pulled, and I pulled. And all of a sudden the whole roof was falling down upon me. "Mathematicians tend to favor restricted definitions, engineers broad definitions, while physicists are somewhere in between" (Turcotte, 1999). What type of definition might fit a networker, or more precisely, a weaklinker, like myself? Natural networks are in continuous growth and remain forever unfinished. I do ask your forgiveness if you feel that the definitions of their links or stability also remain unfinished. Notwithstanding, I hope at least the attempts at definition were interesting and useful.

12.3 Prospects and Extensions

Having analysed the central statement that weak links stabilize all complex systems and discussed related definitions, I begin now by listing a few ideas on further studies in the context of weak links. In the second part of the section, I will outline the possible extensions of weak-link-induced stability to smaller and larger networks than those visited in the tours of Netland in Chaps. 5–11.

Spite, it is time for you to listen, as a future scientist. To begin with, I have already listed 29 novel ideas in Sect. 12.1. All of them may be starting points for exciting further research. Let me list here a few more ideas for further studies, which were put forward in the last section:

- Many examples of the stabilizing effects of weak links are only assumptions at present. The general, mathematical proof for weak-link-induced network stabilization is still lacking.
- Although there are numerous examples of the scale-free distribution of link strength in various networks, we do not know what the limitations of this distribution may be, nor how often we may find them, nor how general and how necessary they are.
- We do not know how the link strength and degree distributions are linked, i.e., we do not know the ratio of hubs having mostly strong or weak links. As an initial characterization of this problem, the 'date hubs' of the yeast networks (Han et al., 2004; Luscombe et al., 2004) are in fact weak hubs, since they establish only transient links with their partners. However, we do not know whether the contribution of weak hubs, weak bridges or the presumably overwhelming majority of simple weak nodes are more important in the stabilization of networks.

- We do not know whether weak links only stabilize networks with scale-free degree distribution, or whether they also stabilize random graphs, star graphs or fully-connected subgraphs.
- The network Le Chatelier principle requires a lot more clarification. Examples have to be given and a formal proof made, and the restrictions of its validity have to be explored.
- The evolution of evolvability will require many experimental, modeling and theoretical studies. Its connections with increasing system complexity seem to be especially important.
- Game theory will certainly go through a renaissance as it conquers more and more networks beyond social nets; even in its classical field of application, i.e., real, human games, the involvement of weak links and their possible connections with the assortativity and clustering of social networks deserve a great deal of attention in the future.

"These are nice ideas, but mostly for mathematicians or physicists. I am neither of these. However, spending such a long time with you and with this book, I have become really interested in weak links. What can I do now?" Spite, first let me thank you for putting aside your reservations and growing to like weak links in the end. In answer to your question, if I had not known you, I would have said: write an email to the address in the Preface of this book with your idea or area of interest. Since we have already been working together for quite a while, I am sure you will find a nice project from those started by the LINK group, after your marriage and the canoe trip, of course!

Do you remember Spite? When we made our trips into Netland, we started with molecules and finished with Gaia. These networks are so nicely nested inside one another that it is tempting to speculate whether there might be anything beyond? Most probably not. The network theory seems to be a really powerful approach to link vastly different systems, and show the general elements of their organization, dynamism and stability. However, like everything we have figured out here on Earth, networks are probably also anthropocentric and may well prove to be useless if we get too far away from human dimensions. In any case, I will make an attempt to extend the network approach both towards the bottom and towards the top – imploring your patience once again.

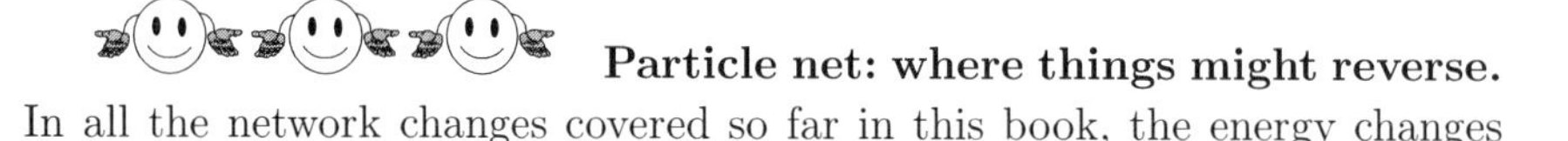

Particle net: where things might reverse. In all the network changes covered so far in this book, the energy changes were modest. Their transformation to matter was negligible. What happens if the energy is so high that we cannot neglect this phenomenon any more? If we speak about elementary particles, can matter serve as a governing force behind transitions in the energy net in the same way as energy served as a governing force behind the transitions in particle nets? If a network is formed from particles, weak links between them correspond to a small probability of interaction between them, i.e., the particle to energy transition is small. If the network is formed from energy levels, weak links between them mean a small probability of transition between these energy levels, i.e., the energy to particle transition is small. In summary, weak links are those which least disturb the dual energy/particle behavior of the given system. Could this also contribute to the stability of the system?

Going to the other extreme, the Solar System and beyond, we find a scale-free density distribution in the hierarchy of stars, galaxies, intergalactic gas clouds, galaxy clusters and probably galaxy superclusters. The fractal property of the Universe was first predicted by Mandelbrot in 1977. By now it is a widely accepted fact.[15] However, it is still an open question as to where we will find the threshold between this scale-free distribution and the presumably isotropic and homogeneous Universe on the 'large scale', which is called the cosmological principle. The fractal property below 100 megaparsec can be explained by the action of gravity invoking a preferential attachment, or self-organization again (Baryshev and Teerikorpi, 2004; Gaite, 2005; Mahdavi and Geller, 2004; Pietronero, 1987; Sarkar, 2005; Wu et al., 1999).

??? ?

What are the weak links in the Universe? Are the weak links in the Universe represented by the dark matter or the dark energy? Do we have something in the Universe which is currently unimaginable but provides stabilizing weak links? Do weak links really stabilize our Universe? Or is the whole network logic no longer valid here, so that the Universe does not have any weak links at all?

Laughlin and Pines wrote in their essay entitled *The Theory of Everything* (Laughlin and Pines, 2000): 'It is impossible to miss the similarity

[15] Let me note here that the network structure of the Universe suggested by its scale-free density distribution should not be confused with a potential top network of the Gaia ecosystem. I have not found the faintest hint of the existence of this latter network.

between the large-scale structure recently discovered in the density of galaxies and the structure of Styrofoam, popcorn or puffed cereals." They also note that: "The central task of theoretical physics in our time is no longer to write down the ultimate equations, but rather to catalogue and understand emergent behavior in its many guises, including potentially life itself." I hope the present book has been able to show that networks are excellent vehicles for this cataloging task. Jiddu Krishnamurti (1992) made a clear distinction between knowledge and wisdom. In his teachings, knowledge is information about the world, whereas wisdom is our knowledge about the processes. I believe networks teach us wisdom in Krishnamurti's sense:

> Mental paradigms are models that simplify complexity, ideas that help to make sense out of the infinitely complex continuum which is reality. No idea can represent reality as it is. A paradigm merely represents a fragment of reality in a way that allows the mind to deal with it: it encodes a part of the world to the mind's specifications. A paradigm is formed and retained because it is useful, not because it is real. A scientific paradigm marks out a conceptual territory for exploration by observation and experimentation. It establishes a world view that defines which questions are worth studying and what answers might be expected. (Cohen, 1992a)

The extent of the usefulness of the network approach in general and the multiple stabilizing roles of weak links in particular remains an open question:

> When a scholar publishes a paper, it is a letter sent to unknown recipients. If the job has been well done, then with luck it may be found and read, perhaps years later .... (Myerson, 2001)

However, at least for me, networks and weak links already "establish a world view that defines which questions are worth studying and what answers might be expected". I owe a great deal to all those I have quoted along the way and also to those I have not quoted due to some regrettable oversight.

12.4 Weak Links and Our Lives

Networks and links not only organize our thoughts and stabilize both ourselves and our environment, but they also help us to develop. Our personal life behaves as a self-organizing network. Our capacity for trust is a signal of our inner stability, showing that we are able to further extend our own life net and universe net. As a result, we feel happy and enriched as our inner or outer networks extend.[16] If the development of our nets is continuously arrested, we start to die. This concluding section will give some advice on how to accomplish this lifetime commitment for our self- and universe-organizing task.

Life net. Life itself, with all its decision points and multi-landscapes, and all the possible but mostly *not* followed pathways, can be imagined as a network. We have very few major decision points, and a large number of small ones. At the big decision points, we might go in a thousand directions, while at the small decision points, we may only select from a very few options. Unfortunately, in this case a control experiment is not possible. We cannot start our life again to explore the choices we have not made. I am afraid the question as to whether the life net is scale-free will never be answered.

In Sects. 8.4 and 10.3, I listed several benefits of links. Links lead you to success. Top performers are distinguished by their well-developed social network. More importantly, links preserve your health (Cross and Parker, 2004). An appropriate social net helps you to control blood pressure, decrease your chances of dying from a heart attack, and improve your immune response (Putnam, 2000). So how should we develop and maintain our links? In fact, this is an inappropriate question. Link development on its own is not enough. A link always has and needs another end besides yourself. Moreover, stability can only be reached and preserved together. So an equally important question arises: how should we help others to develop and maintain their links? Here are some ideas:

- **Leave your gadgets.** If you feel isolated and alone, you may think you do not have links. This assumption is wrong. No one is linkless. Lonely people have their own set of links, but these links are misdirected. It is time to make your link inventory. You may actually play with that as you analyze your coming week. How will you spend your time? What is around you, where is your attention focused, and what is the target of your acts? What are the nodes to which your links lead you? Do you care about your car repair, new mobile, and the pool in the garden, or do you prefer to play with your children, or discover yourself and the wisdom within? If

[16] I am grateful to Csaba Sőti for this idea.

your link inventory is gadget-oriented, it is high time you changed. If you are linked to gadgets, your main direction is downwards and not upwards. Your gadgets will not help your self-organization.

- **Get stabilized.** Unstable subnetworks cannot be part of a top network. Do not expect links if the only help you can offer is to destabilize your neighbors. You will be segregated as part of the self-healing process of the top network. Your life will be changed if you are able to get yourself just a little more stable. Why is this moment so important? Self-stabilization is a highly rewarding project. The smallest step towards your own stabilization will open new links to you, and these will help all the further steps ahead of you.
- **Develop trust.** Once you have improved your stability, you are ready to enjoy the benefits of the immense social net around you. But not only that. You will also enjoy all the other nets above and below. Obviously these indirect effects will be weaker and weaker as the distance between the nested networks grows. But your links are not only your gates into society. They also open the whole Universe to you. Cross-network stabilization may help to improve your health and make life smoother around you. How should you start to make links? Develop trust. How should you start to develop trust? Improve your stability.
- **Select and use your links.** The more links we have, the better our life will become. In fact, this statement is incorrect. If your links lead you everywhere, then you have no links. Moreover, you will have no time to maintain your links. If your links lead you everywhere, your links will become useless. Remember the instability of the over-connected system. The random network is a primitive net with low complexity. You deserve more than that. You need a careful balance of a few strong and a lot of weak links. Treasure all your links appropriately. Make special efforts to be a bridging person between groups with different habits and cultures. Give quality time to your links and use them. You should feel the joy of being linked. This is how you can initiate novelty, achieve success and stability, and be a part of the joyful synchrony these links allow us to develop.
- **Help others to get their links.** Cross and Parker (2004) call the really good link-makers energizing people. What is needed for this? (a) A vision that people can follow your ideas and not you yourself; (b) integrity that validates your vision; (c) a chance for contribution that leaves the path flexible to reach the common goal; (d) positive thinking that binds rather than separates. Relating these to the

terms listed above, (a) and (b) both correspond to your stability, (c) is your trust, and (d) is the joy of using your links.

You may have noticed that a few of the link-related pieces of advice are actually duplicates of those in Sect. 11.3. The stylistic error was intentional. Person- and wisdom-oriented links, a proper inventory of various link strengths and bridging links with the related trust, tolerance and diversity are the most important personal messages of this book. Let me ask you for the last time to take a deep breath, to drink a glass of crystal clear water, to relax, and most importantly: to think. Link management is not only a key for your personal well-being and success. It is the evolutionary duty of us all.

We have come close to the end of the book and it is time to say good-bye to you Spite. Let me thank you for all your criticism, help and companionship. *"That is very nice of you, Peter. I promise to try out the Levy flights on my honeymoon canoe tour with Pity. From now on, whenever we listen to music, we will also think about pink noise, and I have taken good note of all your advice to us and our child."*

This book was actually a link itself. A link to all the wide variety of networks we have visited together and a link for me to say something about the world – as seen from this head. I hope the book is an interesting mixture of 'true science', one- or two-smiley hypotheses, and more conjectural three-smiley beliefs. I have an excuse for the latter two, which are actually related to weak links:

Science and the weak cognitive links. Science wants to give clear answers and thus provides strong cognitive links. When modern science started with Copernicus, Galileo, Newton and Descartes, it increased our inner stability, since it provided a few strong cognitive links which made the plethora of previous links weak, thereby stabilizing the system. However, one has to warn about the other extreme. Extremist positivism may give too much space to science. If thoughts such as 'what is not scientific is not valid and is not part of our world explanation' begin to emerge, we start to reduce our cognitive links towards the outside world to strong ones. Too many strong links destabilize us. Science should be modest, restrict herself, and always leave some ground free for the weaker cognitive links of religion, art and simple beliefs.

This book was about weak links. Here I want to repeat once more that weak links alone do not mean anything. Stable systems most probably need a scale-free distribution in link strength. Stable systems may try to approach a scale-free distribution in all dimensions. A perfect scale-free distribution in all imaginable dimensions and measures looks the same from all possible points on all possible scales. Scale-free then becomes free from any scales. The concept to name this uncompromising scale-freeness was figured out a rather long time ago: it is called God.[17] We are part of a 'background network' providing a net of weak links which are everywhere, and are based on affection, goodwill and love.

The final message: we should not worry, once we learn to accept affection, goodwill and love, we will remain stable. The weak links will stay with us until

THE END ...

[17] I am grateful to Bálint Pató for this idea.

. . . AND BEYOND.

A Useful Links

Note that URL addresses were checked as of 31 August 2005.

Groups Working on Networks and Complex Systems

Adam P. Arkin	http://gobi.lbl.gov/~aparkin
László A. Barabási	http://www.nd.edu/~alb/
Yaneer Bar-Yam	http://necsi.org/publications/dcs/
Eric L. Berlow	http://www.wmrs.edu/people/BIOs/ericberlow/EricBerlow.htm
György Buzsáki	http://osiris.rutgers.edu/frontmid/indexmid.html
Gerald M. Edelman	http://www.nsi.edu/public/scientists/index.php
Jennifer H. Fewell	http://sfi.cyberbee.net/jennifer.html
Murray Gell-Mann	http://www.santafe.edu/sfi/People/mgm/
John Gerhart	http://mcb.berkeley.edu/faculty/CDB/gerhartj.html
Mark Granovetter	http://www.stanford.edu/dept/soc/people/faculty/granovetter/granovet.html
Stuart A. Kauffman	http://www.santafe.edu/sfi/People/kauffman/
János Kertész	http://newton.phy.bme.hu/~kertesz/
Marc W. Kirschner	http://sysbio.med.harvard.edu/faculty/kirschner/
Susan Lindquist	http://www.wi.mit.edu/research/faculty/lindquist.html
Kevin S. McCann	http://www.uoguelph.ca/zoology/department/people/faculty/k_mccann.htm
Mark J. Newman	http://www.santafe.edu/~mark/
Zoltán N. Oltvai	http://path.upmc.edu/people/faculty/oltvai-lab/
Sándor Pongor	http://www.icgeb.trieste.it/RESEARCH/TS/Pongor.htm
Nikos A. Salingaros	http://www.math.utsa.edu/sphere/salingar/
Ricard V. Solé	http://complex.upf.es/~ricard/
Steven H. Strogatz	http://tam.cornell.edu/Strogatz.html
Tamás Vicsek	http://angel.elte.hu/~vicsek/ http://www.angel.elte.hu/clustering/
Duncan J. Watts	http://smallworld.columbia.edu/watts.html

Libraries and General Databases

E-libraries	http://www.arxiv.org
	http://www.santafe.edu/research/publications.php
	http://cogprints.org
	http://comdig.com
	http://www.nslij-genetics.org/wli/1fnoise/

Mathematical Tools

Network map construction	http://vlado.fmf.uni-lj.si/pub/networks/pajek
	http://biodata.mshri.on.ca/osprey/servlet/Index
	http://discover.nci.nih.gov/kohnk/interaction_maps.html
	http://strc.herts.ac.uk/bio/maria/NetBuilder/
	http://paup.csit.fsu.edu
	http://taxonomy.zoology.gla.ac.uk/rod/treeview.html
Network modules	http://www.angel.elte.hu/clustering/
Digital organism development programs	http://physis.sourceforge.net

Protein Networks

Protein interaction databases (mostly yeast)	http://www-unix.mcs.anl.gov/compbio
	http://yeast.cellzome.com
	http://www.bind.ca
	http://dip.doe-mbi.ucla.edu/
Human protein reference database	http://hprd.org
Subnetwork analysis	http://www.genecensus.org/TopNet/

Metabolic Networks

Modeling of biochemical reactions	http://www.gepasi.org/
Whole cell simulation	http://www.nrcam.uchc.edu/
Metabolic networks (general)	http://metacyc.org
Metabolic networks (yeast)	http://www.genome.ad.jp/kegg/
	http://systemsbiology.ucsd.edu/organisms/yeast.html
Metabolic networks (*Escherichia coli*)	http://gcrg.ucsd.edu/organisms/ecoli.html
Red blood cell	http://systemsbiology.ucsd.edu/organisms/rbc.html

Transcriptional Networks

Gene interactions and pathways	http://www.biocarta.com/genes/allpathways.asp http://www.cifn.unam.mx/Computational_Genomics/regulondb http://strc.herts.ac.uk/bio/maria/NetBuilder/
Escherichia coli transcriptional network	http://www.weizmann.ac.il/mcb/UriAlon
Yeast sporulation network	http://cmgm.stanford.edu/pbrown/sporulation
Gene expression databases	http://www.ebi.ac.uk/arrayexpress http://www.ncbi.nih.gov/geo http://transcriptome.ens.fr http://www.gene-regulation.com

Specific Biological Databases

Bacterial chemotaxis simulation program	http://info.anat.cam.ac.uk/groups/comp-cell/Chemotaxis.html
Homepage of complex physiological signals	http://www.physionet.org
Neuronal networks	http://www.cocomac.org/databases.htm http://www.ncl.ac.uk/biol/research/psychology/nsg/neuroinformatics.htm

Social Nets

Social network analysis	http://www.sfu.ca/~insna/
Business networks	http://www.theyrule.net
Dark networks	http://www.orgnet.com/hijackers.html
Zachary friendship network	http://vlado.fmf.uni-lj.si/pub/networks/data/UciNet/zachary.dat
Internet	http://moat.nlanr.net/AS/
Textual networks	www.textarc.org
Network visualizations	www.visualcomplexity.com

B Glossary

This glossary explains a few of the key words used in the book. I would like to apologize if a specific meaning is sometimes given for a word to define it in a way which is used only in this book and slightly restricts or modifies the definition of the same word used in other contexts. Glossary items are marked with *italics* in the explanations for cross-reference.

Allometric scaling laws. Allometric scaling laws cover a wide variety of empirical scaling relationships where the given property is a power law function of the mass. The defining equation of allometric scaling laws is $P = cM^{\alpha}$, where P means the property, c is a constant, M is the mass of the organism or organelle, and α is a scaling exponent which varies depending on the nature of P. In the most studied example, viz., Kleiber's law (1932), P is the basal metabolic rate and the scaling exponent α is $3/4$. In other examples, the value of the exponent may be different, e.g., the dependence of heart rate ($\alpha = -1/4$), life span ($\alpha = 1/4$), the radii of aortas and tree trunks ($\alpha = 3/8$), the unicellular genome length ($\alpha = 1/4$), and RNA concentration ($\alpha = -1/4$), on the mass all have different exponents in their scaling relationship.

Assortativity. We call a *network* assortative, if similar *elements* of the network are selectively and preferentially linked. The elements may be similar by their *degree* or by any other property.

Attractor. The attractor is a set of *network* states on the *stability* mono- or multi-*landscapes* which behaves as a focus where members of a much larger set of network states converge as the network undergoes dynamical changes.

Betweenness centrality. The betweenness centrality of an *element* i is the fraction of shortest paths between any pair of elements in the network which pass through the i th element. (For the definition of the shortest path, see *path length*.)

Canalization. Canalization refers to a reduced sensitivity of an organism to noise from the environment or towards changes in its genotype. In Waddington's formulation, canalization is "the capacity to produce a particular definite end-result in spite of a certain variability both in the initial situation from which development starts and in the conditions met with during its course."

Chaperone. A chaperone, or molecular chaperone, is a protein which helps the folding of other proteins by preventing their aggregation or by partially unfolding them to give them a new opportunity to refold. Chaperones may be RNAs. Both protein- and RNA-chaperones may help the folding of RNAs, besides their help in protein folding. A chaperone is often a stress protein or heat shock protein, which means that its synthesis is induced by *stress* or heat shock, respectively.

Clustering (coefficient). Clustering occurs if neighbors of an *element* have a good chance of being connected. The clustering coefficient is the probability that two neighbors of a given element are also neighbors of each other. The clustering coefficient C for an element is the number of links between all the neighbors of the element (n) divided by the number of links that could possibly exist between all the neighbors of the element (N), i.e., $C = n/N$ (Watts, 1999). Clustering is also often called network transitivity. The range of the clustering coefficient varies between 0 and 1, and the average of all clustering coefficients gives a general measure of the cluster (triangle) formation in a network (Barabasi and Oltvai, 2004).

Cognitive dimensions. In the context of the present book, cognitive dimensions refer to the number of different views (personalities) a person can simultaneously accommodate and evaluate. This process requires the internalization, relative separation and internal conflicts of the intentions, drives, words and deeds of the real or imagined persons, groups of persons or value sets. A typical sentence to reflect this complexity is the following: I believe that A supposes that B intends to guess how C understands what D thinks. Dunbar (2005) shows that the typical average cognitive limit is around 5 cognitive dimensions. However, exceptional minds can think to the 6th or higher order. The cognitive dimensions probably reflect the number of stable oscillation sets a certain person's brain sections can simultaneously accommodate and process.

Degeneracy. A *network* is degenerate if structurally different elements of the network show a functional identity under special circumstances. (See also *redundancy*.)

Degree (distribution). The degree of a network element corresponds to the number of *links* of this element. The degree distribution is the histogram of the total number of elements of the network with a given degree. The degree distribution is a Poissonian distribution for the Erdős–Rényi *random graph* and exponential for single-scaled graphs. It follows a power law for *scale-free* graphs (Barabasi and Oltvai, 2004). The average degree is usually called the coordination number of the network. The origin of the expression 'coordination number' refers to regular lattices where all elements have the same degree.

Element. The element is a single building block of a *network*. The element is also called a vertex in graph theory, site in physics, or actor in sociology. The number of elements is called the order of the graph.

Emergent property. The emergent property of a *network* is derived from the interaction of the network *elements*, and is not observable or inherent in any element of the network considered separately.

Epigenetic inheritance. We call inheritance epigenetic if the inheritable property is not transmitted via a DNA sequence but is inherited by means of other molecular mechanisms. Such a mechanism may use the modulation of DNA accessibility by DNA methylation or histone modification. Epigenetic molecular mechanisms also include RNA- and protein-based inheritance.

Evolvability. Evolvability is the ability of random genetic variation to produce phenotypic changes that can increase *fitness* (intrinsic evolvability), or the ability of a population to respond to selection (extrinsic evolvability). Extrinsic evolvability depends on intrinsic evolvability as well as on external variables such as the history, size and structure of the population (Rutherford, 2003). In the present book, evolvability is mostly used in a broader context, accommodating all mechanisms which determine the ability of a network to respond to changes in the environment.

Exponential cutoff (decay). Most natural scale-free distributions lose their scale-free pattern after a few orders of magnitude. Beyond the scale-free distribution, the probability of the extreme value decays very sharply, usually in an exponential manner.

Fitness. The survival value and reproductive capacity of a given phenotype as compared with the average of the population or other genotypes of the population.

Fractal. Fractal objects are generated by a recursive process in which self-similar objects of different size are repeated. The self-similar objects have a *scale-free* size distribution. The distribution is characterized by the *fractal dimension*. Fractal behavior can be defined in time intervals as well (see *multifractal*).

Fractal dimension (Hausdorff dimension, Hurst exponent). Elements of self-similar, fractal objects have a scale-free size distribution. If we try to fill a larger object with smaller objects, we get the equation $N = (L/l)^d$, where N is the number of smaller objects fitted into the larger object, L/l is the ratio of the characteristic measure of the two objects of different size, and d is the exponent, which is called the fractal dimension. The fractal dimension in space is also called the Hausdorff dimension, and in time the Hurst exponent. In fractal objects, d is not an integer.

Fringe area. The fringe area is an overlap between two *modules* or two independent *networks*. It may either facilitate or prevent communication between the two connected modules or networks. This property of the fringe area can be tightly regulated and may change from time to time (Agnati et al., 2004).

Genetic drift. Genetic drift is a random change in allele frequency within a population. If the population is isolated and genetic drift continues for long enough, it may lead to specitation.

Genome cleansing. The cleansing of the genome occurs, when *stress* makes the consequences of otherwise *silent mutations* visible at the level of the phenotype. Under stress, organisms that have mutations with unfavorable consequences will be sorted out by natural selection. As a consequence the average genome of the whole population will be more uniform, and will 'shed' many of the unfavorable silent mutations.

Giant component. The giant component is the largest part of the *network*, where all *elements* are connected to each other. The giant component contains most of the network elements and appears after the *percolation* threshold.

Hub. A hub is a highly connected *element* of the *network*. A hub usually has more than 1% of total connections.

Keystone species. The keystone species is an important *hub* of an ecosystem whose removal triggers many secondary extinctions and may cause the fragmentation of the whole *network*.

Le Chatelier principle. The Le Chatelier–Brown principle describes the behavior of a system when its equilibrium is perturbed. If a system in equilibrium suffers an effect which changes its conditions, the system will adjust itself to minimize this change.

Link. A link is a connection between two *network elements*. In graph theory, the link is called an edge of the graph. In molecular networks, the link is a bond, and in social networks, the word 'tie' is more often used. The number of links is the size of the graph.

Module. Modules are groups of network *elements* that are relatively isolated from the rest of the network and are functionally or physically linked to each other. Modules may arise from parcellation of a larger network, or from integration of several smaller networks.

Molecular crowding. Molecular crowding occurs if a significant volume of a solution, e.g., the cytoplasm, is occupied by macromolecules. Under such conditions, a large amount of water molecules are transiently bound to the macromolecules present and several phenomena will be drastically changed. As an example of this, protein–protein interactions will be grossly favored in crowded solutions.

Motif. Motifs, network blocks, or patterns are small groups of network *elements* with characteristic linkage patterns. Typical motifs are the feed-forward loops and feedback loops. (See also *negative feedback*.)

Multifractal. We call a distribution in time or space multifractal, when it displays a complex *scale-free* pattern with several scale-free distributions superimposed upon each other and in which the overall distribution has more than one scaling exponent. Multifractal behavior is usually found in time series. More precisely, time dependence in mathematical analysis is typically continuous with continuous derivatives. It can thus be approximated in the vicinity of a time t_{i} by a so-called Taylor series or power series:

$$f(t) = a_0 + a_1(t - t_{\mathrm{i}}) + a_2(t - t_{\mathrm{i}})^2 + \cdots + a_h(t - t_{\mathrm{i}})^h + \cdots ,$$

where h is an integer. In contrast, most time series found in 'real' experiments cannot be approximated by the above formula. If a non-integer number of h is enough to quantify a local singularity in a noisy time series, we call the time series a fractal series. If we find a single value $h = H$ for all singularities t_{i} in the signal, then the signal is a monofractal. If we need several distinct values to describe the time series, than the signal is multifractal.

Negative feedback. Negative feedback is a common regulatory *network motif*, in which an increase in the quantity or function of a network *element* provokes an inhibition of the network elements which caused the increase.

Nested sync. This expression is used in the present book to denote the highly hypothetical induction of synchrony between oscillations at different *network* levels. In other words, nested sync occurs if an element of a top network synchronizes its oscillations with the oscillation of the whole top network, and this phenomenon continues through at least three hierarchical levels of networks. (See also *syntalansis*.)

Nestedness. The *elements* of most real *networks* are not points, but complex networks themselves. This means that real networks are often nested. However, the elements of abstract mathematical networks, known as graphs, are points with no internal structure.

Netquake. A netquake occurs if a *network* has a restricted relaxation and, after the gradual build-up of a tension, the network reaches a state of *self-organized criticality*. Restricted relaxation means that a perturbation of the network is not easily dissipated in the network. In the self-organized critical state, the probability and extent – size and duration – of netquakes often follow a *scale-free* distribution. (See also *pink noise*.)

Netsistance. The netsistance of a *network* refers to its stability against the removal of its *elements* or *links*. Whilst the network is able to preserve its *giant component* and *percolation*, it can be said to have maintained its netsistance. Loss of netsistance implies the death of the network if it is a biological system like a cell or other living organism.

Network. A network is a set of *elements* which are connected to each other by *links*. The elements of most real networks are not points, but complex networks themselves. This means that real networks are often *nested*. However, the elements of abstract mathematical networks, called graphs, are points with no internal structure.

Network diameter. The *network* diameter is the maximal number of *links* in the shortest *path* between any pair of network *elements*.

Node. A node is a *network element* with more than three *links*.

Path length. The path length is the number of *links* we have to pass along when we travel between two *network elements*. The shortest path length is the length of the shortest route between the two elements. The characteristic path length is the average of all the shortest path lengths in the network, and gives a good measure of the navigability of the network (Barabasi and Oltvai, 2004). (See also *network diameter*.)

Percolation (threshold). Percolation is the status of the *network* when it has a *giant component*, so that most of the network *elements* are connected with each other and can therefore communicate. The percolation threshold is the number of *links* when the network reaches percolation.

Pink noise. Noise is usually characterized as a sum of sinusoidal waves. The distribution of the constituent sinusoidal waves follows the equation $P = cD^{-\alpha}$, where P is the contribution of the sinusoidal wave, c is a constant, D is the frequency, and α is a scaling exponent. We call the noise pink if α lies between zero and two. The zero and two exponents correspond to white and brown noise, respectively. Pink noise is also called colored noise, flicker noise, crackling noise or Barkhausen noise. The names $1/f$, $1/t$ or $1/\tau$ noise refer to the situation where α is exactly unity. In pink noise, rare events have a greater effect on the noise than frequent events. This is the reason why this noise is said to be pink. The spectrum of pink noise is biased towards the low frequencies, which correspond to the red light in the spectral analogy with visible light. Therefore the spectrum of pink noise is 'reddened' compared to white noise, i.e., it is pink. Pink noise contains disturbances equally on all time scales, which means that pink noise is *scale-free*. In other words, *netquakes* of *self-organized critical* events are pink-noise processes.

Punctuated equilibrium. Punctuated equilibrium originally referred to a model of evolution in which change occurs in relatively rapid bursts followed by longer periods of stasis (Gould and Eldredge, 1993). In the present book, the expression is used in a more general context. Here, punctuated equilibrium refers to changes of a network on a 'rough' *stability landscape*, where the probability of transition between local minima is relatively low. This gives the same rapid burst and stasis dynamics that characterize the original description, but makes the expression useful to describe changes on the protein energy landscape, the evolutionary fitness landscape, the innovation landscape, software design landscape, scientific progress landscape, the landscape of economic markets, the diegetic landscape of plays, films and novels and any other 'rough' landscapes.

Random graph. The random graph is a mathematical representation of a *network* in which network *elements* are connected at random. The random graph has a Poissonian ('single scale') degree distribution, in which nodes with *degree* deviating significantly from the average degree are extremely rare (Barabasi and Oltvai, 2004).

Redundancy. Two structurally identical *network elements* are redundant. These elements double a certain function. Redundancy is different from *degeneracy*, where the functional identity is displayed only under special circumstances and arises from structurally different elements.

Regular lattice. The regular lattice is a *network* in which all elements have the same *degree*, and are arranged in highly periodical manner.

Resilience. The resilience of a *network* refers to its resistance against the removal of its *elements* or *links*. Resilience is usually measured by the disturbance of network communication (*percolation*). The expression 'resilience' is mainly used for ecological networks. In the present book, this form of network stability is generalized to all networks and called *netsistance*.

Robustness. A *network* is said to be robust if it displays a low sensitivity to environmental fluctuations (network perturbations). In ecosystems, robustness against the removal of *links* and *elements* is called *resilience* or (in the present book) *netsistance*.

Scale-free. A *network* is scale-free if its the *degree distribution* follows a power law. Generally, the distribution of scale-free systems can be written as $P = cM^{-\alpha}$, where P is the probability, c a constant, M the measure, and α a scaling exponent. The names Hurst exponent or *fractal dimension* are used for the scaling exponent when the scale-free distribution is observed in time or space, respectively. Scale-free distributions are best visualized if we take the logarithm of the above equation to obtain $\log P = \log c - \alpha \log M$, which shows that the logarithm of the probability is a linear function of the logarithm of the measure. If we plot the data with this log–log representation (see Fig. 2.5 of Sect. 2.2), we get a straight line. Exceedingly large numbers have a non-zero probability in scale-free distributions. For an order of magnitude higher value, we always have a probability just an order of magnitude lower (Barabasi and Oltvai, 2004). (See also *pink noise*.)

Self-organized criticality. Self-organized criticality lacks a clear-cut definition. In the present book, the expression refers to *networks* where improper relaxation and an increasing tension spontaneously develop long-range correlations between network *elements*. The increasing correlations lead to a statistical steady state of criticality which is characterized by the occurrence of collective behavior manifested by avalanches. The avalanches display a *scale-free* size distribution and occur with a scale-free probability. (See also *pink noise*.)

Silent mutations (hidden mutations). Mutations of the DNA remain silent if their effect is not exposed at the level of the phenotype. Silent mutations may occur in DNA sequences which do not code proteins and are not involved in gene regulation. These mutations may remain silent forever. Silent mutations may also be conditional. These silent mutations may be revealed under specific (*stressful*) conditions, where the rest of the cellular networks cannot substitute the missing function by *redundant* or *degenerate network* segments. A specific form of silent mutation is hidden by *chaperones*. Here the mutation-induced changes in protein structure are repaired by chaperones. This repair becomes compromised after stress, when chaperones become occupied by damaged proteins. Thus, stress often exposes silent mutations at the level of the phenotype and makes them the subject of natural selection. This may cause a *cleansing of the genome* (where the occurrence of the silent mutation in the genome of the population is decreased) or derail *canalization* and give rise to a new dominant phenotype. In extreme cases, the exposure of silent mutations may even cause a jump in evolution.

Small world. We call a *network* a small world when its characteristic *path length* is close to the rather small path length of an Erdős–Rényi *random graph*, but its *clustering coefficient* is much higher than that of the random graph (Watts, 1999).

Social dimensions. The social dimension is the number of features of a social actor, i.e., a person or an element of a social *network*, which helps the classification of this element by other elements of the network. People often use social dimensions to direct and define efficient search and communication in social networks. Social dimensions are also used as an organizational pattern for *clustering*, *motif* and *module* formation in *assortative* social networks.

Stability landscape. On the stability landscape, each state (parameter set) of the *network* is plotted as a function of a 'goodness value'. The stability landscape may characterize any networks, such as proteins (energy landscape), ecosystems (fitness landscape), social networks (economy, innovation, design, scientific progress, etc., landscapes), informational, textual networks (the diegetic landscape of dramas, films and novels), etc. The 'goodness value' depends on the form of the stability landscape. The goodness value may be energy, fitness, market value, story integrity, etc. 'Rough' landscapes have very high goodness values, making high saddles between their local minima. High saddles make the probability of transition low. In contrast, 'buffed' landscapes have low saddles, which make the transition probability high. Rough landscapes often produce a *punctuated equilibrium* and may give rise to *self-organized criticality*.

Stress. Stress is any unexpected, large and sudden perturbation in the life of the *network*, (1) to which the network does not have a prepared adaptive response, or (2) where the network does not have time to mobilize an adaptive response. Stress in this book is used differently from stress in the usual sense in physics, where it is a force that produces strain on a physical body.

Syntalansis. Syntalansis is the extensive synchronization of the oscillators of *network elements*. The development of syntalansis displays a phase transition. As the difference between the frequencies of different oscillators is decreased below a certain threshold, they will suddenly all become synchronized, thereby achieving syntalansis. This phenomenon is similar to the *percolation threshold*.

Topological phase transition. A topological phase transition occurs if a continuous increase in the number of perturbations provokes a singular change in the global topology of the network. The global topology is best monitored by the measure G/N, where G is the size of the largest connected component of the network and N is the total number of links. Alternatively, the measure $k_{\max}/M$ can also be used, where $k_{\max}$ is the largest degree of the network and M is the number of edges in the network (Derenyi et al., 2004; Palla et al., 2004).

Weak links. A *link* of a *network* is weak if its addition or removal does not change the mean value of a target measure, which is usually an *emergent property* of the network, in a statistically discernible way. Weak links stabilize most networks. The effects of weak links are described in every chapter of the present book.

References

The reference list was finalized on 31 August 2005. Before some of the references you will find one or two asterisks:

* a very interesting piece of work, from which I learned a lot;
** a seminal contribution, which you should read.

I would like to apologize for not including several groundbreaking papers in these categories. I restricted these signs to those papers which are necessary to understand the main message of this book: weak links stabilize complex systems. Moreover, I would like to apologize to those whose papers are not cited in the following list. The reference list is already rather voluminous. If I had inserted more papers, the text of the book would have become an appendix to the reference list and not vice versa.

1. Abrams, P.A. (1983): Arguments in favor of higher order interactions. Am. Nat. **121**, 887–891
2. Acinas, S.G., Klepac-Ceraj, V., Hunt, D.E., Pharino, C., Ceraj, I., Distel, D.L. and Potz, M.F. (2004): Fine-scale phylogenetic architecture of a complex bacterial community. Nature **430**, 551–553
3. Aftanas, L.I. and Golocheikine, S.A. (2001): Human anterior and frontal midline theta and lower alpha reflect emotionally positive state and internalized attention. High-resolution EEG investigation of meditation. Neurosci. Lett. **310**, 57–60
4. Agnati, L.F. and Fuxe, K. (2000): Volume transmission as a key feature of information handling in the central nervous system. Possible new interpretative value of the Turing B-type machine. Prog. Brain Res. **125**, 3–19
5. Agnati, L.F., Fuxe, K., Zoli, M., Ozini, I., Toffano. G. and Ferraguti, F. (1986): A correlation analysis of the regional distribution of central enkephalin and beta-endorphin immunoreactive terminals and of opiate receptors in adult and old male rats. Evidence for the existence of two main types of communication in the central nervous system: The volume transmission and the wiring transmission. Acta Physiol. Scand. **128**, 201–207
6. Agnati, L.F., Santarossa, L., Genedani, S., Canela, E.I., Leo, G., Franco, R., Woods, A., Lluis, C., Ferré, S. and Fuxe, K. (2004): On

the nested hierarchical organization of CNS: Basic characteristics of neuronal molecular networks. In: Lecture Notes in Computer Science, ed. by P. Érdi, A. Esposito, M. Marinaro and S. Scarpetta, Springer Verlag, pp. 24–54

7. Agoston, V., Csermely, P. and Pongor, S. (2005): Multiple, weak hits confuse complex systems. www.arxiv.org/q-bio.MN/0410026, Phys. Rev. E **71**, 051909
8. Aiello, L.C. and Wheeler, P. (1995): The expensive tissue hypothesis: The brain and the digestive system in human evolution. Curr. Anthropology **36**, 199–221
9. Albert, R. and Barabasi, L. (2002): Statistical mechanics of complex networks. Rev. Mod. Phys. **74**, 47–97
10. ** Albert, R., Jeong, H. and Barabasi, L. (2000): Attack and error tolerance of complex networks. Nature **406**, 378–382
11. Aldana, M. and Cluzel, P. (2003): A natural class of robust networks. Proc. Natl. Acad. Sci. U.S.A. **100**, 8710–8714
12. Alessandro, B., Beatrice, C., Bertotti, G. and Montorsi, A. (1990): Domain wall dynamics and Barkhausen effect in metallic ferromagnetic materials. I. Theory. J. Appl. Phys. **68**, 2901–2907
13. Alexander, C. (1965): A city is not a tree. Architectural Forum **122** (1,2) 58–62, www.rudi.net/bookshelf/classics/city/alexander/alexander.shtml
14. * Almaas, A., Kovács, B., Vicsek, T., Oltvai, Z.N. and Barabasi, A.-L. (2004): Global organization of metabolic fluxes in the bacterium *Escherichia coli*. Nature **427**, 839–843
15. Allesina, S. and Bodini, A. (2004): Who dominates whom in the ecosystem? Energy flow bottlenecks and cascading extinctions. J. Theor. Biol. **230**, 351–358
16. Alon, U. (2003): Biological networks: The tinkerer as an engineer. Science **301**, 1866–1867
17. Alon, U., Surette, M.G., Barkai, N. and Leibler, S. (1999): Robustness in bacterial chemotaxis. Nature **397**, 168–171
18. Alvarez, L.W, Alvarez, W., Asaro, F. and Michel H.V. (1980): Extraterrestrial cause for the cretaceous–tertiary extinction. Science **208**, 1095–1108
19. Amaral, L.A.N., Scala, A., Barthelemy, M. and Stanley, H.E. (2000): Classes of small-world networks. Proc. Natl. Acad. Sci. U.S.A. **97**, 11149–11152
20. Ansari, A., Berendzen, J., Bowne, S.F., Frauenfelder, H., Iben, I.E.T., Sauke, T.B., Shyamsunder, E. and Young, R.D. (1985): Protein states and proteinquakes. Proc. Natl. Acad. Sci. U.S.A. **82**, 5000–5004
21. Antonovsky, A. (1985): *Health, Stress and Coping*, Jossey-Bass Publ. San Francisco, USA

22. Aoki, M. (1998): The subjective game form and institutional evolution as punctuated equilibrium. Stanford University Working Papers No. 98011 (http://www-econ.stanford.edu/faculty/workp/swp98011.pdf)
23. Aon, M.A., Cortassa, S. and O'Rourke, B. (2004a): Percolation and criticality in a mitochondrial network. Proc. Natl. Acad. Sci. U.S.A. **101**, 4447–4452
24. Aon, M.A., O'Rourke, B. and Cortassa, S. (2004b): The fractal architecture of cytoplasmic organization: Scaling kinetics and emergence in metabolic networks. Mol. Cell. Biochem. **256/257**, 169–184
25. Aranda-Anzaldo, A. and Dent, M.A. (2003): Developmental noise, ageing and cancer. Mech. Aging Dev. **124**, 711–720
26. Arendt, H. (1973): *The Origins of Totalitarianism*, Harcourt, Orlando FL USA
27. Argollo de Menezes, M. and Barabasi, A.-L. (2004): Separating internal and external dynamics of complex systems. Phys. Rev. Lett **93**, 068701
28. Ariaratnam, J.T. and Strogatz, S.H. (2001): Phase diagram for the Winfree model of coupled nonlinear oscillators. Phys. Rev. Lett. **86**, 4278–4281
29. Arita, M. (2004): The metabolic world of *Escherichia coli* is not small. Proc. Natl. Acad. Sci. U.S.A. **101**, 1543–1547
30. Arkin, A., Ross, J. and McAdams, H.H. (1998): Stochastic kinetic analysis of developmental pathway bifurcation in phage lambda-infected *Escherichia coli* cells. Genetics **149**, 1633–1648
31. Arrow, K. (1974): *The Limits of Organization*, Norton and Co., New York NY, USA
32. Atkinson, R.P.D., Rhodes, C.J., Macdonald, D.W. and Anderson, R.M. (2002): Scale-free dynamics in the movement patterns of jackals. Oikos **98**, 134–140
33. Avnir, D., Biham, O., Lidar, D. and Malcai, O. (1998): Is the geometry of nature fractal? Science **279**, 39–40
34. Axelrod, R. (1997): *The Complexity of Cooperation*, Princeton University Press, Princeton NJ, USA
35. Axtell, R.L. (2001): Zipf distribution of U.S. firm sizes. Science **293**, 1818–1820
36. Azbel, M.Y. (1999): Empirical laws of survival and evolution: Universality and implications. Proc. Natl. Acad. Sci. U.S.A. **96**, 15368–15373
37. Baars, B.J. (2002): The conscious access hypothesis: Origins and recent evidence. Trends Cognit. Sci. **6**, 47–52
38. Bacon, F. (1620): *Novum Organum*, Cambridge University Press, Cambridge, UK (2000)
39. Bagler, G. and Sinha, S. (2005): Network properties of protein structures. Physica A **346**, 27–33

40. Baish, J.W. and Jain, R.K. (2000): Fractals and cancer. Cancer Res. **60**, 3683–3688
41. * Bak, P. (1996): *How Nature Works. The Science of Self-Organized Criticality*, Springer-Verlag, New York
42. Bak, P. and Paczuski, M. (1995): Complexity, contingency and criticality. Proc. Natl. Acad. Sci. U.S.A. **92**, 6689–6696
43. Bak, P. and Sneppen, K. (1993): Punctuated equilibrium and criticality in a simple model of evolution. Phys. Rev. Lett. **71**, 4083–4086
44. * Bak, P., Tang, C. and Wiesenfeld, K. (1987): Self-organized criticality: An explanation of $1/f$ noise. Phys. Rev. Lett. **59**, 381–384
45. Bak, P., Paczuski, M. and Shubik, M. (1997): Price variations in a stock market with many agents. Physica A **246**, 430–453
46. Baker, S.N., Spinks, R., Jackson, A. and Lemon, R.N. (2001): Synchronization in monkey motor cortex during a precision grip task. I. Task-dependent modulation in single-unit synchrony. J. Neurophysiol. **85**, 869–885
47. * Ball, K.D., Berry, R.S., Kunz, R.E., Li, F.-Y., Proykova, A. and Wales, D.J. (1996): From topographies to dynamics on multidimensional potential energy surfaces of atomic clusters. Science **271**, 963–966
48. Banavar, J.R., Maritan, A. and Rinaldo, A. (1999): Size and form in efficient transportation networks. Nature **399**, 130–132
49. Bandura, A. (1997): *Self-Efficacy: The Exercise of Control*, W.H. Freeman & Co., San Francisco, USA
50. * Barabasi, A.L. (2002): *Linked: The New Science of Networks*, Perseus Press
51. Barabasi, A.L. (2005): The origin of bursts and heavy tails in human dynamics. Nature **435**, 207–211
52. ** Barabasi, A.L. and Albert, R. (1999): Emergence of scaling in random networks. Science **286**, 509–512
53. Barabasi, A.L. and Oltvai, Z.N. (2004): Network biology: Understanding the cell's functional organization. Nature Rev. Gen. **5**, 101–114
54. Barabasi, A.L., Buldyrev, S.V., Stanley, H.E. and Suki, B. (1996): Avalanches in the lung: A statistical mechanical model. Phys. Rev. Lett. **76**, 2192–2195
55. Barbosa, L.A., Castro e Silva, A. and Kamphorst Leal da Silva, J. (2005): On the universal scaling relations in food webs. `Cond-mat/0507184`
56. Barrahona, M. and Pecora, L.M. (2002): Synchronization in small-world systems. Phys. Rev. Lett. **89**, 054101
57. Barrat, A., Barthélemy, M., Pastor-Satorras, R. and Vespignani, A. (2004a): The architecture of complex weighted networks. Proc. Natl. Acad. Sci. U.S.A. **101**, 3747–3752

58. Barrat, A., Barthélemy, M. and Vespignani, A. (2004b): Weighted evolving networks: Coupling topology and weight dynamics. Phys. Rev. Lett. **92**, 228701
59. * Barron, L.D., Hecht, L. and Wilson, G. (1997): The lubricant of life: A proposal that solvent water promotes extremely fast conformational fluctuations in mobile heteropolypeptide structure. Biochemistry **36**, 13143–13147
60. Bartumeus, F., Catalan, J., Fulco, U.L., Lyra, M.L. and Visnawathan, G.M. (2002): Optimizing the encounter rate in biological interactions: Lévy versus Brownian strategies. Phys. Rev. Lett. **88**, 097901
61. Bar-Yam, Y. and Epstein, I.R. (2004): Response of complex networks to stimuli. Proc. Natl. Acad. Sci. U.S.A. **101**, 4341–4345
62. Baryshev, Y. and Teerikorpi, P. (2002): *Discovery of Cosmic Fractals*, World Scientific, Singapore
63. Bascompte, J., Melián, C.J. and Sala, E. (2005): Interaction strength combinations and the overfishing of a marine food web. Proc. Natl. Acad. Sci. U.S.A. **102**, 5443–5447
64. * Bateson, P., Barker, D., Clutton-Brock, T., Deb, D., d'Udine, B., Foley, R.A., Gluckman, P., Godfrey, K., Kirkwood, T., Lahr, M.M., McNamara, J., Metcalfe, N.B., Monaghan, P., Spencer, H.G. and Sultan, S.E. (2004): Developmental plasticity and human health. Nature **430**, 419–421
65. Batty, M. and Longley, P. (1994): *Fractal Cities*, Academic Press, London
66. Bazant, Z.P. (2004): Scaling theory for quasibrittle structural failure. Proc. Natl. Acad. Sci. U.S.A. **101**, 13400–13407
67. Becskei, A. and Serrano, L. (2000): Engineering stability in gene networks by autoregulation. Nature **405**, 590–593
68. Beck, K. (1999): *Extreme Programming Explained: Embrace Change*, Addison-Wesley Professional, Boston MA, USA
69. Beilock, S.L. and Carr, T.H. (2005): When high-powered people fail. Psych. Science **16**, 101–105
70. Benhamou, C.L., Poupon, S., Lespessailles, E., Loiseau, S., Jennane, R., Siroux, V., Ohley, W. and Pothuaud, L. (2001): Fractal analysis of radiographic trabecular bone texture and bone mineral density: Two complementary parameters related to osteoporotic fractures. J. Bone Miner. Res. **16**, 697–704
71. Bentley, R.A. and Maschner, H.D. (2000): A growing network of ideas. Fractals **8**, 227–237
72. Bentley, R.A. and Maschner, H.D. (2001): Stylistic change as a self-organized critical phenomenon: An archaeological study in complexity. J. Arch. Meth. Theor. **8**, 35–66
73. Benzi, R., Sutera, A. and Vulpiani, A. (1981): The mechanism of stochastic resonance. J. Phys. A **14**, L453–L457

74. Bergendahl, M., Iranmanesh, A., Mulligan, T. and Veldhuis, J.D. (2000): Impact of age on cortisol secretory dynamics basally and as driven by nutrient-withdrawal stress. J. Clin. Endocrinol. Metab. **85**, 2203–2214
75. * Bergman, A. and Siegal, M.L. (2003): Evolutionary capacitance as a general feature of complex gene networks. Nature **424**, 549–551
76. Bergmann, S., Ihmels, J. and Barkai, N. (2004): Similarities and differences in genome-wide expression data of six organisms. PLoS Biology **2**, 85–93
77. ** Berlow, E.L. (1999): Strong effects of weak interactions in ecological communities. Nature **398**, 330–334
78. Bernoulli, D. (1738): Specimen theoriae novae de mensura sortis. Papers Imp. Acad. Sci. St. Petersburg **5**, 175–192
79. Bezrukov, S.M. and Vodyanoy, I. (1995): Noise-induced enhancement of signal transduction across voltage-dependent ion channels. Nature **378**, 362–364
80. Bianconi, G. and Barabasi, A.L. (2001): Competition and multiscaling in evolving networks. Europhys. Lett. **54**, 436–442
81. Biely, C. and Thurner, S. (2005): Statistical mechanics of scale-free networks at a critical point: Complexity without irreversibility? `Cond-mat/0505670`
82. Blake, W.J., Kaern, M., Cantor, C.R. and Collins, J.J. (2003): Noise in eukaryotic gene expression. Nature **422**, 633–637
83. * Blasius, B., Huppert, A. and Stone, L. (1999): Complex dynamics and phase synchronization in spatially extended ecological systems. Nature **399**, 354–359
84. Boccignone, G. and Ferraro, M. (2004): Modelling gaze shift as a constrained random walk. Physica A **331**, 207–218
85. Bollobas, B. (2001): *Random Graphs*, Academic Press, New York
86. Bona, C.A., Heber-Katz, E. and Paul, W.E. (1981): Idiotype–anti-idiotype regulation. I. Immunization with a levan-binding myeloma protein leads to the appearance of auto-anti-(anti-idiotype) antibodies and to the activation of silent clones. J. Exp. Med. **153**, 951–967
87. Bonabeau, E., Theraulaz, G. and Deneubourg, J.L. (1998a): The synchronization of recruitment-based activities in ants. Biosystems **45**, 195–211
88. * Bonabeau, E., Theraulaz, G. and Deneubourg, J.L. (1998b): Group and mass recruitment in ant colonies: The influence of contact rates. J. Theor. Biol. **195**, 157–166
89. Bonabeau, E., Dagorn, L. and Freon, P. (1999): Scaling in animal group-size distributions. Proc. Natl. Acad. Sci. U.S.A. **96**, 4472–4477
90. Bonn, D. and Kegel, W.K. (2003): Stokes–Einstein relations and the fluctuation–dissipation theorem in a supercoiled colloidal fluid. J. Chem. Phys. **118**, 2005–2009

91. Booth, I.R. (2002): Stress and the single cell: Intrapopulation diversity is a mechanism to ensure survival upon exposure to stress. Int. J. Food Microbiol. **78**, 19–30
92. Borisy, A.A., Elliott, P.J., Hurst, N.W., Lee, M.S., Lehár, J., Price, E.R., Serbedzija, G., Zimmermann, G.R., Foley, M.A., Stockwell, B.R. and Keith, C.T. (2003): Systematic discovery of multicomponent therapeutics. Proc. Natl. Acad. Sci. U.S.A. **100**, 7977–7982
93. Borrvall, C., Ebenman, B. and Jonsson, T. (2000): Biodiversity lessens the risk of cascading extinction in model food webs. Ecol. Lett. **3**, 131–136
94. Bortoluzzi, S., Romualdi, C., Bisognin, A. and Danieli, G.A. (2003): Disease genes and intracellular protein networks. Physiol. Genomics **15**, 223–227
95. Bourrin, S., Palle, S., Genty, C. and Alexandre, C. (1995): Physical exercise during remobilization restores a normal bone trabecular network after tail suspension-induced osteopenia in young rats. J. Bone Miner. Res. **10**, 820–828
96. Bovill, C. (1995): *Fractal Geometry in Architecture and Design*, Birkhauser Verlag
97. Braun, T. (2004): Hungarian priority in network theory. Science **304**, 1744b
98. Brede, M. and Behn, U. (2002): Architecture of idiotypic networks: Percolation and scaling behavior. Phys. Rev. E **64**, 011908
99. Brede, M. and Sinha, S. (2005): Assortative mixing by degree makes a network more unstable. `Cond-mat/0507710`
100. Bremner, F.D., Baker, J.R. and Stephens, J.A. (1991): Variation in the degree of synchronization exhibited by motor units lying in different finger muscles in man. J. Physiol. **432**, 381–399
101. Bressloff, P.C. and Coombes, S. (1998): Traveling waves in a chain of pulse-coupled oscillators. Phys. Rev. Lett. **80**, 4815–4818
102. Brinker, A., Pfeifer, G., Kerner, M.J., Naylor, D.J., Hartl, F.U. and Hayer-Hartl, M. (2001): Dual function of protein confinement in chaperonin-assisted protein folding. Cell **107**, 223–233
103. Brooks, R.F. (1985): In: *Temporal Order*, ed. by Rensing, L. and Jaeger, N.I., Springer, Berlin, pp. 304–314
104. Brown, W.H., Malveau, R.C., McCormick, H.W. ('Skip') and Mowbray, T.J. (1998): *Antipatterns: Refactoring Software, Architectures, and Projects in Crisis*, John Wiley & Sons, New York NY, USA
105. Brunk, G.G. (2003): Why are so many important events unpredictable? Self-organized criticality as the 'engine of history'. Japanese J. Political Sci. **3**, 25–44
106. Bryngelson, J.D. and Wolynes, P.G. (1987): Spin glasses and the statistical mechanics of protein folding. Proc. Natl. Acad. Sci. U.S.A. **84**, 7524–7528

107. * Bryngelson, J.D., Onuchic, J.N., Socci, N.D. and Wolynes, P.G. (1995): Funnels, pathways, and the energy landscape of protein folding: A synthesis. Proteins **21**, 167–195
108. Buchanan, M. (2000): *Ubiquity: The Science of History ... Or Why the World Is Simpler Than We Think*, Weidenfeld & Nicholson, London, UK
109. * Buchanan, M. (2002): *Nexus, Small Worlds and the Groundbreaking Science of Networks*, W.W. Norton and Co. New York NY, USA
110. Buck, J.B. (1938): Synchronous rhythmic flashing of fireflies. Quart. Rev. Biol. **13**, 301–314
111. Bukau B. and Horwich A.L. (1998): The Hsp70 and Hsp60 chaperone machines. Cell **92**, 351–366
112. Burgos, J.D. (1996): Fractal representation of the immune B cell repertoire. BioSystems **39**, 19–24
113. Burnet, F.M. (1959): *The Clonal Selection Theory of Acquired Immunity*, Cambridge University Press
114. Buzsáki, G. and Draguhn, A. (2004): Neuronal oscillations in cortical networks. Science **304**, 1926–1929
115. Buzsáki, G., Geisler, C., Henze, D.A. and Wang, X.-J. (2004): Interneuron diversity series: Circuit complexity and axon wiring economy of cortical interneurons. Trends Neurosci. **27**, 186–193
116. Caldarelli, G., Capocci, A., De Los Rios, P. and Munoz, M.A. (2002): Scale-free networks from varying vertex intrinsic fitness. Phys. Rev. Lett. **89**, 258702.1–4
117. Caldarelli, G., Coccetti, F. and De Los Rios, P. (2004): Preferential exchange: Strengthening connections in complex networks. (Cond-mat/0312236) Phys. Rev. E **70**, 027102
118. Camacho, J. and Arenas, A. (2005): Universal scaling in food-web structure? Nature **435**, E3
119. Carlson, J.M. and Doyle, J. (2002): Complexity and robustness. Proc. Natl. Acad. Sci. U.S.A. **99**, 2538–2545
120. Carney, J.M., Starke-Reed, P.E., Oliver, C.N., Landum, R.W., Cheng, M.S., Wu, J.F. and Floyd, R.A. (1991): Reversal of age-related increase in brain protein oxidation, decrease in enzyme activity, and loss in temporal and spatial memory by chronic administration of the spin-trapping compound N-tert-butyl-alpha-phenylnitrone. Proc. Natl. Acad. Sci. U.S.A. **88**, 3633–3636
121. Carreras, B.A., Newman, D.E., Dobson, I. and Poole, A.B. (2004): Evidence for self-organized criticality in a time series of electric power system blackouts. IEEE Trans. Circuits Systems **51**, 1733–1740
122. Cattell, R.B. (1978): *The Scientific Use of Factor Analysis*, Plenum, New York
123. Chalet, D. and Zhang, Y.-C. (1997): Emergence of cooperation and organization in an evolutionary game. Physica A **246**, 407–418

124. Chalet, D. and Zhang, Y.-C. (1998): On the minority game: Analytical and numerical studies. Physica A **256**, 514–532
125. Chan, H.S. and Dill, K.A. (1996): A simple model of chaperonin-mediated protein folding. Proteins **24**, 345–351
126. Changizi, M.A., McDannald, M.A. and Widders, D. (2002): Scaling of differentiation in networks: Nervous systems, organisms, ant colonies, ecosystems, businesses, universities, cities, electronic circuits and LEGOs. J. Theor. Biol. **218**, 215–237
127. Chen, J.Y., Sivachenko, A.Y., Bell, R., Kurschner, C., Ota, I. and Sudhir, S. (2003): Initial large-scale exploration of protein–protein interactions in human brain. CSB 03 Proceedings, IEEE 0-7695-2000-6/03
128. Choi, Y.-M. and Kim, H-J. (2005): A directed network of Greek and Roman mythology. `www.arxiv.org/physics/0506142`
129. Chomsky, N. (1957): *Syntactic Structures*, Mouton, The Hague
130. Chomsky, N. (1968): *Language and Mind*, Harcourt, Brace and World, New York NY, USA
131. Chomsky, N. (1975): *Reflections on Language*, Pantheon Books, New York NY, USA
132. Clauset, A. and Moore, C. (2005): Accuracy and Scaling phenomena in internet mapping. Phys. Rev. Lett. **94**, 018701
133. Clauset, A. and Young, M. (2005): Scale invariance in global terrorism. `www.arxiv.org/physics/0502014`
134. Compiani, M., Fariselli, P., Martelli, P.L and Casadio, R. (1998): An entropy criterion to detect minimally frustrated intermediates in native proteins. Proc. Natl. Acad. Sci. U.S.A. **95**, 9290–9294
135. Cohen, I.R. (1992a): The cognitive principle challenges clonal selection. Immunol. Today **13**, 441–444
136. Cohen, I.R. (1992b): The cognitive paradigm and the immunological homunculus. Immunol. Today **13**, 490–494
137. Cohen, I.R. (2002): Peptide therapy for Type I diabetes: The immunological homunculus and the rational for vaccination. Diabetologia **45**, 1468–1474
138. Cohen, M.N. and Armelagos, G.J. (1984): *Paleopathology at the Origin of Agriculture*, Academic Press, New York NY, USA
139. Cohen, I.R. and Young, D.B. (1991): Autoimmunity, microbial immunity and the immunological homunculus. Immunol. Today **12**, 105–110
140. Cole, B.J. (1995): Fractal time in animal behaviour: The movement activity of *Drosophila*. Animal Behav. **50**, 1317–1324
141. Connor, R.C., Heithaus, M.R. and Barre, L.M. (1999): Superalliance of bottlenose dolphins. Nature **397**, 571–572
142. Cook, D.L., Gerber, A.N. and Tapscott, S.J. (1998): Modeling stochastic gene expression: Implications for haploinsufficiency. Proc. Natl. Acad. Sci. U.S.A. **95**, 15641–15646

143. Cote, P.J. and Meisel, L.V. (1991): Self-organized criticality and the Barkhausen effect. Phys. Rev. Lett. **67**, 1334–1337
144. Coulomb, S., Bauer, M., Bernard, D. and Marsolier-Kergoat, M.-C. (2005): Gene essentiality and the topology of protein interaction networks. Proc. Roy. Soc. B **272**, 1721–1725
145. Crutchfield, J.P. (1994): The calculi of emergence: Computation, dynamics and induction. Physica D **75**, 11–54
146. $*$ Cross, R. and Parker, A. (2004): *The Hidden Power of Social Networks*, Harvard Business School Press, Boston MA, USA
147. Csányi, V. (2005): *If Dogs Could Talk*, North Point Press
148. Csermely, P. (1997): Proteins, RNAs and chaperones in enzyme evolution: A folding perspective. Trends in Biochem. Sci. **22**, 147–149
149. Csermely, P. (1999): The 'chaperone-percolator' model: A possible molecular mechanism of Anfinsen-cage type chaperone action. BioEssays **21**, 959–965
150. Csermely, P. (2001a): Water and cellular folding processes. Cell. Mol. Biol. **47**, 791–800
151. Csermely, P. (2001b): Chaperone-overload as a possible contributor to 'civilization diseases': Atherosclerosis, cancer, diabetes. Trends Genet. **17**, 701–704
152. $*$ Csermely, P. (2004): Strong links are important, but weak links stabilize them. Trends Biochem. Sci. **29**, 331–334
153. $*$ Csermely, P. (2005): *The Strength of Hidden Networks. What Helps the Stability of the World?* (in Hungarian). Vince Publishers, Budapest, Hungary
154. Csermely P., Schnaider T., Sőti C., Prohászka Z. and Nardai G. (1998): The 90 kDa molecular chaperone family: Structure, function and clinical applications. A comprehensive review. Pharmacol. Therap. **79**, 129–168
155. Csermely, P., Ágoston, V. and Pongor, S. (2005): The efficiency of multi-target drugs: The network approach might help drug design. (`www.arxiv.org/abs/q-bio.BM/0412045`) Trends Pharmacol. Sci. **26**, 178–182
156. Csete, M.E. and Doyle, J.C. (2002): Reverse engineering of biological complexity. Science **295**, 1664–1669
157. Csikszentmihalyi, M. (1990): *Flow: The Psychology of Optimal Experience* Harper and Row, New York NY, USA
158. da Fontoura Costa, L. and Palhares Viana, M. (2005): Hierarchy, fractality, small-world and resilience of Haversian bone structure: A complex network study. `Cond-mat/0506019`
159. Dahrendorf, R. (1968): *Essays in the Theory of Society*, Stanford University Press, Stanford CA, USA
160. Damasio, A. (1994): *Descartes' Error: Emotion, Reason, and the Human Brain*, Avon Books

161. Darwin, C. (1859): *On the Origin of Species by Means of Natural Selection or the Preservation of Favored Races in the Struggle for Life*, Murray, London
162. Datta, A.K., Farmer, S.F. and Stephens, J.A. (1991): Central nervous pathways underlying synchronization of human motor unit firing studied during voluntary contractions. J. Physiol. **432**, 401–425
163. Deacon, T. (1997): *The Symbolic Species*, W. Norton and Co. New York NY, USA
164. De Boer, R.J. and Perelson, A.S. (1991): Size and connectivity as emergent properties of a developing immune network. J. Theor. Biol. **149**, 381–424
165. Deeds, E.J., Ashenberg, O. and Shakhnovich, E.I. (2005): A simple physical model for scaling in protein–protein interaction networks. `www.arxiv.org/q-bio.MN/0509001` (Proc. Natl. Acad. Sci. U.S.A. **103**, 311–316)
166. * Degenne, A. and Forsé, M. (1999): *Introducing Social Networks*, SAGE Publications, London
167. De Montis, A., Barthélemy, M., Chessa, A. and Vespignani, A. (2005): The structure of inter-urban traffic: A weighted network analysis. `physics/0507106`
168. DePace, A.H. and Weissman, J.S. (2002): Origins and kinetic consequences of diversity in Sup35 yeast prion fibers. Nature Struct. Biol. **9**, 389–396
169. Derégnaucourt, S., Mitra, P.P., Fehér, O., Pytte, C. and Tchernichovski, O. (2005): How sleep affects the developmental learning of bird song. Nature **433**, 710–716
170. ** Derenyi, I., Farkas, I., Palla, G. and Vicsek, T. (2004): Topological phase transitions of random networks. Physica A **334**, 583–590
171. de Ruiter, P.C., Neutel, A.-M. and Moore, J.C. (1995): Energetics, patterns of interaction strengths, and stability in real ecosystems. Science **269**, 1257–1260
172. de Solla Price, D.J. (1965): Network of scientific papers. Science **149**, 510–515
173. * de Visser, J.A.G.M., Hermisson, J., Wagner, G.P., Meyers, L.A., Bagheri-Chaichian, H., Blanchard, J.L., Chiao, L., Cheverud, J.M., Elena, S.F., Fontana, W., Gibson, G., Hansen, T.F., Krakauer, D., Lewontin, R.C., Ofria, C., Rice, S.H., von Dassow, G., Wagner, A. and Whitlock, M.C. (2003): Perspective: Evolution and detection of genetic robustness. Evolution **57**, 1959–1972
174. Dewey, T.G. and Bann, J.G. (1992): Protein dynamics and $1/f$ noise. Biophys. J. **63**, 594–598
175. Dial, K.P. (2003): Wing-assisted incline running and the evolution of flight. Science **299**, 402–404

176. di Bernardo, M., Garofalo, F. and Sorrentino, F. (2005): Synchronizability of degree correlated networks. Cond-mat/0504335
177. Dill, K.A. (1985): Theory for the folding and stability of globular proteins. Biochemistry **24**, 1501–1509
178. Dill, K.A. (1999): Polymer principles and protein folding. Protein Sci. **8**, 1166–1180
179. Dill, K.A., Bromberg, S., Yue, K., Fiebig, K.M., Yee, D.P., Thomas, P.D. and Chan, H.S. (1995): Principles of protein folding – A perspective from simple exact models. Protein Sci. **4**, 561–602
180. Dobson, C.M., Evans, P.A. and Radford, S.E. (1994): Understanding how proteins fold: The lysozyme story so far. Trends Biochem. Sci. **19**, 31–37
181. Dodds, P.S., Rothman, D.H. and Weitz, J.S. (2001): Re-examination of the '3/4 law' of metabolism. J. Theor. Biol. **209**, 9–27
182. * Dodds, P.S., Muhamad, R. and Watts, D.J. (2003a): An experimental study of search in global social networks. Science **301**, 827–829
183. Dodds, P.S., Watts, D.J. and Sabel, C.F. (2003b): Information exchange and the robustness of organizational networks. Proc. Natl. Acad. Sci. U.S.A. **100**, 12516–12521
184. D'Odorico, P., Laio, F. and Ridolfi, L. (2005): Noise-induced stability in dryland plant ecoysystems. Proc. Natl. Acad. Sci. U.S.A. **102**, 10819–10822
185. Dokholyan, N.V., Li, L., Ding, F. and Shakhnovich, E.I. (2002): Topological determinants of protein folding. Proc. Natl. Acad. Sci. U.S.A. **99**, 8637–8641
186. Dorogovtsev, S.N. and Mendes, J.F.F. (2001): Scaling properties of scale-free evolving networks: Continuous approach. Phys. Rev. E **63**, 056125.1–19
187. Dorogovtsev, S.N. and Mendes, J.F.F. (2002): Evolution of networks. Adv. Phys. **51**, 1079–1187
188. Doye, J.P.K. (2002): The network topology of a potential energy landscape: A static scale-free network. Phys. Rev. Lett. **88**, 238701
189. Doye, J.P.K. and Massen, C.P. (2005): Characterizing the network topology of the energy landscapes of atomic clusters. (www.arxiv.org/cond-mat/0411144) J. Chem. Phys. **122**, 084105
190. Dressel, R., Grzeszik, C., Kreiss, M., Lindemann, D., Herrmann, T., Walter, L. and Günther, E. (2003): Differential effect of acute and permanent heat shock protein 70 overexpression in tumor cells on lysability by cytotoxic T lymphocytes. Cancer Res. **63**, 8212–8220
191. Duarte, M. and Zatsiorsky, V.M. (2000): On the fractal properties of natural human standing. Neurosci. Lett. **283**, 173–176
192. Duch, J. and Arenas, A. (2005): Community detection in complex networks using extremal optimization. Phys. Rev. E **72**, 027104

193. * Dunbar, R.I.M. (1998): *Grooming, Gossip and the Evolution of Language*. Harvard University Press
194. Dunbar, R.I.M. (2003): The origin and subsequent evolution of language. In: *Language Evolution*, ed. by M. Christiansen and S. Kirby, Oxford Univ. Press, Oxford UK
195. ** Dunbar, R.I.M. (2005): Why are good writers so rare? An evolutionary perspective on literature. J. Cult. Evol. Psychol. **3**, 7–22
196. Dunbar, R.I.M., Duncan, N. and Nettle, D. (1994): Size and structure of freely forming conversational groups. Human Nature **6**, 67–78
197. Dunker, A.K., Brown, C.J., Lawson, J.D., Iakoucheva, L.M. and Obradovic, Z. (2002): Intrinsic disorder and protein function. Biochemistry **41**, 6573–6582
198. Dunne, J.A., Williams, R.I. and Martinez, N.D. (2002a): Network structure and biodiversity loss in food webs: Robustness increases with connectance. Ecol. Lett. **5**, 558–567
199. Dunne, J.A., Williams, R.I. and Martinez, N.D. (2002b): Food-web structure and network theory: The role of connectance and size. Proc. Natl. Acad. Sci. U.S.A. **99**, 12917–12922
200. Dupouy-Camet, J. (2002): Trichinellosis is unlikely to be responsible for Mozart's death. Arch. Intern. Med. **162**, 946. Hirschmann's reply: 946–947
201. Durkheim, E. (1933): *The Divison of Labor in Society*, MacMillan, New York
202. * Earl, D.J. and Deem, M.W. (2004): Evolvability is a selectable trait. Proc. Natl. Acad. Sci. U.S.A. **101**, 11531–11536
203. Earnshaw, W.C., Honda, B.M., Laskey, R.A. and Thomas, J.O. (1980): Assembly of nucleosomes: The reaction involving *X. laevis* nucleoplasmin. Cell **21**, 373–383
204. ** Edelman, G.M. and Gally, J.A. (2001): Degeneracy and complexity in biological systems. Proc. Natl. Acad. Sci. U.S.A. **98**, 13763–13768
205. Edman, L. and Rigler, R. (2000): Memory landscapes of single-enzyme molecules. Proc. Natl. Acad. Sci. U.S.A. **97**, 8266–8271
206. Egghe, L. (2005): A characterization of the law of Lotka in terms of sampling. Scientometrics **62**, 321–328
207. Eigen, M. (2002): Error catastrophe and antiviral strategy. Proc. Natl. Acad. Sci. U.S.A. **99**, 13374–13376
208. Eguíluz, V.M., Chialvo, D.R., Cecchi, G.A., Baliki, M. and Apkarian, A.V. (2005): Scale-free brain functional networks. Phys. Rev. Lett. **94**, 018102
209. Eke, A., Herman, P., Kocsis, L. and Kozak, L.R. (2002): Fractal characterization of complexity in temporal physiological signals. Physiol. Meas. **23**, R1–R38
210. Eldredge, N. (1985): *Unfinished Synthesis. Biological Hierarchies and Modern Evolutionary Thought*. Oxford University Press, Oxford

211. Elliott, J.I., Fesenstein, R., Tolaini, M. and Kioussis, D. (1995): Random activation of a transgene under the control of a hybrid hCD2 locus control region/lg enhancer regulatory element. EMBO J. **14**, 575–586
212. Elowitz, M.B., Levine, A.J., Siggia, E.D. and Swain, P.S. (2002): Stochastic gene expression in a single cell. Science **297**, 1183–1186
213. Enoka, R.M., Christou E.A., Hunter, S.K., Kornatz, K.W., Semmler, J.G., Taylor, A.M. and Tracy, B.L. (2003): Mechanisms that contribute to differences in motor performance between young and old adults. J. Electromyography Kinesiol. **13**, 1–12
214. Enright, J.T. (1980): Temporal precision of circadian systems. A reliable neuronal clock from unreliable components? Science **209**, 1542–1545
215. Erdős, P. and Rényi, A. (1959): On random graphs. Publicationes Mathematicae Debrecen **6**, 290–297
216. Erdős, P. and Rényi, A. (1960): On the evolution of random graphs. Magyar Tud. Akad. Mat. Kutató Int. Közl. **5**, 17–61
217. Estrada, E. (2005): Virtual identification of essential proteins within the protein interaction network of yeast. `Cond-mat/0505005`. Proteomics (Proteomics **6**, 35–40)
218. Eysenck, H.J. (1970): *The Structure of Human Personality*, Methuen, London
219. Fagan, W.F. (1997): Omnivory as a stabilizing feature of natural communities. Am. Nat. **150**, 554–567
220. Fares M.A., Ruiz-Gonzalez M.X., Moya A., Elena S.F. and Barrio E. (2002): GroEL buffers against deleterious mutations. Nature **417**, 398
221. Farkas, I., Helbing, D. and Vicsek, T. (2002): Mexican waves in an excitable medium. Nature **419**, 131–132
222. Farmer, J.D. and Lo, A.W. (1999): Frontiers of finance: Evolution and efficient markets. Proc. Natl. Acad. Sci. U.S.A. **96**, 9991–9992
223. Feder, J.H., Rossi, J.M., Solomon, J., Solomon, N. and Lindquist, S. (1992): The consequences of expressing hsp70 in *Drosophila* cells at normal temperatures. Genes Dev. **6**, 1402–1413
224. Feenstra, R. (1996): Trade and uneven growth. J. Dev. Econ. **49**, 229–256
225. Feibleman, J.K. (1954): The integrative levels in nature. Br. J. Phylos. Sci. **5**, 59–66
226. Fell, J., Klaver, P., Lehnertz, K., Grunwald, T., Schaller, C., Elger, C.E. and Fernandez, G. (2001): Human memory formation is accompanied by rhinal-hippocampal coupling and decoupling. Nature Neurosci. **4**, 1259–1264
227. Fenimore, P.W., Fraunefelder, H., McMahon, B.H. and Parak, F.G. (2002): Slaving: Solvent fluctuations dominate protein dynamics and functions. Proc. Natl. Acad. Sci. U.S.A. **99**, 16047–16051

228. Ferrarini, L., Bertelli, L., Feala, J., McCulloch, A.D. and Paternostro, G. (2005): A more efficient search strategy for aging genes based on connectivity. Bioinformatics **21**, 338–348
229. Ferrer Cancho, R. and Sole, R.V. (2001): The small-world of human language. Proc. Roy. Soc. B **268**, 2261–2265
230. Ferrer Cancho, R. and Sole, R.V. (2003): Least effort and the origins of scaling in human language. Proc. Natl. Acad. Sci. U.S.A. **100**, 788–791
231. ** Fewell, J.H. (2003): Social insect networks. Science **301**, 1867–1870
232. Feynman, R.P., Leighton, R.B. and Sands, M. (1965): *The Feynman Lectures on Physics*, Vol. 3. *Quantum Mechanics*. Addison-Wesley Professional, Boston MA, U.S.A.. Sect. 21.9
233. Fink, L.H. (1991): Proceedings of bulk power system voltage phenomena II. Voltage stability and security. ECC Inc.
234. Finnegan, E.J. (2001): Epialleles – A source of random variation in times of stress. Curr. Op. Plant Biol. **5**, 101–106
235. Flomenbom, O., Velonia, K., Loos, D., Masuo, S., Cotlet, M., Engelborghs, Y., Hofkens, J., Rowan, A.E., Nolte, R.J.M., van der Auweraer, M. and de Schryver, F.C. (2005): Stretched exponential decay and correlations in the catalytic activity of fluctuating single lipase molecules. Proc. Natl. Acad. Sci. U.S.A. **102**, 2368–2372
236. Fowler, M., Beck, K., Brant, J. and Opdyke, W. (1999): *Refactoring*, Addison-Wesley Professional, Boston MA, USA
237. Fox, J.C. and Keaveny T.M. (2001): Trabecular eccentricity and bone adaptation. J. Theor. Biol. **212**, 211–221
238. Fox, J.J. and Hill, C.C. (2001): From topology to dynamics in biochemical networks. Chaos **11**, 809–813
239. Fox, J.C., Snyder, A.Z., Vincent, J.L., Corbetta, M., van Essen, D.C. and Raichle, M.E. (2005): The human brain is intrinsically organized into dynamic, anticorrelated functional networks. Proc. Natl. Acad. Sci. U.S.A. **102**, 9673–9678
240. Frank, A. (1885): Über die physiologische Bedeutung der Mycorrhiza. Ber. Dt. Bot. Ges. **6**, 248–269
241. Freud, S. (1915): Observations on transference-love. In: *The Standard Edition of the Complete Works of Sigmund Freud*, ed. by J. Strachey, Vol. XII, pp. 157–173, Hogarth Press, London UK
242. Freund, T.F. (2003): Interneuron diversity series: Rhythm and mood in perisomatic inhibition. Trends Neurosci. **26**, 489–495
243. Fronczak, P., Fronczak, A. and Holyst, J.A. (2005): Interplay between network structure and self-organized criticality. `Cond-mat/0509043`
244. Fukuyama, F. (1992): *The End of History and the Last Man*, Free Press New York NY, USA
245. * Fukuyama, F. (1995): *Trust: The Social Virtues and the Creation of Prosperity*, Free Press, New York NY, USA

246. Gaite, J. (2005): The fractal distribution of haloes. Europhys. Lett. **71**, 332–338, www.arxiv.org/astro-ph/0505607
247. Gallese, V. and Goldman, A. (1998): Mirror neurons and the simulation theory of mind-reading. Trends Cogn. Sci. **2**, 493–501
248. Gamma, E., Helm, R., Johnson, R. and Vlissides, J. (1994): *Design Patterns Elements of Reusable Object-Oriented Software*, Addison-Wesley Professional, Boston MA, USA
249. Ganopolski, A. and Rahmstorf, S. (2002): Abrupt glacial climate changes due to stochastic resonance. Phys. Rev. Lett. **88**, 038501
250. Gao, Z., Hu, B. and Hu, G. (2001): Stochastic resonance of small-world networks. Phys. Rev. E **65**, 016209
251. Gardner, M.R. (1978): Mathematical games – White and brown music, fractal curves and one-over-f fluctuations. Sci. Am. **238**, 16–32
252. Gardner, M.R. and Ashby, W.R. (1970): Connectance of large dynamic (cybernetic) systems: Critical values for stability. Nature **228**, 784
253. $*$ Garlaschelli, Caldarelli, G., and Pietronello, L. (2003): Universal scaling relations in food webs. Nature **423**, 165–168
254. Garlaschelli, D., Battison, S., Castri, M., Servedio, V.D.P. and Caldarelli, G. (2005a): The scale-free topology of market investments. (Cond-mat/0310503) Physica A **350**, 491–499
255. Garlaschelli, Caldarelli, G., and Pietronello, L. (2005b): Universal scaling in food-web structure? Reply. Nature **435**, E4
256. Geisel, T., Nierwetberg, J. and Zacherl, A. (1985): Accelerated diffusion in Josephson junctions and related chaotic systems. Phys. Rev. Lett. **54**, 616–619
257. $*$ Gell-Mann, M. (1994): *The Quark and the Jaguar. Adventures in the simple and the complex*, Little Brown, London
258. Gell-Mann, M. (1995): What is complexity? Complexity **1**, 16–19
259. Gfeller, D., Chappelier, J.-C. and De Los Rios, P. (2005): Finding instabilities in the community structure of complex networks. Cond-mat/0503593 (Phys. Rev. E **72**, 056135)
260. Gheorghiu, S. and Coppens, M.O. (2004): Heterogeneity explains features of 'anomalous' thermodynamics and statistics. Proc. Natl. Acad. Sci. U.S.A. **101**, 15852–15856
261. Ghim, C-M., Oh, E., Goh, K.-I., Khang, B. and Kim, D. (2004): Packet transport along the shortest pathways in scale-free networks. Eur. Phys. J. B **38**, 193–199
262. Gibson, G. and van Helden, S. (1997): Is function of the *Drosophila* homeotic gene Ultrabithorax canalized? Genetics **147**, 1155–1168
263. Gibson, G., and Wagner, G. (2000): Canalization in evolutionary genetics: A stabilizing theory? BioEssays **22**, 372–380
264. Gilden D.L. (2001): Cognitive emissions of $1/f$ noise. Psychol. Rev. **108**, 33–56

265. Gilden, D.L., Thornton, T. and Mallon, M.W. (1995): $1/f$ noise in human cognition. Science **267**, 1837–1839
266. Girvan, M. and Newman, M.E.J. (2002): Community structure in social and biological networks. Proc. Natl. Acad. Sci. U.S.A. **99**, 7821–7826
267. Gisiger, T. (2001): Scale invariance in biology: Coincidence or footprint of universal mechanism? Biol. Rev. **76**, 161–209
268. Goetze, T. and Brickmann, J. (1992): Self-similarity of protein surfaces. Biophys. J. **61**, 109–118
269. Goh, K-I., Kahng, B. and Kim, D. (2001): Universal behavior of load distribution in scale-free networks. Phys. Rev. Lett. **87**, 278701
270. Goh, K-I., Salvi, G., Kahng, B. and Kim, D. (2005): Skeleton and fractal scaling in complex networks. `Cond-mat/0508332` (Phys. Rev. Lett., in press)
271. Goldberger, A.L., Amaral, L.A.N., Hausdorf, J.M., Ivanov, P.C., Peng, C.-K. and Stanley, H.E. (2002): Fractal dynamics in physiology: Alterations with disease and aging. Proc. Natl. Acad. Sci. U.S.A. **99**, 2466–2472
272. Goodwin, D.W., Powell, B., Bremer, D., Hoine, H. and Stern, J. (1969): Alcohol and recall: State-dependent effects in man. Science **163**, 1358–1360
273. * Gould, S.J. and Eldredge, N. (1993): Punctuated equilibrium comes of age. Nature **366**, 223–227
274. * Gould, S.J. and Lewontin, R.C. (1979): The spandrels of San Marco and the Panglossian Paradigm: A critique of the adaptationist programme. Proc. Roy. Soc. London B **205**, 581–598
275. Grabher, G. (1993): *The Embedded Firm. On the Socioeconomics of Industrial Networks*, Routledge, London and New York
276. ** Granovetter, M. (1973): The strength of weak ties. Am. J. Sociology **78**, 1360–1380
277. * Granovetter, M. (1983): The strength of weak ties: A network theory revisited. Sociological Theory **1**, 202–233
278. Grassly, N.G., Fraser, C. and Garnett, G.P. (2005): Host immunity and synchronized epidemics of syphilis across the United States. Nature **433**, 417–421
279. Greene, L.H. and Higman, V.A. (2003): Uncovering network systems within protein structures. J. Mol. Biol. **334**, 781–791
280. Grigera, T.S. and Israeloff, N.E. (1999): Observation of fluctuation–dissipation theorem in a structural glass. Phys. Rev. Lett. **83**, 5038–5041
281. Gross, T.S., Poliachik, S.L., Ausk, B.J., Sanford, D.A., Becker, B.A. and Srinivasan, S. (2004): Why rest stimulates bone formation: A hypothesis based on complex adaptive phenomenon. Exerc. Sport Sci. Rev. **32**, 9–13

282. Gu, Z., David, L., Petrov, D., Jones, T., Davis, R.W. and Steinmetz, L.M. (2005): Elevated evolutionary rates in the laboratory strain of *Saccharomyces cerevisiae*. Proc. Natl. Acad. Sci. U.S.A. **102**, 1092–1097
283. Guclu, H. and Korniss, G. (2004): Extreme fluctuations in small-worlds with relaxational dynamics. (Cond-mat/0311575) Phys. Rev. E **69**, 065104
284. Guimera, R. and Amaral, L.A.N. (2005): Functional cartography of complex networks. Nature **433**, 895–900
285. Guimera, R., Sales-Pardo, M. and Amaral, L.A.N. (2004): Modularity from fluctuations in random graphs and complex networks. Phys. Rev. E **70**, 025101
286. Gülow, K., Bienert, D. and Haas, I.G. (2002): BiP is feed-back regulated by control of protein translation efficiency. J. Cell Sci. **115**, 2443–2452
287. Gulyas, B. (2001): Neural networks for internal reading and visual imagery of reading: A PET study. Brain Res. Bull. **54**, 319–328
288. Gutenberg, B. and Richter, C.F. (1956): Magnitude and energy of earthquakes. Ann. Geofis. **9**, 1–15
289. Hagerhall, C.M., Purcell, T. and Taylor, R. (2004): Fractal dimension of landscape silhouette outlines as a predictor of landscape preference. J. Environm. Psychology **24**, 247–255
290. Hales, C.N. and Barker, D.J. (1992): Type 2 (non-insulin-dependent) diabetes mellitus: The thrifty phenotype. Diabetologia **35**, 595–601
291. Hall, D. and Minton, A.P. (2003): Macromolecular crowding: Qualitative and semiquantitative successes, quantitative challenges. Biochim. Biophys. Acta **1649**, 127–139
292. Hallinan, J. (2003): Self-organization leads to hierarchical modularity in an Internet community. Lect. Notes Artif. Intell. **2773**, 914–920
293. ∗ Han, J.-D.J., Bertin, N., Hao, T., Goldberg, D.S., Berriz, G.F., Zhang, L.V., Dupuy, D., Walhout, A.J.M., Cusick, M.E., Roth, F.P. and Vidal, M. (2004): Evidence for dynamically organized modularity in the yeast protein–protein interaction network. Nature **430**, 88–93
294. Hanahan, D. and Weinberg, R.A. (2000): The hallmarks of cancer. Cell **100**, 57–70
295. Hansen, M.T. (1999): The search-transfer problem: The role of weak ties in sharing knowledge across organization subunits. Adm. Sci. Quart. **44**, 82–111
296. Harsanyi, J.C. and Selten, R. (1988): *A General Theory of Equilibrium Selection in Games*, MIT Press, Cambridge MA, USA
297. Hartl F-U. (1996): Molecular chaperones in cellular protein folding. Nature **381**, 571–580
298. Hartwell, L.H., Hopfield, J.J., Leibler, S. and Murray, A.W. (1999): From molecular to modular cell biology. Nature **402**, C47–C52

299. Hatano, T. and Sasa, S. (2001): Steady state thermodynamics of Langevin systems. Phys. Rev. Lett. **86**, 3463–3466
300. Hayashi, Y. and Miyazaki, T. (2005): Emergent rewirings for cascades on correlated networks. `Cond-mat/0503615`
301. Hayflick, L. (2000): The future of ageing. Nature **408**, 267–269
302. Hawkes, K., O'Connell, J.F., Blurton Jones, N.G., Alvarez, H. and Charnov, E.L. (1989): Grandmothering, menopause, and the evolution of human life histories. Proc. Natl. Acad. Sci. U.S.A. **95**, 1336–1339
303. Hegel, G.W.F. (1989): *Science of Logic*, Prometheus Books, Amherst, NY USA
304. Helbing, D., Farkas, I. and Vicsek, T. (2000): Simulating dynamical features of escape panic. Nature **407**, 487–490
305. Hemelrijk, C.K. (2002): Self-organization and natural selection in the evolution of complex despotic societies. Biol. Bull. **202**, 283–288
306. Hemmingsen, S.M., Woolford, C., van der Vies, S.M., Tilly, K., Dennis, D.T., Georgopoulos, C.P., Hendrix, R.W. and Ellis, R.J. (1988): Homologous plant and bacterial proteins chaperone oligomeric protein assembly. Nature **333**, 330–334
307. Hermisson, J., Hansen, T.F. and Wagner, G.P. (2003): Epistasis in polygenic traits and the evolution of genetic architecture under stabilizing selection. Am. Nat. **161**, 708–734
308. Herndon L.A., Schmeissner, P.J., Dudaronek, J.M., Brown, P.A., Listner, K.M., Sakano, Y., Paupard, M.C., Hall, D.H. and Driscoll, M. (2003): Stochastic and genetic factors influence tissue-specific decline in aging *C. elegans*. Nature **419**, 808–814
309. Halley, J.M. (1996): Ecology, evolution and $1/f$ noise. Trends Ecol. Evol. **11**, 33–37
310. Hill, R. A. and Dunbar, R.I.M. (1994): Social network size in humans. Human Nature **14**, 53–72
311. * Himmelstein, D.U., Levins, R. and Woolhandler, S. (1990): Beyond our means: Patterns of variability of physiological traits. Int. J. Health Serv. **20**, 115–124
312. Hirase, H., Qian, L., Barthó, P. and Buzsáki, G. (2004): Calcium dynamics of cortical astrocytic networks in vivo. PLoS Biology **2**, 494–499
313. Hirschmann, J.V. (2001): What killed Mozart? Arch. Intern. Med. **161**, 1381–1389
314. * Holling, C.S. (1973): Resilience and stability of ecological systems. Annu. Rev. Ecology Systematics **4**, 1–23
315. Holland, H.D. (1997): Evidence for life on Earth more than 3 850 million years ago. Science **275**, 38–39
316. Holyoak, M. and Sachdev, S. (1998): Omnivory and the stability of simple food webs. Oecologia **117**, 413–419
317. Honeyman, C. (1987): In defense of Ambiguity. Negotiation J. **3**, 81–86

318. * Hong, L. and Page, S.E. (2004): Groups of diverse problem solvers can outperform groups of high-ability problem solvers. Proc. Natl. Acad. Sci. U.S.A. **101**, 16385–16389
319. Hruza, Z. and Wachtlova, M. (1969): Diminution of bone blood flow and capillary network in rats during aging. J. Gerontol. **24**, 315–320
320. Hsu, K.J. and Hsu, A. (1991): Self-similarity of the '$1/f$ noise' called music. Proc. Natl. Acad. Sci. U.S.A. **88**, 3507–3509
321. Huang, S. (2002): Rational drug discovery: What can we learn from regulatory networks? Drug Discov. Today **7**, S163–S169
322. Huber, R., Ghilardi, M.F., Massimini, M. and Tononi, G. (2004): Local sleep and learning. Nature **430**, 78–81
323. Huberman, B.A., Pirolli, P.L., Pitkow, J.E. and Lukose, R.M. (1998): Strong regularities in world wide web surfing. Science **280**, 95–97
324. Hughes, A.R. and Stachowicz, J.J. (2004): Genetic diversity enhances the resistance of a seagrass ecosystem to disturbance. Proc. Natl. Acad. Sci. U.S.A. **101**, 8998–9002
325. Huxley, J.S. (1932): *Problems of Relative Growth*, Dial Press, New York NY, USA
326. Huygens, C. (1665): Letter to R. Moray (February 27, 1665) In: *Œuvres Complètes de Christian Huygens*, ed. by M. Nijhoff, Société Hollandaise des Sciences, 1893, The Hague, Vol. 5, pp. 246–249
327. Ivanov, P.C., Amaral, L.A.N., Goldberger, A.L., Havlin, S., Rosenblum, M.G., Struzik, Z.R. and Stanley, H.E. (1999): Multifractality in human heartbeat dynamics. Nature **399**, 461–465
328. Ives, A.R. and Cardinale, B.J. (2004): Food-web interactions govern the resistance of communities after non-random extinctions. Nature **429**, 174–177
329. * Jacob, F. (1977): Evolution and tinkering. Science **196**, 1161–1166
330. Janssen, M.A. and Jager, W. (2003): Simulating market dynamics: Interactions between consumer psychology and social networks. Art. Life **9**, 343–356
331. Jausovec, N. and Jausovec, K. (2000): Differences in resting EEG related to ability. Brain Topogr. **12**, 229–240
332. Jeong, H., Tombor, B., Albert, R., Oltvai, Z.N. and Barabasi, A.L. (2000): The large-scale organization of metabolic networks. Nature **407**, 651–654
333. Jeong, H., Mason, S.P., Barabasi, A.L. and Oltvai, Z.N. (2001): Lethality and centrality in protein networks. Nature **411**, 41–42
334. Jerne, N.K. (1974): Towards a network theory of the immune system. Ann. Immunol. (Inst. Pasteur) **125** C, 373–389
335. Jerne, N.K. (1984): Idiotypic networks and other preconceived ideas. Immunol. Rev. **79**, 5–23

336. ∗ Johansen, A. and Sornette, D. (2001): Finite-time singularity in the dynamics of the world population, economic and financial indices. Physica A **294**, 465–502
337. Johnson, J.B. (1925): The Schottky effect in low frequency circuits. Phys. Rev. **26**, 71–85
338. ∗ Jones, J.C., Myerscough, M.R., Graham, S. and Oldroyd, B.P. (2004): Honey bee nest thermoregulation: Diversity promotes stability. Science **305**, 402–404
339. Jordan, F. and Molnar, I. (1999): Reliable flows and preferred patterns in food webs. Evol. Ecol. Res. **1**, 591–609
340. Jordan, F. and Scheuring, I. (2002): Searching for keystones in ecological networks. Oikos **99**, 607–612
341. Jordan, F., Scheuring, I. and Vida, G. (2002): Species positions and extinction dynamics in simple food webs. J. Theor. Biol. **215**, 441–448
342. Jordano, P., Bascompte, J. and Olesen, J.M. (2003): Invariant properties in coevolutionary networks of plant–animal interactions. Ecol. Lett. **6**, 69–81
343. Jung, C.G. (1969): *The Structure and Dynamics of the Psyche*, Princeton University Press
344. Karinthy, F. (1929): *Minden másképpen van* (Everything is the Other Way, in Hungarian), Atheneum Press, Budapest, Hungary
345. Karsai, I. and Wenzel, J.W. (1998): Productivity, individual-level and colony-level flexibility, and organization of work as consequences of colony size. Proc. Natl. Acad. Sci. U.S.A. **95**, 8665–8669
346. Kauffman, S.A. (1969): Metabolic stability and epigenesis in randomly constructed genetic nets. J. Theor. Biol. **22**, 437–467
347. ∗∗ Kauffman, S.A. (2000): *Investigations*, Oxford University Press, Oxford, UK
348. Kauffman, S.A. and Johnsen, S. (1991): Co-evolution to the edge of chaos: Coupled fitness landscapes, poised states and co-evolutionary avalanches. J. Theor. Biol. **149**, 467–506
349. ∗ Kauffman, S.A. and Levin, S. (1987): Towards a general theory of adaptive walks on rugged landscapes. J. Theor. Biol. **128**, 11–45
350. Kawachi, I. and Berkman, L.F. (2001): Social ties and mental health. J. Urban Health **78**, 458–467
351. Keitt, T.H. and Stanley, H.E. (1998): Dynamics of North American breeding bird populations. Nature **393**, 257–260
352. Kenworthy, A.K. Nichols, B.J. Remmert, C.L., Hendrix, G.M. Kumar, M., Zimmerberg, J. and Lippincott-Schwartz, J. (2004): Dynamics of putative raft-associated proteins at the cell surface. J. Cell Biol. **165**, 735–746
353. Kemkemer, R., Schrank, S., Vogel, W., Gruler, H. and Kaufmann, D. (2002): Increased noise as an effect of haploinsufficiency of the tumor-

suppressor gene neurofibromatosis type 1 in vitro. Proc. Natl. Acad. Sci. U.S.A. **99**, 13783–13788
354. Kennedy, P. (1987): *The Rise and Fall of the Great Powers*, Random House, New York NY, USA
355. Kerckhoff, A., Back, K. and Miller, N. (1965): Sociometric patterns in hysterical contagion. Sociometry **28**, 2–15
356. Kerr, B., Riley, M.A., Feldman, M.W. and Bohannan, B.J. (2002): Local dispersal promotes biodiversity in a real-life game of rock–paper–scissors. Nature **418**, 171–174
357. Keynes, J.M. (1936): *The General Theory of Employment, Interest and Money*, Harcourt, Brace. New York NY, USA
358. * Kiessling, W. (2005): Long-term relationships between ecological stability and biodiversity in Phanerozoic reefs. Nature **433**, 410–413
359. Killworth, P.D. and Bernard, R. (1978/79): The reversal of the small-world experiment? Social Networks **1**, 159–192
360. Kim, P.S. and Baldwin, R.L. (1990): Intermediates in the folding reactions of small proteins. Annu. Rev. Biochem. **59**, 631–660
361. Kirkwood, T.B.L. and Austad, S.N. (2000): Why do we age? Nature **408**, 233–238
362. ** Kirschner, M. and Gerhart, J. (1998): Evolvability. Proc. Natl. Acad. Sci. U.S.A. **95**, 8420–8427
363. Kiss, I.Z., Zhai, Y. and Hudson, J.L. (2002): Emerging coherence in a population of chemical oscillators. Science **296**, 1676–1678
364. Kitami, T. and Nadeau, J.H. (2002): Biochemical networking contributes more to genetic buffering in human and mouse metabolic pathways than does gene duplication. Nature Genetics **32**, 191–194
365. Kleiber, M. (1932): Body size and metabolism. Hilgardia **6**, 315–353
366. Kleinberg, J. (2000): Navigation in a small world. Nature **406**, 845
367. Kleinfield, J.S. (2002): The small world problem. Society **39**, 61–66
368. Klibanov, A.M. (1995): What is remembered and why? Nature **374**, 596
369. Kniffki, K.D., Pavlak, M. and Vahle-Hinz, C. (1993): Scaling behavior of the dendritic branches of thalamic neurons. Fractals **1**, 171–178
370. Koestler, A. and Smythies, J.R. (1969): *Beyond Reductionism: New Perspectives in the Life Sciences*, Hutchinson, London UK
371. Kohlrausch, R. (1854): Theorie des elektrischen Rückstandes in der Leidner Flasche. Pogg. Ann. Phys. **91**, 179–214
372. Kolmogorov, A.N. (1965): Three approaches to the quantitative definition of information. Problems Information Transmission **1**, 1–7
373. Kondoh, M. (2003): Foraging adaptation and the relationship between food-web complexity and stability. Science **299**, 1388–1391
374. Koonin, E.V. and Galperin, M.Y. (2002): *Sequence, Evolution, Function: Computational Approaches in Functional Genomics*, Kluwer Academic Publishers

375. Koonin, E.V., Wolf, Y.I. and Karev, G.P. (2002): The structure of the protein universe and genome evolution. Nature **420**, 218–223
376. Kopp, M. (2000): Cultural transition. In: *Encyclopedia of Stress*, ed. by G. Fink, Academic Press, San Diego, Vol. 1. pp. 611–615
377. Kopp, M. and Réthelyi, J. (2004): Where psychology meets physiology: Chronic stress and premature mortality – the Central-Eastern European health paradox. Brain Res. Bull. **62**, 351–367
378. Korte, C. and Milgram, S. (1970): Acquaintance networks between racial groups. J. Pers. Soc. Psychol. **15**, 101–108
379. Kovacs, I.A., Szalay, M.S. and Csermely, P. (2005): Water and molecular chaperones act as weak links of protein folding networks: Energy landscape and punctuated equilibrium changes point towards a game theory of proteins. (`www.arxiv.org/q-bio/0409030`) FEBS Lett. **579**, 2254–2260
380. Kowner, R. (2001): Psychological perspective on human developmental stability and fluctuating asymmetry: Sources, applications and implications. Br. J. Psychol. **92**, 447–469
381. Krakauer, D.C. and Sasaki, A. (2002): Noisy clues to the origin of life. Proc. Roy. Soc. Lond. B. **269**, 2423–2428
382. Krebs, V.E. (2002): Uncloaking terrorist networks. First Monday 7, April (`www.firstmonday.org`)
383. Kretschmer, E. (1921): *Körperbau und Charakter*, Springer Verlag, Berlin
384. Krishnamurti, J. (1992): *On Relationships*, Harper Collins, New York NY, USA
385. Krugman, P.R. (1996): *The Self-Organizing Economy*, Blackwell Publishers, Oxford, UK
386. Kuhn, T.S. (1962): *The Structure of Scientific Revolutions*, The University of Chicago Press, Chicago IL, USA
387. Kun, F., Wittel, F.K., Herrmann, H.J., Kröplin, B.H. and Máløy, K.J. (2005): Scaling behavior of fragment shapes. `Cond-mat/0506686`. Phys. Rev. Lett. (in press)
388. Kunde, W., Kiesel, A. and Hoffmann, J. (2003): Conscious control over the content of unconscious cognition. Cognition **88**, 223–242
389. Kunin, V., Pereira-Leal, J.B. and Ouzounis, C.A. (2004): Functional evolution of the yeast protein interaction network. Mol. Biol. Evol. **21**, 1171–1176
390. * Kunovich, R.M. and Hodson, R. (1999): Civil war, social integration and mental health in Croatia. J. Health Soc. Behav. **40**, 323–343
391. Kuperman, M. and Zanette, D. (2001): Stochastic resonance in a model of opinion formation on small-world networks. Phys. Rev. E **64**, 050901
392. Kuramoto, Y. (1984): *Chemical Oscillation, Waves, and Turbulence*, Springer Verlag Berlin, Germany

393. Kurkal, V., Daniel, R.M., Finney, J.L., Tehei, M., Dunn, R.V. and Smith, J.C. (2005): Enzyme activity and flexibility at very low hydration. Biophys. J. **89**, 1282–1287
394. Kuznetsov, V.A., Knott, G.D. and Bonner, R.F. (2002): General statistics of stochastic process of gene expression in eukaryotic cells. Genetics **161**, 1321–1332
395. Kydd, A. (2003): Which side are you on? Bias, credibility and mediation. Am. J. Political Sci. **47**, 597–611
396. Lahdenperä, M., Lummaa, V., Helle, S., Tremblay, M. and Russell, A.F. (2004): Fitness benefits of prolonged post-reproductive lifespan in women. Nature **428**, 178–181
397. Lai, Y.-C., Motter, A., Nishikawa, T., Park, K. and Zhao, L. (2005): Complex networks: Dynamics and security. Pramana J. Physics **64**, 483–502
398. Lakhina, A., Byers, J.W., Crovella, M. and Xie, P. (2003): Sampling bias in IP topology measurements. Proc. IEEE INFOCOM '03, San Francisco CA, USA. April 2003
399. Lampl, M., Veldhuis, J.D. and Johnson, M.L. (1992): Saltation and stasis: A model of human growth. Science **258**, 801–803
400. Latora, V. and Marchiori, M. (2003): Economic small-world behavior in weighted networks. Eur. Phys. J. B **32**, 249–263
401. Laughlin, R.B. and Pines, D. (2000): The theory of everything. Proc. Natl. Acad. Sci. U.S.A. **97**, 28–31
402. Laughlin, S.B. and Sejnowski, T.J. (2003): Communication in neuronal networks. Science **301**, 1870–1874
403. Le Comber, S.C., Spinks, A.C., Bennett, N.C., Jarvis, J.U.M. and Faulkes, C.G. (2002): Fractal dimension of African mole-rat burrows. Can. J. Zool. **80**, 436–441
404. Lederman, L. (1993): *The God Particle*, Delta Publishing, New York
405. Lee, E.J., Goh, K.-I., Kahng, B. and Kim, D. (2005a): Robustness of the avalanche dynamics in data packet transport on scale-free networks. (`www.arxiv.org/cond-mat/0410684`) Phys. Rev. E **71**, 056108
406. Lee, S.H., Kim, P-J. and Jeong, H. (2005b): Statistical properties of sampled networks. `Cond-mat/0505232`. (Phys. Rev. E **73**, 016102)
407. Lehman, M.M., Ramil, J.F., Wernick, P.D., Perry, D.E. and Turski, W.M. (1998): Metrics and laws of software evolution – The nineties view. In: *Elements of Software Process Assessment and Improvement*, ed. by K. El Eman and N.H. Madhavji, IEEE CS Press
408. Leland, W.E., Taqqu, M.S., Willinger, W. and Wilson, D.V. (1994): On the self-similar nature of Ethernet traffic. IEEE/ACM Trans. Netw. **2**, 1–15
409. Lenton, T.M. (1998): Gaia and natural selection. Nature **394**, 439–447

410. Levin, M.D. (2003): Noise in gene expression as the source of non-genetic individuality in the chemotactic response of *Escherichia coli*. FEBS Lett. **550**, 135–138
411. * Levin, D.Z. and Cross, R. (2004): The strength of weak ties you can trust: The mediating role of trust in effective knowledge transfer. J. Management Sci. **50**, 1477–1490
412. Levine, M. and Tjian, R. (2003): Transcription regulation and animal diversity. Nature **424**, 147–151
413. Levinthal, C. (1968): Are there pathways for protein folding? J. Chem. Phys. **65**, 44–45
414. Levy, P. (1937): *Théorie de l'addition des variables aleatoires*, Gauthiers-Villars, Paris
415. Levy, J.S. (1983): *War in the Modern Great Power System, 1495–1975*, Kentucky University Press, Lexington KY, USA
416. Levy, Y. (1996): Modularity of language reconsidered. Brain Lang. **55**, 240–263
417. Levy, Y. and Onuchic, J.N. (2004): Water and proteins: A love–hate relationship. Proc. Natl. Acad. Sci. U.S.A. **101**, 3325–3326
418. Lewis, K. (2000): Programmed cell death in bacteria. Microbiol. Mol. Biol. Rev. **64**, 503–514
419. Lewis, M. and Rees, D.C. (1985): Fractal surfaces on proteins. Science **230**, 1163–1165
420. Lewontin, R.C. (1970): The units of selection. Annu. Rev. Ecol. Syst. **1**, 1–18
421. Li, C. and Chen, G. (2004): A comprehensive weighted evolving network model. (Cond-mat/0406299) Physica A **343**, 288–294
422. Li, H., Helling, R., Tang, C. and Wingreen, N. (1996): Emergence of preferred structures in a simple model of protein folding. Science **273**, 666–669
423. Li, X., Jin, Y.Y. and Chen, G. (2003): Complexity and synchronization of the world trade web. Physica A **328**, 287–296
424. Liben-Nowell, D., Novak, J., Kumar, R., Raghavan, P. and Tomkins, A. (2005): Geographic routing in social networks. Proc. Natl. Acad. Sci. U.S.A. **102**, 11623–11628
425. * Lieberman, E., Hauert, C. and Nowak, M.A. (2005): Evolutionary dynamics on graphs. Nature **433**, 312–316
426. Liljeros, F., Edling, C.R., Amaral, L.A.N., Stanley, H.E. and Aberg, Y. (2001): The web of human sexual contacts. Nature **411**, 907–908
427. Lin, N. (1999): Social networks and status attainment. Annu. Rev. Sociol. **25**, 467–487
428. Lin, N., Dayton, P. and Greenwald, P. (1978): Analyzing the instrumental use of relations in the context of social structure. Sociol. Meth. Res. **7**, 149–166

429. Lindner, J.F., Meadows, B.K. and Ditto, W.L. (1995): Array enhanced stochastic resonance and spatiotemporal synchronization. Phys. Rev. Lett. **75**, 3–6
430. Lindner, J.F., Meadows, B.K., Ditto, W.L., Inchiosa, M.E. and Bulsara, A.R. (1996): Scaling laws for spatiotemporal synchronization and array enhanced stochastic resonance. Phys. Rev. E **53**, 2081–2086
431. Linville, P.W. (1987): Self-complexity as a cognitive buffer against stress-related illness and depression. J. Pers. Soc. Psychol. **52**, 663–676
432. Lippiello, E., de Arrangelis, L. and Godano, C. (2005): Memory in self-organized criticality. `Cond-mat/0505129`. Europhys. Lett. **72**, 678–684
433. * Liu, C., Weaver, D.R., Strogatz, S.H. and Reppert, S.M. (1997): Cellular construction of a circadian clock: Period determination in the suprachiasmatic nuclei. Cell **91**, 855–860
434. Liu, Y., Fratini, E., Baglioni, P., Chen, W.-R. and Chen, S.-H. (2005): An effective long-range attraction between protein molecules in solutions studied by small angle neutron scattering. `Cond-mat/0508162`. (Phys. Rev. Lett. **95**, 118102)
435. Loeb, L.A., Loeb, K.R. and Anderson, J.P. (2003): Multiple mutations and cancer. Proc. Natl. Acad. Sci. U.S.A. **100**, 776–781
436. * Lolle, S.J., Victor, J.L., Young, J.M. and Pruitt, R.E. (2005): Genome-wide non-Mendelian inheritance of extra-genomic information in *Arabidopsis*. Nature **434**, 505–509
437. Lorenz, W.E. (2003): Fractals and fractal architecture. Diplom-arbeit, Wienna University of Technology, Austria (`www.iemar.tuwien.ac.at/modul23/Fractals/subpages/10home.html`)
438. Lotka, A.J. (1926): The frequency distribution of scientific productivity. J. Washington Acad. Sci. **16**, 317–323
439. Lovelock, J. (1979): *Gaia: A New Look at Life on Earth*, Oxford University Press, Oxford UK
440. Lovelock, J. (2003): The living Earth. Nature **426**, 769–770
441. Lu, E.T. and Hamilton, R.J. (1991): Avalanches and the distribution of solar flares. Astrophys. J. Lett. **380**, L89–L92
442. Lu, H.P., Xun, L. and Xie, X.S. (1998): Single-molecule enzymatic dynamics. Science **282**, 1877–1882
443. Lundkvist, I., Coutinho, A., Varela, F. and Holmberg, D. (1989): Evidence for a functional idiotypic network among natural antibodies in normal mice. Proc. Natl. Acad. Sci. U.S.A. **86**, 5074–5078
444. Luscombe, N.M., Babu, M.M., Yu, H., Snyder, M., Teichmann, S.A. and Gerstein, M. (2004): Genomic analysis of regulatory network dynamics reveals large topological changes. Nature **431**, 308–312
445. Lusseau, D. (2003): The emergent properties of a dolphin social network. Proc. Roy. Soc. London B **270**, S186–S188
446. Lux, T. and Marchesi, M. (1999): Scaling and criticality in a stochastic multi-agent model of a financial market. Nature **397**, 498–500

447. Ma, H.W. and Zeng, A.P. (2003): Reconstruction of metabolic networks from genome data and analysis of their global structure for various organisms. Bioinformatics **19**, 220–277
448. MacDuffie, J.P. (1997): The road to root cause: Shop floor problem-solving at three auto assembly plants. Management Sci. **43**, 479–502
449. Magnasco, M.O. (2000): The thunder of distant net storms. `Nlin.AO/0010051`
450. Mahdavi, A. and Geller, M.J. (2004): A new redshift survey of galaxies in groups: The shape of the mass density profile. Astrophys. J. **607**, 202–219
451. Makse, H.A., Havlin, S. and Stanley, H.E. (1995): Modelling urban growth patterns. Nature **377**, 608–612
452. Malamud, B.D., Morein, G. and Turcotte, D.L. (1998): Forest fires: An example of self-organized critical behavior. Science **281**, 1840–1842
453. Malcai, O., Lidar, A.D. and Biham, O. (1997): Scaling range and cutoffs in empirical research. Phys. Rev. E **56**, 2817–2828
454. Mandelbrot, B. (1963): The variation of certain speculative prices. J. Business **36**, 394–419
455. Mandelbrot, B. (1967): How long is the coast of Britain? Statistical self-similarity and fractional dimension. Science **156**, 636–638
456. ** Mandelbrot, B. (1977): *Fractal: Form, Chance and Dimension*, Freeman, San Francisco
457. Mantegna, R.N. and Stanley, H.E. (1995): Scaling behaviour in the dynamics of an economic index. Nature **376**, 46–49
458. Margulis, L. (1998): *Symbiotic Planet: A New Look at Evolution*, Basic Books
459. Marmot, M.G. and Smith, G.D. (1989): Why are Japanese living longer? Br. Med. J. **299**, 1547–1551
460. * Maslow, S. and Sneppen, K. (2002): Specificity and stability in topology of protein networks. Science **296**, 910–913
461. Masuda and Konno (2005): VIP-club phenomenon: Inevitable emergence of elites and masterminds in social networks. `Cond-mat/0501129`. Social Networks (in press)
462. May, R.M. (1973): *Stability and Complexity in Model Ecosystems*, Princeton University Press
463. May, R.M. and Lloyd, A.L. (2001): Infection dynamics on scale-free networks. Phys. Rev. E **64**, 066112.1–4
464. Mayani, H., Dragowska, W. and Lansdorf, P.M. (1993): Lineage commitment in human hemopoiesis involves asymmetric cell division of multipotent progenitors and does not appear to be influenced by cytokines. J. Cell Physiol. **157**, 579–586
465. Maynard-Smith, J. (1970): Natural selection and the concept of a protein space. Nature **225**, 563–564

466. Maynard-Smith, J. and Szathmary, E. (1995): *The Major Transitions in Evolution*, W.H. Freeman, Oxford
467. McAdams, H.H. and Arkin, A. (1997): Stochastic mechanisms in gene expression. Proc. Natl. Acad. Sci. U.S.A. **94**, 814–819
468. * McCann, K.S. (2000): The diversity–stability debate. Nature **405**, 228–233
469. McCann, K.S. and Hastings, A. (1997): Re-evaluating the omnivory-stability relationship in food webs. Proc. R. Soc. Lond. B **264**, 1249–1254
470. ** McCann, K.S., Hastings, A. and Huxel, G. (1998): Weak trophic interactions and the balance of nature. Nature **395**, 794–798
471. McClintock, M.K. (1971): Menstrual synchrony and suppression. Nature **229**, 244–245
472. McNamee, J.E. (1991): Fractal perspectives in pulmonary physiology. J. Appl. Physiol. **71**, 1–8
473. Menge, B.A. (1995): Indirect effects in marine rocky intertidal interaction webs: Patterns and importance. Ecol. Monog. **65**, 21–74
474. Merton, R.K. (1968): The Matthew effect in science. Science **159**, 56–63
475. Metzler, R., Klafter, J., Jortner, J. and Volk, M. (1998): Multiple time scales for dispersive kinetics in early events of peptide folding. Chem. Phys. Lett. **293**, 477–484
476. Micolich, A.P., Taylor, R.P., Davies, A.G., Bird, J.P., Newbury, R., Fromhold, T.M., Ehlert, A., Linke, H., Macks, L.D., Tribe, W.R., Linfield, E.H., Ritchie, D.A., Cooper, J., Aoyagi, Y. and Wilkinson, P.B. (2001): Evolution of fractal patterns during a classical–quantum transition. Phys. Rev. Lett. **87**, 036802
477. Mikiten, T.M., Salingaros, N.A. and Yu, H.-S. (2000): Pavements as embodiments of meaning for a fractal mind. Nexus Network Journal **2**, 61–72 (`www.nexusjournal.com/Miki-Sali-Yu.html`)
478. Milgram, S. (1967): The small-world problem. Psych. Today **1**, 62–67
479. Milgrom, P. and Roberts, J. (1990): Rationalizability, learning and equilibrium in games with strategic complementarities. Econometrica **58**, 1255–1277
480. Milo, R., Shen-Orr, S., Itzkovitz, S., Kashtan, N., Chklovskii, D. and Alon, U. (2002): Network motifs: Simple building blocks of complex networks. Science. **298**, 824–827
481. Milotti, E. (2002): $1/f$ noise: A pedagogical review. `Physics/0204033`
482. Milroy, J. and Milroy, L. (1985): Linguistic change, social network and speaker innovation. J. Linguistics **21**, 339–395
483. Milward, H.B. and Raab, J. (2003): Dark networks as problems. J. Publ. Adm. Res. Theor. **13**, 413–439
484. Mirollo, R.E. and Strogatz, S.H. (1990): Synchronization of pulse-coupled biological oscillators. SIAM J. Appl. Math. **50**, 1645–1662

485. Misteli, T. (2001): Protein dynamics: Implications for nuclear architecture and gene expression. Science **291**, 843–847
486. Molnar, I. (2002): The reliability theoretical aspects of the biological continuity principles. Acta Zool. Acad. Sci. Hung. **48** (S1), 177–196
487. Montesquieu (1734): *Considerations on the Causes of the Greatness of the Romans and Their Decline*, W. Innys and R. Manby, London UK
488. Montoya, J.M. and Sole, R.V. (2003): Topological properties of food webs: From real data to community assembly models. Oikos **102**, 614–622
489. Montroll, E.W. and Shlesinger, M.F. (1982): On $1/f$ noise and other distributions with long tails. Proc. Natl. Acad. Sci. U.S.A. **79**, 3380–3383
490. Moreno, J.L. (1934): *Who Shall Survive?*, Beacon House, Beacon NY, USA
491. Moreno, Y., Gomez, J.B. and Pacheco, A.F. (2002): Instability of scale-free networks under node-breaking avalanches. Europhys. Lett. **58**, 630–636
492. Morishita, Y. and Aihara, K. (2004): Noise-reduction through interaction in gene expression and biochemical reaction processes. J. Theor. Biol. **228**, 315–325
493. Morphy, R., Kay, C. and Rankovic, Z. (2004): From magic bullets to designed multiple ligands. Drug Discovery Today **9**, 641–651
494. Mosekilde, L. (2000): Age-related changes in bone mass, structure, and strength-effects of loading. Z. Rheumatol. **59**, S1–S9
495. Motter, A.E. (2004): Cascade control and defense in complex networks. Phys. Rev. Lett. **93**, 098701
496. Motter, A.E., Zhou, C. and Kurths, J. (2005): Network synchronization, diffusion, and the paradox of heterogeneity. Phys. Rev. E **71**, 016116
497. Myers, C.R. (2003): Software systems as complex networks: Structure, function and evolvability of software collaboration graphs. Phys. Rev. E **68**, 046116
498. Myerson, R. (2001): Learning game theory from John Harsanyi. Games Econ. Behav. **36**, 20–25
499. Nash, J. (1950): Equilibrium points in n-person games. Proc. Natl. Acad. Sci. U.S.A. **36**, 48–49
500. Neckers, L. (2002): Hsp90 inhibitors as novel cancer chemotherapeutic agents. Trends Mol. Med. **8**, S55–S61
501. Néda, Z., Ravasz, E., Vicsek, T., Brechet, Y. and Barabasi, A.-L. (2000): The sound of many hands clapping. Nature **403**, 849–850
502. Nedergaard, M., Ransom, B. and Goldman, S.A. (2003): New roles for astrocytes: Redefining the functional architecture of the brain. Trends Neurosci. **26**, 523–530
503. Neel, J.V. (1962): Diabetes mellitus: A ‘thrifty’ genotype rendered detrimental by ‘progress’? Am. J. Hum. Genet. **14**, 353–362

504. ∗ Neutel, A.-M., Heesterbeek, J.A.P. and de Ruiter, P.C. (2002): Stability in real food webs: Weak links in long loops. Science **296**, 1120–1123
505. Newman, M.E.J. (2003a): Mixing patterns in networks. Phys. Rev. E **67**, 026126
506. ∗∗ Newman, M.E.J. (2003b): The structure and function of complex networks. SIAM Rev. **45**, 167–256
507. Newman, M.E.J. (2003c): Ego-centered networks and the ripple effect, or why all your friends are weird. Social Networks **25**, 83–95
508. Newman, E.A. (2003d): New roles for astrocytes: Regulation of synaptic transmission. Trends Neurosci. **26**, 536–542
509. Newman, M.E.J. (2004): Fast algorithm in detecting community structure in networks. Phys. Rev. E **69**, 066133
510. Newman, M.E.J. and Girvan, M. (2004): Finding and evaluating community structure in networks. Phys. Rev. E **69**, 026113
511. ∗ Newman, M.E.J. and Park, J. (2003): Why social networks are different from other types of networks. Phys. Rev. E **68**, 036122
512. Nishiguchi, T. and Beaudet, A. (1998): The Toyota group and the Aisin fire. Sloan Management Rev. **40**, 49–59
513. Nishikawa, T., Motter, A.E., Lai, Y.-C. and Hoppensteadt, F.C. (2003): Heterogeneity in oscillator networks: Are smaller worlds easier to synchronize? Phys. Rev. Lett. **91**, 014101
514. Noe, R. (1994): A model of coalition formation among male baboons with fighting ability as the crucial parameter. Anim. Behav. **47**, 211–224
515. Nowak, M.A. and Sigmund, K. (2004): Evolutionary dynamics of biological games. Science **303**, 793–799
516. Nowicki, S., Searcy, W.A. and Peters, S. (2002): Quality of song learning affects female response to male bird song. Proc. Roy. Soc. Lond. B **269**, 1949–1954
517. Nyíri, K. (2003): Towards a knowledge society. Digicult. Info **6**, 55–58 (`www.digicult.info`)
518. Ochman, H., Lawrence, J.G. and Groisman, E.A. (2000): Lateral gene transfer and the nature of bacterial innovation. Nature **405**, 299–304
519. O'Donnell, A.J. (2003): Soul of the grid: A cultural biography of the California independent system operator. iUniverse
520. Ogle, J.W. (1866): On the diurnal variations in the temperature of the human body in health. St. George's Hosp. Rep. **1**, 220–245
521. Oleinikova, A., Brovchenko, I., Smolin, N., Krukau, A., Geiger, A. and Winter, R. (2005): The percolation transition of hydration water: From planar hydrophilic surfaces to proteins. `Cond-mat/0505564`. (Phys. Rev. Lett. **95**, 247802)
522. Oltvai, Z.N. and Barabasi, A.-L. (1999): Life's complexity pyramid. Science **298**, 763–764

523. * Onnela, J.-P., Hyvönen, J., Saramaki, J., Kaski, K., Kertész, J., Szabó G. and Barabási, A.L. (2005): Weak links and strong cliques in social network. In preparation
524. * Oono, Y. and Paniconi, M. (1998): Steady state thermodynamics. Progr. Theor. Phys. Suppl. **130**, 29–44
525. Orgel, L.E. (1963): The maintenance of the accuracy of protein synthesis and its relevance to ageing. Proc. Natl. Acad. Sci. U.S.A. **49**, 517–521 (correction **67**, 1476)
526. Orme-Johnson, D.W. and Hayes, C.T. (1981): EEG phase coherence, pure consciousness, creativity, and TM – Sidhi experiences. Int. J. Neurosci. **13**, 211–217
527. Ormerod, P. and Roach, A.P. (2004): The medieval inquisition: Scale-free networks and the suppression of heresy. Physica A **339**, 645–652
528. Ostwald, M.J. (2001): Fractal architecture: Late twentieth century connections between architecture and fractal geometry. Nexus Network Journal, Vol. 3 (`www.nexusjournal.com/Ostwald-Fractal.html`)
529. Otterbein, K. (1968): Internal war: A cross-cultural study. American Anthropologist **70**, 277–289
530. Ottino, J.M. (2004): Engineering complex systems. The emergent properties of complex systems are far removed from the traditional preoccupation of engineers with design and purpose. Nature **427**, 399
531. Otzen, D.E. and Oliveberg, M. (1999): Salt-induced detour through compact regions of the protein folding landscape. Proc. Natl. Acad. Sci. U.S.A. **96**, 11746–11751
532. Ozbudak, E.M., Thattai, M., Kurtser, I., Grossman, A.D. and van Oudenaarden, A. (2002): Regulation of noise in the expression of a single gene. Nat. Genetics **31**, 69–73
533. Packer, C., Tatar, M. and Collins, A. (1998): Reproductive cessation in female mammals. Nature **392**, 807–811
534. Page, R.E. Jr. and Erber, J. (2002): Levels of behavioral organization and the evolution of division of labor. Naturwissenschaften **89**, 91–106
535. Pagel, M. and Mace, R. (2004): The cultural wealth of nations. Nature **428**, 275–278
536. Paine, R.T. (1969): A note on trophic complexity and community stability. Am. Nat. **103**, 91–93
537. Paine, R.T. (1992): Food-web analysis through field measurement of per capita interaction strength. Nature **355**, 73–75
538. Pal, C. (2001): Yeast prions and evolvability. Trends Genet. **17**, 167–169
539. Palla, G., Derenyi, I., Farkas, T. and Vicsek, T. (2004): Statistical mechanics of topological phase transitions in networks. Phys. Rev. E **69**, 046117
540. Palla, G., Derenyi, I., Farkas, T. and Vicsek, T. (2005): Uncovering the overlapping community structure of complex networks in nature and society. Nature **435**, 814–818

541. Pankiw, T., Page, R.E. and Fondrk, M.K. (1998): Brood pheromone stimulates pollen foraging in honey bees (*Apis mellifera*). Behav. Ecol. Sociobiol. **44**, 193–198
542. Panksepp, J. and Burgdorf, J. (2000): 50-kHz chirping (laughter?) in response to conditioned and unconditioned tickle-induced reward in rats: Effects of social housing and genetic variables. Behav. Brain Res. **115**, 25–38
543. Papin, J.A., Reed, J.L. and Palsson, B.O. (2004): Hierarchical thinking in network biology: The unbiased modularization of biochemical networks. Trends Biochem. Sci. **29**, 641–647
544. * Papoian, G.A., Ulander, J., Eastwood, M.P., Luthey-Schulten, Z. and Wolynes, P.G. (2004): Water in protein structure prediction. Proc. Natl. Acad. Sci. U.S.A. **101**, 3352–3357
545. Papp, B., Pal, C. and Hurst, L.D. (2004): Metabolic network analysis of the causes and evolution of enzyme dispensability in yeast. Nature **429**, 661–664
546. Pareto, V. (1897): The new theories of economics. J. Pol. Econ. **5**, 485–502
547. Park, R. and Burgess, E.W. (1925): *The City*, University of Chicago Press
548. Park, D., Lee, S., Boiser, D., Schroeder, M., Lappe, M., Oh, D. and Bhak, J. (2005): Comparative interactomics analysis of protein family interaction networks using PSIMAP (protein structural interactome map). Bioinformatics **21**, 3234–3240
549. Park, K., Lai, Y.-C., Zhao, L. and Ye, N. (2005b): Jamming in complex gradient networks. Phys. Rev. E **71**, 065105
550. Parkinson, C.N. (1957): *Parkinson's Law and Other Studies in Administration*, Houghton Mifflin, Boston MA, USA
551. Paton, W.D.M. and Vizi, E.S. (1969): The inhibitory action of noradrenaline and adrenaline on acetylcholine output by guinea-pig ileum longitudinal muscle strip. Br. J. Pharmacol. **35**, 10–28
552. Paul, G., Tanizawa, T., Havlin, S. and Stanley, H.E. (2004): Optimization of robustness of complex networks. Eur. J. Phys. B **38**, 187–191
553. Paulsson, J., Berg, O.G. and Ehrenberg, M. (2000): Stochastic focusing: Fluctuation-enhanced sensitivity of intracellular regulation. Proc. Natl. Acad. Sci. U.S.A. **97**, 7148–7153
554. Peccei, J.S. (2001): Menopause: adaptation or epiphenomenon? Evol. Anthropol. **10**, 43–57
555. Pedrazza, J.M. and van Oudenaarden, A. (2005): Noise propagation in gene networks. Science **307**, 1965–1969
556. Penna ,T.J., de Oliveira, P.M., Sartorelli, J.C., Goncalves, W.M. and Pinto, R.D. (1995): Long-range anticorrelations and non-Gaussian behavior of a leaky faucet. Phys. Rev. E **52**, R2168–R2171

557. Pereira-Leal, J.B., Audit, B., Peregin-Alvarez, J.M. and Ouzounis, C.A. (2005): An exponential core in the heart of the yeast protein interaction network. Mol. Biol. Evol. **22**, 421–425
558. Perrett, D.I., May, K.A. and Yoshikawa, S. (1994): Facial shape and judgements of female attractiveness. Nature **368**, 239–242
559. Pertsemlidis, A., Soper, A.K., Sorenson, J.M. and Head-Gordon, T. (1999): Evidence for microscopic, long-range hydration forces for a hydrophobic amino acid. Proc. Natl. Acad. Sci. U.S.A. **96**, 481–486
560. Peters, O. and Christensen, K. (2002): Rain: Relaxations in the sky. Phys. Rev. E **66**, 036120
561. Peterson, B.S. and Leckman, J.F. (1998): The temporal dynamics of ticks in Gilles de la Tourette syndrome. Biol. Psychiatry **44**, 1337–1348
562. Petsche, H. (1996): Approaches to verbal, visual and musical creativity by EEG coherence analysis. Int. J. Psychophysiol. **24**, 145–159
563. Pettijohn, T.F. (1999): Popularity in environmental context: Facial feature assessment of American movie actresses. Med. Psychol. **1**, 229–247
564. Pfefferkorn, H.W. (2004): The complexity of mass extinction? Proc. Natl. Acad. Sci. U.S.A. **101**, 12779–12780
565. Pietronero, L. (1987): The fractal structure of the universe: Correlations of galaxies and clusters and the average mass density. Physica A **144**, 257–287
566. Pimm, S.L. and Lawton, J.H. (1978): On feeding on more than one trophic level. Nature **275**, 542–544
567. Plank, L.D. and Harvey, J.D. (1979): Generation time statistics of *Escherichia coli B* measured by synchronous culture techniques. J. Gen. Microbiol. **115**, 69–77
568. Pleh, C. (1998): Computers and personality. In: *Sprache und Verstehen*, ed. by K. Neumer, Österreichische Institut für Semiotik, Wien, Austria
569. ∗ Plotkin, S.S. and Wolynes, P.G. (2003): Buffed energy landscapes: Another solution to the kinetic paradoxes of protein folding. Proc. Natl. Acad. Sci. U.S.A. **100**, 4417–4422
570. Poletaev, A. and Osipenko, L. (2003): General network of natural autoantibodies as immunological homunculus (immunculus). Autoimmune Rev. **2**, 264–271
571. Ponzi, A. and Aizawa, Y. (2000): Criticality and punctuated equilibrium in a spin system model of a financial market. Chaos Solitons Fractals **11**, 1739–1746
572. Portugali, J. (1999): *Self-Organizing and the City*, Springer Verlag, Berlin
573. Potanin, A., Noble, J., Frean, M. and Biddle, R. (2005): Scale-free geometry in object-oriented programs. Commun. ACM, **48**, 99–103
574. Poyatos, J.F. and Hurst, L.D. (2004): How biologically relevant are interaction-based mocules in protein networks? Genome Biol. **5**, R93

575. Pressman, R.S. (1992): *Software Engineering: A Practitioner's Approach*, McGraw-Hill, New York NY, USA
576. Putnam, R.D. (1995): Bowling alone: America's declining social capital. J. Democracy **6**, 65–78
577. * Putnam, R.D. (2000): *Bowling Alone: The Collapse and Revival of American Community*, Simon Schuster New York NY, USA
578. Qin, H., Lu, H.H.S., Wu, W.B. and Li, W-H. (2003): Evolution of the yeast protein interaction network. Proc. Natl. Acad. Sci. U.S.A. **100**, 12820–12824
579. Queitsch, C., Sangster, T.A. and Lindquist, S. (2002): Hsp90 as a capacitor of phenotypic variation. Nature **417**, 618–624
580. Radicchi, F., Castellano, C., Cecconi, F., Loreto, V. and Parisi, D. (2004): Defining and identifying communities in networks. Proc. Natl. Acad. Sci. U.S.A. **101**, 2658–2663
581. Radman, M., Taddei, F. and Matic, I. (2000): Evolution-driving genes. Res. Microbiol. **151**, 91–95
582. Ramos-Fernandez, G., Mateos, J.L., Miramontes, O., Cocho, G., Larralde, H. and Ayala-Orozco, B. (2004): Lévy walk patterns in the foraging movements of spider monkeys (*Ateles geoffroyi*). Behav. Ecol. Sociobiol. **55**, 223–230
583. Rao, F. and Caflisch, A. (2004): The protein folding network. J. Mol. Biol. **342**, 299–306
584. ** Rao, C.V., Wolf, D.M. and Arkin, A.P. (2002): Control, exploitation and tolerance of intracellular noise. Nature **420**, 231–237
585. Rapoport, A. and Horvath, W. (1961): A study of a large sociogram. Behav. Sci. **6**, 279–291
586. * Rauch, E.M. and Bar-Yam, Y. (2004): Theory predicts the uneven distribution of genetic diversity within species. Nature **431**, 449–452
587. * Ravasz, R., Somera, A.L., Mongru, D.A., Oltvai, Z.N. and Barabasi, A.L. (2002): Hierarchical organization of modularity in metabolic networks. Science **297**, 1551–1555
588. Reichardt, J. and Bornholdt, S. (2004): Detecting fuzzy community structures in complex networks with a Potts model. Phys. Rev. Lett. **93**, 218701
589. Reish, O., Orlovski, A., Mashevitz, M., Sher, C., Libman, V., Rosenblat, M. and Avivi, L. (2003): Modified allelic replication in lymphocytes of patients with neurofibromatosis type 1. Cancer Genet. Cytogenet. **143**, 133–139
590. Reynolds, O. (1900): On the action of rain to calm the sea. In: *Papers on Mechanical and Physical Subjects*, pp. 86–88, Cambridge University Press, London
591. Richardson L.F. (1948): Variation in the frequency of fatal quarrels with magnitude. Am. Stat. Assoc. **43**, 523–546

592. Ridley, M. (1998): *The Origins of Virtue: Human Instincts and the Evolution of Cooperation*, Penguin Books
593. Ritossa, F. (1962): A new puffing pattern induced by heat shock and DNP in *Drosophila*. Experientia **18**, 571–573
594. Rivera, M.C. and Lake, J.A. (2004): The ring of life provides evidence for a genome fusion origin of eukaryotes. Nature **431**, 152–155
595. Rives, A.W. and Galitski, T. (2003): Modular organization of cellular networks. Proc. Natl. Acad. Sci. U.S.A. **100**, 1128–1133
596. Roberts S.P. and Feder M. (1999): Natural hyperthermia and expression of the heat shock protein Hsp70 affect developmental abnormalities in *Drosophila melanogaster*. Oecologia **121**, 323–329
597. Roberts, D. and Turcotte, D. (1998): Fractality and self-organized criticality of wars. Fractals **6**, 351–357
598. Rocha, E.P.C., Matic, I. and Taddei, F. (2002): Over-representation of repeats in stress response genes: A strategy to increase versatility under stressful conditions? Nucl. Acids Res. **30**, 1886–1894
599. Rogers, E.M. and Shoemaker, F. (1971): *Communication of Innovations: A Cross-Cultural Approach*, Free Press, New York
600. Rolls, E.T. (1999): *The Brain and Emotion*, Oxford University Press, Oxford UK
601. Rook, K.S. (1984): The negative side of social interaction: Impact on psychological well-being. J. Pers. Soc. Psychol. **46**, 1097–1108
602. Rose, S. (1997): *Lifelines*, Oxford University Press
603. ∗ Rosvall, M., Grönlund, A., Minnhagen, P. and Sneppen, K. (2005): Searchability of networks. `Cond-mat/0505400`. Phys. Rev. E **72**, 046117
604. Rubenstein, R.C. and Zeitlin, P.L. (2000): Sodium 4-phenylbutyrate downregulates Hsc70: Implications for intracellular trafficking of deltaF508-CFTR. Am. J. Physiol. **278**, C259–C267
605. Rupley, J.A. and Careri, G. (1991): Protein hydration and function. Adv. Protein Chem. **41**, 37–172
606. Russell, D.F., Wilkens, L.A. and Moss, F. (1999): Use of behavioural stochastic resonance by paddle fish for feeding. Nature **402**, 291–294
607. Ruthen, R. (1993): Adapting to complexity. Sci. Am. **268** (January), 110–117
608. ∗∗ Rutherford, S.L. and Lindquist, S. (1998): Hsp90 as a capacitor for morphological evolution. Nature **396**, 336–342
609. Ryan, B. and Gross, N.C. (1943): The diffusion of hybrid seed corn in two Iowa communities. Rural Sociol. **8**, 15–24
610. Ryan, W. and Pitman, W. (1998): *Noah's Flood: New Scientific Discoveries about the Event that Changed History*, Simon & Schuster
611. ∗ Sabel, C.F. (2002): Diversity, not specialization: The ties that bind the (new) industrial district. In: *Complexity and Industrial Cluster Dynamics Models in Theory and Practice*, ed. by A. Quadrio Curzio, M.

Fortis and A. Quadrio Quadrio, Physica Verlag, Heidelberg, pp. 107–122

612. ** Salingaros, N.A. (2004): Small-world networks and the fractal city. PLANUM – Eur. J. Planning On-line (`www.planum.net/topics/themesonline-salingaros.html`)
613. Saloma, C., Perez, G.J., Tapang, G., Lim, M. and Palmes-Saloma, C. (2003): Self-organized queuing and scale-free behavior in real escape panic. Proc. Natl. Acad. Sci. U.S.A. **100**, 11947–11952
614. Sangster, T.A., Lindquist, S. and Queitsch, C. (2004): Under cover: causes, effects and implications of Hsp90-mediated genetic capacitance. BioEssays **26**, 348–362
615. Sarkar, S. (2005): Measuring the cosmological density perturbation. Nucl. Phys. B Proc. Suppl. **148**, 1–6
616. Scala, A., Amaral, L.A.N. and Barthelemy, M. (2001): Small-world networks and the conformation space of a short lattice polymer chain. Europhys. Lett. **55**, 594–600
617. Scarrott, G.G. (1998): The formulation of a science of information: An engineering perspective on the natural properties of information. Cybernetics Human Knowl. **5**, 7–17
618. Scharloo, W. (1991): Canalization: Genetic and developmental aspects. Annu. Rev. Ecol. Syst. **22**, 65–93
619. Scheffer, M., Carpenter, S., Foley, J.A., Folke, C. and Walker, B. (2001): Catastrophic shifts in ecosystems. Nature **413**, 591–596
620. Scheibel, A.B. (1985): Falls, motor dysfunction, and correlative neurohistologic changes in the elderly. Clin. Geriatr. Med. **1**, 671–677
621. Scheinkman, J.A. and Woodford, M. (1994): Self-organized criticality and economic fluctuations. Am. Econ. Rev. **84**, 417–421
622. Schidlowski, M. (1988): A 3 800-million-year isotopic record of life from carbon in sedimentary rocks. Nature **333**, 313–318
623. Schiermeier, Q. (2004): Noah's flood. Nature **430**, 718–719
624. Schmalhausen, I.I. (1949): *Factors of Evolution. The Theory of Stabilizing Selection*, Blakiston, Philadelphia PA, USA
625. Schmith, J., Lemke, N., Mombach, J.C.M., Benelli, P., Barcellos, C.K. and Bedin, G.B. (2005): Damage, connectivity and essentiality in protein–protein interaction networks. Physica A **349**, 675–684
626. Schrödinger, E. (1935): Die gegenwärtige Situation in der Quantenmechanik. Naturwissenschaften **23**, 807–812; 823–828, 844–849
627. Schumpeter, J. (1947): The creative response to economic history. J. Econ. History **7**, 149–159
628. Sclavi, B., Sullivan, M., Chance, M.R., Brenowitz, M. and Woodson, S.A. (1998): RNA folding at millisecond intervals by synchrotron hydroxyl radical footprinting. Science **279**, 1940–1943
629. Scott, M.R., Will, R., Ironside, J., Nguyen, H.O., Tremblay, P., DeArmond, S.J. and Prusiner, S.B. (1999): Compelling transgenetic evidence

for transmission of bovine spongiform encephalopathy prions to humans. Proc. Natl. Acad. Sci. U.S.A. **96**, 15137–15142

630. Seligman, A.B. (1997): *The Problem of Trust*, Princeton University Press, Princeton NJ, USA
631. Selye, H. (1955): Stress and disease. Science **122**, 625–631
632. Selye, H. (1956): *The Stress of Life*, McGraw-Hill, New York NY, USA
633. Semmler, J.G. (2001): Motor unit synchronization and neuromuscular performance. Exercise Sport Sci. Rev. **30**, 8–14
634. * Semmler, J.G. and Nordstrom, M.A. (1998): Motor unit discharge and force tremor in skill- and strength-trained individuals. Exp. Brain Res. **119**, 27–38
635. Semmler, J.G., Kornatz, K.W. and Enoka, R.M. (2003): Motor-unit coherence during isometric contractions of a hand muscle are greater in older adults. J. Neurophysiol. **90**, 1346–1349
636. Sergio, F., Newton, I. and Marchesi, L. (2005): Top predators and biodiversity. Nature **436**, 192
637. Serrano, A. and Boguna, M. (2003): Topology of the world trade web. Phys. Rev. E **68**, 015101
638. ** Sethna, J.P., Dahmen, K. and Myers, C.R. (2001): Crackling noise. Nature **410**, 242–250
639. Shanley, D.P. and Kirkwood, B.L. (2001): Evolution of the human menopause. BioEssays **23**, 282–287
640. Shargel, B., Sayama, H., Epstein, I.R. and Bar-Yam, Y. (2003): Optimization of robustness and connectivity in complex networks. Phys. Rev. Lett. **90**, 068701
641. Sharom, J.R., Bellows, D.S. and Tyers, M. (2004): From large networks to small molecules. Curr. Op. Chem. Biol. **8**, 81–90
642. Shenkav, B., Solomon, A., Lancet, D. and Kafri, R. (2005): Early systems biology and prebiotic networks. In: *Transactions on Computational Systems Biology*, ed. by C. Priami et al., LNBI 3380, pp. 14–27, Springer-Verlag, Heidelberg, Germany
643. Sherman, P.W. (1998): The evolution of menopause. Nature **392**, 759–760
644. Shlesinger, M.F. (1987): Fractal time and $1/f$ noise in complex systems. Ann. NY Acad. Sci. **504**, 214–228
645. Shockley, W. (1957): On the statistics of individual variations of productivity in research laboratories. Proc. Inst. Radio Engineers **45**, 279–290
646. Shoichet, B.K., Baase, W.A., Kuroki, R. and Matthews, B.W. (1995): A relationship between protein stability and protein function. Proc. Natl. Acad. Sci. U.S.A. **92**, 452–456
647. Siljak, D.D. (1978): *Large-Scale Dynamic Systems: Stability and Structure*, North Holland, New York NY, USA

648. * Silk, J.B., Alberts, S.C. and Altmann, J. (2003): Social bonds of female baboons enhance infant survival. Science **302**, 1231–1234
649. Simmel, G. (1990): *A Philosophy of Money*, Routledge, Oxford UK
650. Simon, H. (1955): On a class of skew distribution functions. Biometrika **42**, 425–440
651. Simon, S.M., Peskin, C.S. and Oster, G.F. (1992): What drives the translocation of proteins? Proc. Natl. Acad. Sci. U.S.A. **89**, 3770–3774
652. Skinner, B.F. (1937): The distribution of associated words. Psychol. Record **1**, 71–76
653. Skrabski, À., Kopp, M. and Kawachi, I. (2003): Social capital in a changing society: Cross sectional associations with middle-aged female and male mortality rates. J. Epidemiol. Community Heatlh **57**, 114–119
654. Skrabski, À., Kopp, M. and Kawachi, I. (2004): Social capital and collective efficacy in Hungary: Cross-sectional associations with middle-aged female and male mortality rates. J. Epidemiol. Community Heatlh **58**, 340–345
655. Skvoretz, J. and Fararo, T.J. (1989): Connectivity and the small-world problem. In: *The Small World*, ed. by M. Kochen, pp. 296–326, Ablex, Norwood, NJ, U.S.A.
656. Slezak, M. and Pfrieger, F.W. (2003): New roles for astrocytes: Regulation of CNS synaptogenesis. Trends Neurosci. **26**, 531–535
657. * Smith, S.M. and Vela, E. (2001): Environmental context-dependent memory: a review and meta-analysis. Psychon. Bull. Rev. **8**, 203–220
658. Smith, M.J., Kulkarni, S. and Pawson, T. (2004): FF domains of CA150 bind transcription and splicing factors through multiple weak interactions. Mol. Cell. Biol. **24**, 9274–9285
659. Sneppen, K., Bak, P., Flyvbjerg, H. and Jensen, M.H. (1995): Evolution as a self-organized critical phenomenon. Proc. Natl. Acad. Sci. U.S.A. **92**, 5209–5213
660. Sole, R.V. and Fernandez, P. (2005): Modularity 'for free' in genome architecture? `q-bio.GN/0312032`
661. Sole, R.V., Pastor-Satorras, R., Smith, E.D. and Kepler, T. (2002): A model of large-scale proteome evolution. Adv. Complex Syst. **5**, 43–54
662. ** Sole, R.V., Ferrer Cancho, R., Montoya, J.M. and Valverde, S. (2003a): Selection, tinkering and emergence in complex networks. Complexity **8**, 20–33
663. Sole, R.V., Fernandez, P. and Kauffman, S.A. (2003b): Adaptive walks in a gene network model of morphogenesis: Insights into the Cambrian explosion. Int. J. Dev. Biol. **47**, 693–701
664. Solomon, T.H., Weeks, E.R. and Swinney, H.L. (1993): Observation of anomalous diffusion and Levy flights in a two-dimensional rotating flow. Phys. Rev. Lett. **71**, 3975–3978

665. Soma, R., Nozaki, D., Kwak, S. and Yamamoto, N. (2003): $1/f$ noise outperforms white noise in sensitizing baroreflex function in the human brain. Phys. Rev. Lett. **91**, 078101
666. * Song, C., Havlin, S. and Makse, H.A. (2005a): Self-assembly of complex networks. Nature **433**, 392–395
667. Song, C., Havlin, S. and Makse, H.A. (2005b): Fractal growth of complex networks: Repulsion between hubs. `Cond-mat/0507216`
668. * Sornette D. (2002): Predictability of catastrophic events: Material rupture, earthquakes, turbulence, financial crashes, and human birth. Proc. Natl. Acad. Sci. U.S.A. **100**, 12123–12128
669. Sornette, D. (2003): Critical market crashes. Physics Rep. **378**, 1–98
670. Sőti, C., Sreedhar, A.S. and Csermely, P. (2003): Apoptosis, necrosis and cellular senescence: Chaperone occupancy as a potential switch. Aging Cell **2**, 39–45
671. Sőti, C., Pal, C., Papp, B. and Csermely, P. (2005): Molecular chaperones as regulatory elements of cellular networks. Curr. Opin. Cell Biol. **17**, 210–215
672. Spearman, C. (1931): *The Creative Mind*, D. Appleton and Co., The University Press, New York
673. Spehar, B., Clifford, C.W.G., Newell, B.R. and Taylor, R.P. (2003): Universal aesthetic of fractals. Computers Graphics **27**, 813–820
674. Spirin, V. and Mirny, L.A. (2003): Protein complexes and functional modules in molecular networks. Proc. Natl. Acad. Sci. U.S.A. **100**, 12123–12128
675. Sporns, O. (2003): Network analysis, complexity and brain function. Complexity **8**, 56–60
676. Sporns, O. and Edelman, G.M. (1998): Bernstein's dynamic view of the brain: The current problems of modern neurophysiology (1945). Motor Control **2**, 283–305
677. Spudich, J.L. and Koshland, D.E. Jr. (1976): Non-genetic individuality: Chance in the single cell. Nature **262**, 467–471
678. Sreedhar, A.S. and Csermely, P. (2004): Heat shock proteins in the regulation of apoptosis. A comprehensive review. Pharmacology and Therapeutics **101**, 227–257
679. Stam, C.J., Montez, T., Jones, B.F., Rombouts, S.A.R.B., van der Made, Y., Pijnenburg, Y.A.L. and Scheltens, P. (2005): Disturbed fluctuations of resting state EEG synchronization in Alzheimer's disease. Clin. Neurophysiol. **116**, 708–715
680. Stanley, M.H.R., Amaral, L.A.N., Buldyrev, S.V., Havlin, S., Leschhorn, H., Maass, P., Salinger, M.A. and Stanley, H.E. (1996): Scaling behaviour in the growth of companies. Nature **379**, 804–806
681. Stark, D. (1996): Recombinant property in East European capitalism. Am. J. Sociol. **101**, 993–1027

682. Stark, D. and Vedres, B. (2002): Pathways of property transformation: Enterprise network careers in Hungary, 1988–2000. Outline of an analytic strategy. Santa Fe Institute Working Papers No. 200112081
683. Steinglass, P., Weisstub, E. and De-Nour, A.K. (1988): Perceived personal networks as mediators of stress reactions. Am. J. Psychiatry **145**, 1259–1264
684. Stern, W. (1911): *Die differenzielle Psychologie in ihren methodischen Grundlagen*, Fischer, Jena
685. Stewart, J., Varela, F.J. and Coutinho, A. (1989): The relationship between connectivity and tolerance as revealed by computer simulation of the immune network: Some lessons for an understanding of autoimmunity. J. Autoimmun. **2**, S15–S23
686. Steyvers, M. and Tennenbaum, J.B. (2005): The large-scale structure of semantic networks: Statistical analyses and model of semantic growth. Cognitive Sci. **29**, 41–78
687. Stickgold, R., Hobson, J.A., Fosse, R. and Fosse, M. (2001): Sleep, learning, and dreams: Off-line memory reprocessing. Science **294**, 1052–1057
688. * Stiller, J. and Hudson, M. (2005): Weak links and scene cliques within the small world of Shakespeare. J. Cult. Evol. Psychol. **3**, 57–73
689. Stiller, J., Nettle, D. and Dunbar, R.I.M. (2003): The small world of Shakespeare's plays. Human Nature **14**, 397–408
690. Stoycheva, K. (2003): Talent, science and education: How do we cope with uncertainty and ambiguities? In: *Science Education: Talent Recruitment and Public Understanding*, ed. by P. Csermely and L. Lederman, NATO Science Series V/38, IOS press, Amsterdam, pp. 31–43
691. ** Strogatz, S.H. (2003): *Sync*, Hyperion, New York
692. * Strogatz, S.H., Mirollo, R.E. and Matthews, P.C. (1992): Coupled nonlinear oscillators below the synchronization threshold: Relaxation by generalized Landau damping. Phys. Rev. Lett. **68**, 2730–2733
693. Stuart, J.M., Segal, E., Koller, D. and Kim, S.K. (2003): A gene-coexpression network for global discovery of conserved genetic modules. Science **302**, 249–255
694. Stumpf, M.P.H. and Ingram, P.J. (2005): Probability models for degree distributions of protein interaction networks. `www.arxiv.org/q-bio.MN/0507005`. Europhys. Lett. **71**, 152–158
695. Stumpf, M.P.H., Wiuf, C. and May, R.M. (2005): Subnets of scale-free networks are not scale-free: Sampling properties of networks. Proc. Natl. Acad. Sci. U.S.A. **102**, 4221–4224
696. Suki, B., Barabasi, A.L., Hantos, Z., Petak, F. and Stanley, H.E. (1994): Avalanches and power-law behaviour in lung inflation. Nature **368**, 615–618

697. Swaddle, J.P. and Cuthill, I.C. (1995): Asymmetry and human facial attractiveness: Symmetry may not always be beautiful. Proc. Roy. Soc. Lond. B Biol. Sci. **261**, 111–116
698. Swain, P.S., Elowitz, M.B. and Siggia, E.D. (2002): Intrinsic and extrinsic contributions to stochasticity in gene expression. Proc. Natl. Acad. Sci. U.S.A. **99**, 12795–12800
699. Swanson, K.A., Kang, R.S., Stamenova, S.D., Hicke, L. and Radhakrishnan. I. (2003): Solution structure of Vps27 UIM–ubiquitin complex important for endosomal sorting and receptor downregulation. EMBO J. **22**, 4597–4606
700. Szvetelszky, Zs. (2002): *The Gossip*, Gondolat, Budapest, Hungary (in Hungarian)
701. Szvetelszky, Zs. (2003): Ways and transformations of gossip. J. Cult. Evol. Psychol. **2**, 109–122
702. Tanaka, R. (2005): Scale-rich metabolic networks. Phys. Rev. Lett. **94**, 168101
703. Tanaka, S.M., Alam, I.M. and Turner, C.H. (2003): Stochastic resonance in osteogenic response to mechanical loading. FASEB J. **17**, 313–314
704. Tanaka, R., Yi, T.-M. and Doyle, J. (2005): Some protein interaction data do not exhibit power law statistics. www.arxiv.org/q-bio.MN/0506038. FEBS Lett. **579**, 5140–5144
705. Tautz, J., Maier, S., Groh, C., Rössler, W. and Brockmann, A. (2003): Behavioral performance in adult honey bees is influenced by the temperature experienced during their pupal development. Proc. Natl. Acad. Sci. U.S.A. **100**, 7343–7347
706. Taylor, S.E. (2002): *The Tending Instinct: How Nurturing Is Essential to Who We Are and How We Live*, Times Books, New York NY, USA
707. Tegano, D. (1990): Relationship of tolerance of ambiguity and playfulness to creativity. Psychol. Rep. **66**, 1047–1056
708. * Tewari, S. and Toner, J. (2005): Dissipate locally, couple globally: A sharp transition from decoupling to infinite range coupling in Josephson arrays with on-line dissipation. Cond-mat/0508138
709. Thalange, N.K., Foster, P.J., Gill, M.S., Price, D.A. and Clayton, P.E. (1996): Model of normal prepubertal growth. Arch. Dis. Child. **75**, 427–431
710. Theraulaz, G., Bonabeau, E., Solé, R.V., Schatz, B. and Deneubourg, J.-L. (2002): Task partitioning in a ponerine ant. J. Theor. Biol. **215**, 481–489
711. Thirumalai, D. and Guo, Z. (1994): Nucleation mechanism for protein folding and theoretical predictions for hydrogen-exchange labeling experiments. Biopolymers **35**, 137–140
712. Thomas, C.D., Cameron, A., Green, R.E., Bakkenes, M., Beaumont, L.J., Collingham, Y.C., Erasmus, B.F.N., Ferreira de Siqueira, M.,

Grainger, A., Hannah, L., Hughes, L., Huntley, B., van Jaarsveld, A.S., Midgey, G.F., Miles, L., Ortega-Huerta, M.A., Peterson, A.T. and Williams, S.E. (2003): Extinction risk from climate change. Nature **427**, 145–148
713. Tiana, G., Shakhnovich, B.E., Dokholyan, N.V. and Shaknovich, E.I. (2004): Imprint of evolution on protein structures. Proc. Natl. Acad. Sci. U.S.A. **101**, 2846–2851
714. Todorov, A., Mandisodza, A.N., Goren, A. and Hall, C.C. (2005): Inferences of competence from faces predict election outcomes. Science **308**, 1623–1626
715. Tompa, P. (2002): Intrinsically unstructured proteins. Trends Biochem. Sci. **27**, 527–533
716. Tompa, P. and Csermely, P. (2004): The role of structural disorder in RNA and protein chaperone function. FASEB J. **18**, 1169–1175
717. * Tononi, G. and Edelman, G.M. (1998): Consciousness and complexity. Science **282**, 1846–1851
718. Tononi, G., Sporns, O. and Edelman, G.M. (1992): Reentry and the problem of integrating multiple cortical areas: Simulation of dynamic integration in the visual system. Cerebral Cortex **2**, 310–335
719. Tononi, G., Edelman, G.M. and Sporns, O. (1998): Complexity and coherency: Integrating information in the brain. Trends Cognitive Sci. **2**, 474–484
720. Topkis, D.M. (1979): Equilibrium points in nonzero-sum n-person submodular games. SIAM J. Contr. Optim. **17**, 773–787
721. * Toroczkai, Z. and Bassler, K.E. (2004): Jamming is limited in scale-free systems. Nature **428**, 716
722. Trepalingner, E.H., Jarzynski, C., Ritort, F., Crooks, G.E., Bustamante, C.J. and Liphardt, J. (2005): Experimental test of Hatano and Sasa's nonequilibrium steady-state equality. Proc. Natl. Acad. Sci. U.S.A. **101**, 15038–15041
723. True, H.L. and Lindquist, S. (2000): A yeast prion provides a mechanism for genetic variation and phenotypic diversity. Nature **407**, 477–483
724. True, H.L., Berlin, I. and Lindquist, S. (2004): Epigenetic regulation of translation reveals hidden genetic variation to produce complex traits. Nature **431**, 184–187
725. Tsigelny, I.F. and Nigam, S.K. (2004): Complex dynamics of chaperone–protein interactions under cellular stress. Cell. Biochem. Biophys. **40**, 263–276
726. Turcotte, D.L. (1999): Self-organized criticality. Rep. Prog. Phys. **62**, 1377–1429
727. Tyre, M. and Orlikowski, W. (1994): Windows of opportunity: Temporal patterns of technological adaptation in organization. Organization Sci. **5**, 98–118

728. Uptain, S.M. and Lindquist, S. (2002): Prions as protein-based genetic elements. Annu. Rev. Microbiol. **56**, 703–741
729. Usher, M. and Feingold, M. (2000): Stochastic resonance in the speed of memory retrieval. Biol. Cybern. **83**, L11–L16
730. Utasi, Á. (2002): *A bizalom hálója* (The Net of Trust, Hungarian), Új Mandátum Publishers, Budapest, Hungary
731. Uversky, V.N. (2002): Natively unfolded proteins: A point where biology waits for physics. Protein Sci. **11**, 739–756
732. Valente, A. X.C.N., Sarkar, A. and Stone, H.A. (2004): 2-peak and 3-peak optimal complex networks. Phys. Rev. Lett. **92**, 118702
733. Valverde, S., Ferrer Cancho, R. and Sole, R.V. (2002): Scale-free networks from optimal design. Europhys. Lett. **60**, 512–517
734. van Beek, J.H.G.M., Roger, S.A. and Bassingthwaighte, J.B. (1989): Regional myocardial flow heterogeneity explained with fractal networks. Am. J. Physiol. **257**, H1670–H1680
735. van der Stelt, O., Belger, A. and Lieberman, J.A. (2004): Macroscopic fast neuronal oscillations and synchrony in schizophrenia. Proc. Natl. Acad. Sci. U.S.A. **101**, 17567–17568
736. van Galen, G.P. and van Huygevoort, M. (2000): Error, stress and the role of neuromotor noise in space oriented behaviour. Biol. Psychol. **51**, 151–171
737. van Oudenaarden, A. and Theriott, J.A. (1999): Cooperative symmetry-breaking by actin polymerization in a model for cell motility. Nature Cell Biol. **1**, 493–499
738. Varela, F.J. and Coutinho, A. (1991): Second generation immune networks. Immunol. Today **12**, 159–166
739. Varela, F., Andersson, A., Dietrich, G., Sundblad, A., Holmberg, D., Kazatchkine, M. and Coutinho, A. (1991): Population dynamics of natural antibodies in normal and autoimmune individuals. Proc. Natl. Acad. Sci. U.S.A. **88**, 5917–5921
740. Vazquez, A., Flammini, A., Maritan, A. and Vespignani, A. (2002): Modeling of protein interaction networks. Complexus **1**, 38–44
741. Veiel, H.O. (1993): Detrimental effects of kin support networks on the course of depression. J. Abnorm. Psychol. **102**, 419–429
742. Vendruscolo, M., Dokholyan, N.V., Paci, E. and Karplus, M. (2002): Small-world view of the amino acids that play a key role in protein folding. Phys. Rev. E **65**, 061910.1–4
743. Venegas, J.G., Winkler, T., Musch, G., Vidai Melo, M.F., Layfield, D., Tgavalekos, N., Fischman, A.J., Callahan, R.J., Bellani, G. and Harris, R.S. (2005): Self-organized patchiness in asthma as a prelude to catastrophic shifts. Nature **434**, 777–782
744. Vicsek, T. (1989): *Fractal Growth Phenomena*, World Scientific, Singapore

745. Viswanathan, G.M., Afanasyev, V., Buldyrev, S.V., Murphy, E.J., Prince, P.A. and Stanley, H.E. (1996): Levy-flight search patterns of wandering albatrosses. Nature **381**, 413–415
746. * Viswanathan, G.M., Buldyrev, S.V., Havlin, S., da Luz, M.G.E., Raposo, E.P. and Stanley, H.E. (1999): Optimizing the success of random searches. Nature **401**, 911–914
747. Vizi, E.S. (1979): Presynaptic modulation of neurochemical transmission. Prog. Neurobiol. **12**, 181–290
748. Vizi, E.S. (1984): *Non-Synaptic Interactions Between Neurons: Modulation of Neurochemical Transmission*, Wiley, New York NY, USA
749. Vogelstein, B., Lane, D. and Levine, A.J. (2000): Surfing the p53 network. Nature **408**, 307–310
750. Vohradsky, J. (2001): Neural model of the genetic network. J. Biol. Chem. **276**, 36168–36173
751. von Bertalanffy, L. (1950): The theory of open systems in physics and biology. Science **111**, 23–29
752. von Mering, C., Krause, R., Snel, B., Cornell, M., Oliver, S.G., Fields, S. and Bork, P. (2002): Comparative assessment of large-scale data sets of protein–protein interactions. Nature **417**, 399–403
753. von Mering, C., Zdobnov, E.M., Tsoka, S., Cicarelli, F.D., Pereira-Leal, J.B., Ouzounis, C.A. and Bork, P. (2003): Genome evolution reveals biochemical networks and functional modules. Proc. Natl. Acad. Sci. U.S.A. **100**, 15428–15433
754. Voss, R.F. and Clarke, J. (1975): $1/f$ noise in music and speech. Nature **258**, 317–318
755. Vrba, E.S. and Gould, S.J. (1986): The hierarchical expansion of sorting and selection: Sorting and selection cannot be equated. Paleobiology **12**, 217–228
756. Waddington, C.H. (1942): Analization of development and the inheritance of acquired characters. Nature **150**, 563–565
757. Waddington, C.H. (1953): Genetic assimilation of an acquired character. Evolution **7**, 118–126
758. Waddington, C.H. (1959): Evolutionary systems – animal and human. Nature **183**, 1634–1638
759. Wagner, A. (2000): Robustness against mutations in genetic networks of yeast. Nature Genetics **24**, 355–361
760. Wakeham, D.E., Chen, C.-Y., Greene, B., Hwang, P.K. and Brodsky, F.M. (2003): Clathrin self-assembly involves coordinated weak interactions favorable for cellular regulation. EMBO J. **22**, 4980–4990
761. Walker, T.J. (1969): Acoustic synchrony: Two mechanisms in the snowy tree cricket. Science **166**, 891–894
762. Warwick, R.M. and Clarke, K.R. (1993): Increased variability as a symptom of stress in murine communities. J. Exp. Mar. Biol. Ecol. **172**, 215–226

763. Wasserman, S. and Faust, K. (1994): *Social Network Analysis*, Cambridge University Press, Cambridge, UK
764. Watts, D.J. (1999): *Small Worlds. The Dynamics of Networks Between Order and Randomness*, Princetown University Press
765. * Watts, D.J. (2002): A simple model of global cascades on random networks. Proc. Natl. Acad. Sci. U.S.A. **99**, 5766–5771
766. * Watts, D.J. (2003): *Six Degrees. The Science of a Connected Age*, W.W. Norton & Company
767. ** Watts, D.J. and Strogatz, S.H. (1998): Collective dynamics of 'small-world' networks. Nature **393**, 440–442
768. * Watts, D.J., Dodds, P.S. and Newman, M.E.J. (2002): Identity and search in social networks. Science **296**, 1302–1305
769. Weinert, D. (2000): Age-dependent changes of the circadian system. Chronobiology Int. **17**, 261–283
770. Weisbuch, G., De Boer, R.J. and Perelson, A.S. (1990): Localized memories in idiotypic networks. J. Theor. Biol. **146**, 483–499
771. Wellman, B. (2001): Computer networks as social networks. Science **293**, 2031–2034
772. West, B.J. (1990): Physiology in fractal dimensions: Error tolerance. Ann. Biomed. Eng. **18**, 135–149
773. West, B.J. and Deering, W. (1994): Fractal physiology for physicists: Lévy statistics. Phys. Rep. **246**, 1–100
774. West, G.B. and Brown, J.H. (2004): Life's universal scaling laws. Physics Today **57**, 36–42
775. West, G.B., Brown, J.H. and Enquist, B.J. (1997): A general model for the origin of allometric scaling laws in biology. Science **276**, 122–126
776. West, G.B., Woodruff, W.H. and Brown, J.H. (2002): Allometric scaling of metabolic rate from molecules and mitochondria to cells and mammals. Proc. Natl. Acad. Sci. U.S.A. **99**, 2473–2478
777. * White, D.R. and Houseman, M. (2003): The navigability of strong ties: Small worlds, tie strength, and network topology. Complexity **8**, 72–81
778. Wiemken, V. and Boller, T. (2002): *Ectomycorrhiza*: Gene expression, metabolism and the wood-wide web. Curr. Opin. Plant Biol. **5**, 1–7
779. Wiesenfeld, K. and Jaramillo, F. (1998): Minireview of stochastic resonance. Chaos **8**, 539–548
780. Wilhelm, T. and Hanggi, P. (2003): Power-law distributions resulting from finite resources. Physica A **329**, 499–508
781. Wilkins, A. (1997): Canalization: A molecular genetic perspective. BioEssays **19**, 257–262
782. Wilkins, A. (2004): Interview with Gerald M. Edelman, Part II. BioEssays **26**, 326–335

783. Wilkinson, D.M. and Huberman, B.A. (2004): A method for finding communities of related genes. Proc. Natl. Acad. Sci. U.S.A. **101**, 5241–5248
784. Williams, P. (1998): The nature of drug-trafficking networks. Curr. History **97**, 154–159
785. Williams, R.J., Berlow, E.L., Dunne, J.A., Barabasi, A.-L. and Martinez, N.D. (2002): Two degrees of separation in complex food webs. Proc. Natl. Acad. Sci. U.S.A. **99**, 12913–12916
786. Willmore, K.E., Klingenberg, C.P. and Hallgrimsson, B. (2005): The relationship between fluctuating asymmetry and environmental variance in rhesus macaque skulls. Evolution **59**, 898–909
787. Wilson, M.A., Meaux, S., Parker, R. and van Hoof, A. (2005): Genetic interactions between [PSI^+] and nonstop mRNA decay affect phenotypic variation. Proc. Natl. Acad. Sci. U.S.A. **99**, 10244–10249
788. Winfree, A.T. (1967): Biological rhythms and the behavior of populations of coupled oscillators. J. Theor. Biol. **16**, 15–42
789. Woese, C. (1998): The universal ancestor. Proc. Natl. Acad. Sci. U.S.A. **95**, 6854–6859
790. Wright, S. (1932): The roles of mutation, inbreeding, crossbreeding, and selection in evolution. In: Proc. Sixth International Congress on Genetics pp. 355–366
791. Wright, P.E. and Dyson, H.J. (1999): Intrinsically unstructured proteins: Re-assessing the protein structure-function paradigm. J. Mol. Biol. **293**, 321–331
792. Wu, K.K.S., Lahav, O. and Rees, M.J. (1999): The large-scale smoothness of the Universe. Nature **397**, 225–230
793. Yan, G., Zhou, T., Hu, B., Fu, Z.-Q. and Wang, B.-H. (2005): Efficient routing on complex networks. `Cond-mat/0505366`
794. Yang, H., Luo, G., Karnchanaphanurach, P., Louie, T.-M., Rech, I., Cova, S., Xun, L. and Xie, X.S. (2003): Protein conformational dynamics probed by single-molecule electron transfer. Science **302**, 262–266
795. * Yao, W., Fuglevand, A.J. and Enoka, R.M. (2000): Motor-unit synchronization increases EMG amplitude and decreases force steadiness of stimulated contractions. J. Neurophysiol. **83**, 441–452
796. Yashin, A.I., Cypser, J.W., Johnson, T.E., Michalski, A.I., Boyko, S.I. and Novosletsev, V.N. (2002): Heat shock changes the heterogeneity distribution in populations of *Caenorhabditis elegans*: Does it tell us anything about the biological mechanism of stress response? J. Gerontol. **57** A, B83–B92
797. Yazdanbakhsh, M., Kremsner, P.G. and van Ree, R. (2002): Allergy, parasites, and the hygiene hypothesis. Science **296**, 490–494
798. Yip, A.M. and Horvath, S. (2005): The generalized topology overlap matrix for detecting modules in gene networks. `www.genetics.ucla.edu/labs/horvath/GTOM`

799. Yodzis, P. (1981): The stability of real ecosystems. Nature **289**, 674–676
800. Yook, S.H., Jeong, H., Barabasi, A.-L. and Tu, Y. (2001): Weighted evolving networks. Phys. Rev. Lett. **86**, 5835–5838
801. Youdim, M.B.H. and Buccafusco, J.J. (2005): Multi-functional drugs for various CNS targets in the treatment of neurodegenerative disorders. Trends Pharmacol. Sci. **26**, 27–36
802. Zanette, D. H. and Manrubia, S.C. (1997): Role of intermittency in urban development: A model of large-scale city formation. Phys. Rev. Lett. **79**, 523–526
803. Zeldovich, K.B., Berezovsky, I.N. and Shakknovich, E.I. (2005): Physical origins of protein superfamilies. `q-bio.GN/0508036`
804. Zhang, Y., Kihara, D. and Skolnick, J. (2002): Local energy landscape flattening: Parallel hyperbolic Monte Carlo sampling of protein folding. Proteins **48**, 192–201
805. Zhao, M., Zhou, T., Wang, B.-H. and Wang, W.-X. (2005): Enhance synchronizability by structural perturbations. `Cond-mat/0507221` (Phys. Rev. E **72**, 057102)
806. Zheng, W., Buhlmann, P. and Jacobs, H.O. (2004): Sequential shape-and-solder-directed self-assembly of functional microsystems. Proc. Natl. Acad. Sci. U.S.A. **101**, 12814–12817
807. Zhou, S. and Mondragon, R.J. (2003): The rich club phenomenon in internet topology. IEEE Commun. Lett. (`www.arxiv/cs.NI/0308036`)
808. Zipf, G.K. (1949): *Human Behaviour and the Principle of Least Effort: An Introduction to Human Ecology*, Addison-Wesley, Cambridge MA, USA

Index

absolutism, 252
activation energy, 104, 116–118, 121, 284, 287
  lowering, 286
adaptive response, 48, 59, 77, 129, 184
adolescence, 171, 172
Afghanistan, 195, 196, 279
Africa, 253
African buffalo, 183
Afro-American, 2, 193
age relativism, 214
aggregation, 114, 115, 128, 130, 135, 136
aging, 19, 152–156, 165, 279
  of cellular network, 153
  weak-link theory, 155
air-conditioning revolution, 216
Aisin crisis, 202, 204
Al Qaeda, 208
Albania, 196
albatross, 27
alcohol, 169
alienation, 13, 193, 232
allergy, 160
allometric scaling law, 22, 162, 264, *see Glossary*
alpha male, 78
alternative medicine, 152
altruism, VII, 210, 217, 231, 256
Alzheimer's disease, 136, 168
ambiguity, 110, 205, 221–224
  of individual, 257
  tolerance of, 175
ambiguity-induced stabilization, 223
amplitude synchrony, 86
anarchy, 251
angiogenesis, 150
animal communities, 181–184, 278
ant, 27, 182, 192, 198, 232
antibody, 158, 229
antigen, 110, 158, 159
  hunger, 159
antipattern, 239
aorta, 24
apoptosis, 78, 141, 279
appraiser, 300
Arabidopsis thaliana, 93, 132
arbitration, 223
arbitrator, 300
architecture, 232, 233, 279
  Gothic, 234
  human-friendly, 235
argon, 118
arms dealing, 207
arrogance, 196
art, 30, 176, 224 231, 278, 310
arterial tree, 162
assimilation, 255
assortative network, 39, 84
assortativity, 69, 178, *see Glossary*
  in social network, 189, 192, 199, 301, 305
  negative, 301
astrocyte as weak link, 168
astrology, 84
asymmetry, 149
atherosclerosis, 254
atom, XI, 33
ATP, 11, 78

hydrolysis, 120
attack tolerance, 69
attention-deficit disorder, 168
attractor, 103, *see Glossary*
authoritarianism, 88
autism, 168, 171
autoantibody, 158, 161
autoimmune disease, 159, 161
autoimmunity, 160, 161
avalanche, 61, 66, 96, 122, 188, 287

baboon, 183–185, 193
baby quake, 65
Bach, J.S., 30
Balkans, 195, 279
Barkhauser effect, 61
bearish market, 244
beauty, 30, 64, 149, 176, 233, 234
bee, 27, 181–183, 192, 198, 232
belief, 310
benevolence, 199, 260
Bernoulli law, 25
bet-hedging, 26, 247
fund, 26
betweenness centrality, 37, 38, 70, *see Glossary*
big phenotype, 173, 196, 254, 279
binary evolution game, 301
biological network, 16, 77, 301
bird lifespans, 264
birdsong, 149, 294
black noise, 51
blogs, 214
blood vessel, 143, 150, 162
Boeing 777, 91, 289, 290
bone, 110, 154
growth, 54
Boolean network, 158
bottom network, *see* network, bottom
brain, 9, 12, 22, 111, 159, 166, 183, 279
cloned, 128
functions, 167
oscillations, 168
plasticity, 188
weak links, 169
brainstorming, 294
breathing, 62
brown noise, 49–51
Brownian
motion, 49, 51, 61, 120
music, 30
search strategy, 28
Budapest, 231
buffering, IX, 131, 132, 134–136, 141, 144, 279, 289
weak-link-induced, 145
bullish market, 244, 245
business network, 201

cafeteria, 204, 205
call-ins, 212
canalization, 144, *see Glossary*
cancer, X, 134, 150–151, 279
Cantor Fitzgerald, 205
carbon monoxide, 122, 168
caspase, 141
catharsis, 226, 230, 280
Catholic Church, 200
cell, VII, 28, 33, 44, 70, 72, 110–115, 290, 299
cycle, 141
death, 78, 106, 140, 278
division, 41
immune, 110, 157
pacemaker, 83
stability, 125–156
synchrony, 83
cellular network, XI, 33, 77, 125–156, 279, 301
chain of command, 204
chaperone, VII–IX, XI, 78, 120, 121, 128–156, 279, *see Glossary*
as hub, 133
overload, 155
chat party, 214
chatting, 195, 216
cheats, 256
chemical bond, 113

child exploitation, 256
childbirth, 165, 279
China, 216, 273
Chinese, 193
Chinese medicine, 152
chronic resentment, 273
circadian rhythm, 81, 86
circulation, 150
cities, 16, 39, 189, 210, 231–237
civilization disease, 155, 254
clapping audience, 81, 88, 188, 280
climate, 53
  warming, 271
clone selection theory, 160
cluster analysis model, 235
clustering, 10, 11, 27, 36, 37, *see Glossary*
  fuzzy, 37
  in language network, 219
  in social network, 189, 192, 195, 199, 301, 305
  superparamagnetic, 37
cognition, 52, 106
cognitive
  dimensions, 10, 175, *see Glossary*
  flexibility, 196, 197
  isolation, 197
  links, 310
cohesion, 193, 198, 210, 261
coin tossing, 25
collective behavior, 244–249
color, 169
combination therapy, 152
communism, 253
competition, 79, 148, 216, 258
competitiveness, 259
complex network, 37, 108, 142
  aging, 154
  relaxation, 291
  stability, 289–292
complexity, 41, 42, 52, 59, 61, 91, 97, 106–108, 111, 148, 155, 288, 290
  in literature, 225, 227
  in neural network, 170
  in software networks, 239
  of language, 220
complicated system, 91, 107, 111
condensation, 75, 78, 79
confidence, 260
conflict, 192, 195, 196
consciousness, 169
  in game theory, 299
conservative parties, 197
consumerism, 249
cooperation, 256
correlated percolation model, 235
cortisol, 153
cosmological principle, 306
coughing, 62, 278
crackling noise, 49
creativity, 168, 176, 223, 290
Creutzfeldt–Jakob disease, 135
cricket, 53
Croatia, 88, 258
crying, 153, 278, 280
cultural diversity, 272
culture quake, 65
cytokine, 157
cytoskeletal network, 125, 139

dark energy and matter, 306
dark network, 207–209
data bases, 201
death, 57, 66, 76, 78, 89, 93, 106, 142, 252, 308
decision-making, 41, 52
deductive approach, 97, 142
deer, 27, 232
degeneracy, 108–109, 145, 162, 183, *see Glossary*
  in language network, 224
  in organizational networks, 202
  in software networks, 239
degree, 14, 44, *see Glossary*
  distribution, 76, 127, 304
  maximal, 20
democracy, 191, 251, 253
  exportation of, 253, 254, 279

dendogram, 36
depression, 176
design, 66, 90, 92, 240
  of software, 238–240
designability, 92, 289, 292
diabetes, 156, 173, 254
diamond, 57
dictatorship, 75, 78, 184, 251, 256
diegetic landscape, 241, 279
  rugged, 289
diffuse problem, 207
diffusion, 61
diffusion-limited aggregation, 235
dilution effect, 159, 199
dinosaur, 293
disassortative network, 40, 58, 70
disassortativity of world markets, 243
disease, 151–152
  civilization, 155, 254
disintegration, 105, 106, 140, 279
  of society, 251
disposable
  contacts, 213
  grooming, 212
  love, 213
  soma theory, 152
distance in social network, 192
distribution, 13
disulfide bridge, 114, 123
divergence, 39, 84, 192
diversity, VIII, 26, 97, 104, 105, 122, 132, 133, 139, 184, 217, 273, 279
  as stabilizer, 193, 198
  cultural, 256, 272
  developmental, 132, 134
  genetic, 266
  hidden, 289, 290, 292
  in ecological networks, 265, 266, 271
  in modern world, 256
  management, 149
  morphological, IX
  optimization, 136, 156
  phenotypic, 132
  portfolio, 247
  stress-induced, 141–143
  tolerance, 279
diversity–stability debate, 266, 267
division of labor, 71, 192, 217, 260
DNA, 22, 72, 93, 126, 135, 141, 146, 151
  lateral transfer, 265
dog, 191, 221, 248
dolphin, 183, 185
domino effect, 68
Down syndrome, 184
dreams, 84, 169
Drosophila melanogaster, 131, 150
Drosophila melanogaster, VIII, 128, 130, 133, 134, 141, 143
drug trafficking, 207
duplication, 71
duplication and divergence, 20

earthquake, 24, 61, 68
eco-landscape, 269
ecological network, 16, 263–274
ecology, 102
economic collapse, 42, 68, 243, 246
economics, 52, 77, 243–247, 284
ecosystem, 154, 174, 301
  management, 272
80–20 rule, 18, 101
elections, 150, 177, 197
electrical synctium, 169
electroencephalogram (EEG), 85, 168
electronic devices, 52
element, 98, 282, *see Glossary*
  degenerate, 108
elementary particle, 305
embargos, 247
embryogenesis, 130, 149
emergent property, 4, 90, 104, *see Glossary*
emotional trauma, 172
emotions, 61, 169

empathy, 231
energy landscape, 115–119, 241, 279, 284, 288
  buffed, 118
energy network, 116–118, 306
engineering, 90–92, 166, 176, 240, 279
  as an art, 237, 240
  network-driven, 241
enthalpy, 112
entropy, 67, 112, 113
environment, 105
enzyme, 70, 110, 121, 123
epigenetic inheritance, 135, 141, 153, *see Glossary*
epilepsy, 170
error
  catastrophe, 151
  tolerance, 21, 69, 289
Escherichia coli, 16, 38, 43, 132
Ethernet traffic, 43
ethology, 78
evolution, 34, 38, 55, 69, 91, 108, 120, 144, 147, 148, 162, 202, 267, 284, 288, 302
  as tinkerer, 91, 166
  cultural, 45
  molecular, 148
  of evolvability, 293, 305
  reversed, 93
evolutionary
  continuity, 148
  failure, 202
  heritage, 278
  jump, 143, 144, 148, 156
  network, 92, 301
  response, 173
evolvability, 144–146, 266, 305, *see Glossary*
  in complex network, 292–294
  of society, 258
  of software networks, 238, 240
  optimal, 148
exponential cutoff, 15, 22, 63, *see Glossary*
extinction, 42, 270, 271
  of dinosaur, 293
  partial, 266
eye development, 149

factor analysis, 170
facultative essentiality, 265
fairy tales, 153
fasting stress, 153
feed-forward loop, 238
female genital mutilation, 256
fertility, 184, 185
firefly, 81
firm quake, 203
firms, 201–207, 278
fish, 53, 54
fitness, 19, 37, 55, 148, *see Glossary*
  landscape, 279, 284, 288, 290
  of society, 198
fitting data, 15
flash mob, 214
flexibility, 260
flicker noise, 49
fluctuation–dissipation theorem, 61
folding trap, 115, 118
food web, 16, 72, 264, 265, 268
Ford, 205
forest, 265, 267
  fire, 61, 273
fractal, 21, 25, 36, 43, 137, 153, 162, 232–234, 248, 306, *see Glossary*
  architecture, 234
  cities, 235, 236
  dimension, 21, *see Glossary*
fracture, 61, 73
fragmentation, 203, 210
free energy, 67, 112, 114
French absolutism, 259
Freudian revolution, 170
friends, 187, 189, 196, 199, 201, 300
friendship, 100
  network, XII

fringe area, 35, 37, 39, 167, 279, *see Glossary*
  in business, 202
  in software networks, 238
  in towns, 237
fundamentalist strategy, 244

Gaia, 42, 154, 270–274, 293, 295, 306
game theory, 256, 282, 296–302
  renaissance, 305
games, 31, 278
gangs, 215
gays, 193
gender relativism, 214, 280
gene, 72, 102, 133, 134, 137, 151, 233
  duplication, 131
  network, 49, 59
  transcription, 49, 110, 126, 127, 142
General Motors, 205
genetic
  diversity, 182, 266
  drift, *see Glossary*
  information, 146
  malformation, 184
  network, 109, 138, 139
  stability, 150
genius, 168
genome cleansing, 141, *see Glossary*
giant component, 48, 56, 69, 70, 76, 80, 103, 282, *see Glossary*
  disappearance, 76
  in organization, 203
  of social network, 188
gifted people, 153, 168, 172, 193
Girvan–Newman method, 38
glial cell, 166
gliovascular unit, 168
global
  communication, 58, 59, 66
  connectedness, 87, 118
  village, 243, 247
globalization, 191, 253, 255, 258
glycolysis, 150
goal-setting, 206
God, 295, 311
gossip, 105, 209, 210, 278, 280, 300
  quake, 63
Gothic architecture, 234
grandmother effect, 184, 185, 279
gravity, 306
Greek and Roman mythology, 228
grooming, 183, 184, 193
growth
  arrest, 92
  quake, 65
  rate, 22
Gutenberg–Richter law, 24
gypsies, 193

hamster, 86
handicapped people, 193
hare and lynx populations, 81, 264, 268
heart
  failure, 163, 259
  rate, 22, 65, 83, 90, 162
herding behavior, 61, 187, 244, 245, 247–249, 261, 273
heresy, 200
hierarchy, 33, 39, 59, 123, 127
  age-related, 214
  in insect networks, 182
  in organizational networks, 202
  in social networks, 188
  in software networks, 238
high blood pressure, 254, 259
highly optimized tolerance, 90, 241
history, 247–258
  end of, 253
  turning points, 278
HIV, 188, 200
holism, 280
holon, 34
homunculus, 158
horizontal contacts, 203
hospitals, 105, 152
housekeeping heat, 55, 93
Hsp60, 132
Hsp70, 132

Hsp90, VIII, 130, 132, 141, 144, 148
  inhibition, 130
hub, 12, 15, 20, 24, 39, 69, 70, 84, 133, 153, 304, *see Glossary*
  date, 44, 304
  party, 44
humanities, 176
Hungary, 187, 193
hunter–gatherer societies, 250
hydrogen bond, 113, 121, 123
hydrophobic collapse, 113
hypoxia, 150

idiotypic network, 157, 159
immigration, 255
immunculus, 158, 161
immune
  cell, *see* cell
  deficiency, 160
  response, 110, 143, 150, 157, 161, 308
  system, 157–161, 273
immunological
  disease, 159
  network, 157–161, 230
India, 216, 273
Indian medicine, 152
indirect effect, 102
induction, 97
infection, 57
inflammation, 150, 160
information network, 16
information relativism, 215
innovation, 66, 202, 254, 260, 284, 290
  landscape, 241, 279, 284, 289, 290
  Schumpeterian clustering, 66
  spread of, 199, 222, 259
innovative potential, 205, 206
innovativeness, 259
Inquisition, 200
insect networks, 181
institutions, 285
integration, 255
interdisciplinary research, 176, 200
Internet, 16, 42, 49, 166, 210
ion channels, 53
irregularity, 97, 98
Islamites, 193
isolation, 52, 202, 210, 308
  in modern society, 216

jackal, 27, 232
Jewish people, 193
joint stock companies, 257
Joseph effect, 26
Josephson effect, 86
jump length, 291
Jung, C., 84, 168, 170

keystone species, 264, *see Glossary*
Kleiber law, 23, 162
Kleinberg model, 12, 13
knowledge network, 31
Koch curve, 23
Kuramoto model, 86
Kurds, 193

landslide, 61
language network, 219–224
lattice-type network, 11, 12, 20, 57
laughter, 278, 280
  quake, 88, 226
law-abidance, 191
laws, 105
Le Chatelier principle, 103, 272, *see Glossary*
  for networks, 283, 293, 305
leadership collaboration, 202
learning, 54, 82, 169, 278
Lehman's second law, 239
lesbians, 193
Levinthal paradox, 111, 112
Levy flight, 27–29, 123, 175, 232, 233, 278, 291, 292
liberal parties, 197
life, 27, 34, 38, 67, 102, 272
  network, 308
lifespan, 22, 162
lightning, 62, 278

link, 56, *see Glossary*
  management, 310
  relativism, 216, 302
  strength, 3, 44, 92, 100, 137, 207, 212, 228, 243, 300, 304
lion, 185
London Millennium bridge, 81
loneliness, 176, 259, 308
long-range contacts, 10, 11, 43, 96, 120, 168, 173, 191, 198, 199, 217, 259, 287
longevity, 259
Lotka law, 16
love, 311
lung, 162
  quake, 62
lyophilization, 121

macaque, 183
madhouses, 105
magnetization, 61
mall kiss, 212
marginal utility, 26
market
  behavior, 212
  diversity, 247
  dynamics, 244
  guru, 245, 246, 280
  investment, 43, 244, 246
  quake, 244–246
  stability, 245
Markov process, 51
Marx, K., 252
masterpieces, 226, 227, 230
  nested sync, 230
mathematics, 172, 176, 200
Matthew effect, 18, 235
meaning, 97, 110, 219–221
mediator, 300, 302
medieval society, 257
meditation, 85, 168
membrane, 125
memory, 54, 81, 82, 169
  immunological, 158
  landscape, 123
men, 100, 186, 194
  under stress, 259
menopause, 184–186, 279
menstrual cycle, 81, 88
meritocracy, 251
metabolic
  enzyme, 38
  network, 39, 70, 78, 110, 126, 127, 139
  rate, 22, 162
metastasis, 143, 151
Mexican wave, *see* waves in stadiums
microchip, 49, 241
microdiversity, 265
Middle East, 195, 196, 279
Milgram experiment, 7, 12
minorities, 193, 256
minority game, 301
mistakes, 167
mitochondria, 22, 126, 138, 141, 300
mobile phone, 211
moderator, 300
modern life, 216, 261
modernity, 257
modular network, 35
modularity, 59, 66, 73, 83, 91, 127
  in language network, 221–222
  in organizational networks, 202
  in social networks, 188
  in software networks, 238
  of energy landscape, 118
  of immune system, 158
  of towns, 235
module, 3, 17, 35–43, 55, 108, 109, 127, 146, *see Glossary*
  formation, 36, 39
  in neural network, 167
  in software networks, 238
mole-rat, 232, 233
molecular
  crowding, 136, *see Glossary*
  evolution, 148
  network, 16, 111–124
molten globule, 113, 115, 116

Mona Lisa, 149
money, 257
monkey, 27, 194, 232, 248
mono-landscape, 297, 300
monofractal, 25
monomaniac, 171
motif, 55, 108, 109, 127, 146, 228, *see Glossary*
  in software networks, 238
motor unit, 110, 163, 279
  synchrony, 164, 165
mouse-to-elephant curve, 23
movie stars, 176
Mozart, W.A., 30, 54, 85, 230
multi-landscape, 296–303
multi-talented people, 171
multicellular organism, 148
multifractal, 25, 65, 137, 163, *see Glossary*
multilevel selection, 146
multiscale distribution, 17
multitarget drug, 151, 152, 279
muscle, 110
  network, 163–166
music, 30, 54, 82, 220, 278
  atonal, 30
musician, 164, 165
mutation, IX, 38, 130, 131, 150
  accumulation theory, 152
mycorrhiza, 265, 267
mythology network, 228

nanotechnology, 241
Nash equilibrium, 282, 296
nature conservation, 264
Neanderthal, 269
Nebuchadnezzar, 62
necrosis, 78
negative feedback, 55, 90, 138, 270, *see Glossary*
negotiator, 300
neighbors, 309
nested sync, 84, 85, 230, 279, *see Glossary*
nestedness, 29, 32–44, 66, 80, 82, 105, 286, 291, *see Glossary*
  of towns, 236
netquake, 60–65, 273, 285, *see Glossary*
netsistance, 103, 140, 203, 210, 282, *see Glossary*
network, 5, *see Glossary*
  bottom, 32, 33, 40, 42, 80, 92, 96, 104, 105, 146, 258, 286, 287, 296
  communication, 96
  diameter, *see Glossary*
  element, *see* element
  failure, 68–74, 91
  frozen, 293
  integrity, 107, 133, 282
  motif, *see* motif
  navigation, 12, 20
  plastic, 293
  properties, 6
  resilience, 48, 104
  skeleton, 22
  stability, 47, 49, 55, 58, 66, 96, 103, 118, 282, 286, 287
  stabilization, 90–93
  survival, 80
  top, 32, 33, 38, 40, 42, 84, 105, 258, 286, 296
  topology, 20, 146
  traffic, 40
  transition, 20
neural network, 10, 33, 110, 166–170
neuro-glial network, 166–170
neurodegenerative disease, 135
neuron, 110, 166, 168
  death, 111
  mirror, 170
nitric oxide, 168
nitrogen, 113
Noah effect, 24
node, 15, 56, *see Glossary*
noise, 47–55, 59, 96, 97, 104, 133–135, 146, 155, 279, 287

buffer, 156
extrinsic, 49
generator, 156
in muscle network, 163
intrinsic, 49
management, 148, 156
of aging, 153
optimal, 55, 136, 149, 156
spectrum, 50
traders, 244
non-equilibrium systems, 55
norms, 105
novel quake, 228
nucleoplasmin, 130
nutrition, 255

obesity, 173, 254
Occam's razor, 294
ogling quake, 63
omnivory, 267–270, 288
opinion, 199
formation, 52, 256
organelle, 110, 125, 138
diversity, 138, 279
network, 139
organizational network, 201, 202
oscillator, 52, 86
chemical, 81
frequency distribution, 88
network, 82, 83
osteocyte, 154
osteoporosis, 154
overconnectedness, 72, 176, 199, 266, 269
in immune system, 159
ownership, 206
oxidation, 300
oxygen, 113

p53, 134, 138
painting, 30
Paleolithic, 173, 250
Panglossian paradigm, 92
panic, 61
quake, 244

parcellation, 35, 42
Pareto law, 16, 18, 27, 101
Paris, 236
Parkinson law, 41
Parkinson's disease, 136, 165
parsimony principle, 77, 79
particle network, 306
path length, 219, *see Glossary*
patience, 196, 254, 256, 273
pendulum clock, 80
percolation, 48, 57, 66, 103, *see Glossary*
in hydrogen-bond network, 121
in immune system, 159
in insect networks, 182
in social network, 188
peristaltic movements, 90
Permian catastrophe, 271
persecution, 200
personal network, 201, 217
personality disorder, 172
perspective-taking ability, 201
perturbation, 21, 39, 47–55, 70, 129, 283
dissipation, 56–67, 87, 148
phase synchrony, 86
pheromones, 184
phone calls, 16, 179, 210
physicians, 105
pidgin English, 221, 222
pidgin formalization, 238, 240, 279
pink noise, 50, 51, 53, 54, 60, 63, 123, 151, 159, 278, 279, *see Glossary*
in novels, 228
pit bull, 263
pleiotropy theory, 152
Poisson distribution, 17, 20, 120
political movements, 215
politicians, 149, 187, 195, 200
positivism, 310
postmodern tribes, 215
potassium chloride, 118
pottery, 66

Potts model, 37
poverty, 153, 254
power law, 14, 16, 22, 24, 28, 29, 50, 76, 162
power network, 16, 19, 72
  failure, 68
  quake, 68
prebiotic network, 79
predator, 263, 266, 268
preferential attachment, 19, 27, 44
prey, 263
price fluctuations, 243
primes, 170
prion, 111, 135, 265
prisons, 105
probability, 24, 27, 43
problem solving, 205
protein, VII, 17, 33, 44, 70, 110, 241, 279, 299
  backbone, 113
  complex, 124, 125
  conformation, 111, 112, 115–117
  designability, 289
  domain, 16, 120
  dynamics, 52, 123
  evolution, 148
  folding, 39, 111–115, 121, 134, 284, 288
  folding pathway, 118
  heat shock, 128
  hydrophobic core, 114, 115, 120
  memory, 121
  mobility, 121
  network, XI, 16, 33, 133, 138, 139, 144, 150
  quake, 61, 122
  stability, 284
  stabilization, 123
  stress, 128
  unstructured, 138
proton, 113
pseudo-grooming, 194, 195, 209–217, 280
pseudo^2-grooming, 210, 211
pseudo^3-grooming, 212
pseudo-strong link, 215
psychic recovery, 88
psycho quake, 171, 278
psychological network, 170–179
psychology, 105, 170–179
psychotherapy, 178
punctuated equilibrium, 91, 122, 144, 148, 241, 250, 284, *see Glossary*

quantum mechanics, 17, 113
quasar emissions, 52, 61
queen bee, 182
queuing, 26

racist pressure, 153
radio, 212
rain, 24, 52, 61, 174
random graph, 10, 11, 17, 18, 20, 75–78, 147, 171, 176, *see Glossary*
  in history, 250
  relaxation, 291
rapid eye movement, 169
rat, 53
  laughter, 88
recognition, 190
recombination, 146
redirected growth, 273
reductionism, 280
redundancy, 145, 146, *see Glossary*
  in dark networks, 209
  in language network, 221, 224
  in organizational networks, 202
reentry, 170
refactoring, 239
regular lattice, *see Glossary*
Reims Cathedral, 234
relativization, 261
  of link strength, 216, 280
relaxation, 28, 56–67, 80, 88, 89, 96, 104, 118, 129, 133, 175, 250, 272, 273, 280
  in complex network, 291

in immune system, 159
in market dynamics, 244
in random network, 291
psychic, 226, 230
religion, 176, 257, 310
religious sects, 193, 215
replacement, 146
replication, 150
research institutes, 105
resilience of ecosystems, 270
resourcefulness, 259
resources, 20, 41, 75, 78, 79, 103, 182, 250, 251, 254
abundance, 254, 264
revolution, 251
French, 252
ribosome, 115
rich club, 39, 178
Richardson law, 248
Richter scale, 24
RNA, 24, 49, 93, 110, 115, 126, 133
interference, 72
stabilization, 123
robustness, 55, 91, *see Glossary*
of democracy, 253
of ecological networks, 267, 270
Rome, 236
Romeo and Juliet, 195
rules, 191
rumor, 57

saddle, *see* activation energy
sailing-ship effect, 206
saint, 168
salt bridge, 123
saltatoric growth, 65
sampling, 17
sardinella, 183
scale-freeness, 5, 12–32, 43, 49–51, 57, 59, 60, 66, 69, 75–78, 80, 84, 100, 123, 147, 176, 278, *see Glossary*
in architecture, 233, 280
in art, 228
in history, 251
in immune system, 158, 159, 161
in language network, 219
in muscle network, 163
in neural network, 167
in organizational networks, 202
in social networks, 188
in towns, 235
of genetic diversity, 266
of link strength, 304
of terrorist attacks, 209
of wars, 248
of world trade, 243
scaling exponent, 14, 16, 22, 28, 50, 76, 162
schizophrenia, 168
school, 31
scientific
collaboration, 16, 43, 200
discovery, 293
language, 221
method, 97, 106, 310
policy, 290
productivity, 16
progress, 285
publications, 5, 15, 16
reductionism, 280
scrapie, 135
sculpture, 30
segmentation, 193, 195
segmented society, 192, 195
segregation, 105, 158, 193, 255, 287
in modern society, 215
self-assembly, 241
self-efficacy, 259
self-organization, 18–24, 27, 35, 40, 41, 55, 59, 105, 289, 294
self-organized criticality, 52, 59–67, 83, 88, 219, 278, 285, *see Glossary*
in firms, 203
in market dynamics, 244
in social networks, 188
in war, 248
self-regulation, 270

self-similarity, 21, 30, 43
self-stabilization, 309
senses, 110
sex, 16, 34, 278
in primates, 212
quake, 64
sexual contact network, 16, 188
Shakespeare, W., 30, 224, 225, 245
shark, 266
signal, 48, 59
threshold, 53, 54
transduction, VIII, 114
signal-to-noise ratio, 52, 157, 163
signaling network, 110, 126
silent mutation, VIII, IX, 128, 130–132, 136, 140, 141, 147, 148, *see Glossary*
sixth-order thinking, 227, 245, 280, 296
slander, 210, 280
slang, 221
slaved process, 121
slavery, 256
sleep, 169, 170
small phenotype, 173, 196, 197, 254, 255, 279
small-talk, 209, 224
small-worldness, 5, 7–13, 43, 66, 80, 84, 96, 119, 127, *see Glossary*
in literature, 225
of energy landscape, 117
of language network, 219
of neural network, 167
of organizational networks, 202
of social networks, 187
of world trade, 243
smell, 169
SMS
messages, 179, 211
votes, 212
social
capital, 198, 258–262, 280
circles, 13, 189, 202
classes, 252
cohesion, 2
dimensions, 10, 191, 199, 221, *see Glossary*
network, 10, 16, 21, 33, 41, 44, 75, 154, 177–179, 181–217, 258, 301–302
search, 199
stability, 100
socialization, 297
society quake, 248, 251, 255
socioeconomic status, 12
software, 16, 279, 285
network, 238–240
solar flares, 52, 61
solidarity, 198, 261
South Sea Company scandal, 243
specialization, 253, 260
specitation, 265
speculative bubble, 244
speech, 220
spending habits, 173, 176
spirituality, 215
St Petersburg paradox, 25, 26
St. Peter's Cathedral, 237
St. Stephen, 193, 255
stability landscape, 116, 122, 148, 279, 282, 284, 294–310, *see Glossary*
accessibility, 286–288
rugged, 284, 288–292
smooth, 286, 291
star network, 20, 39, 75–78, 147, 171, 176, 251
in history, 251
stasis, 284, 290, 291
sterility of modern world, 160
stochastic resonance, 52–54, 86, 136, 172, 278
Stokes–Einstein relation, 61
stress, 39, 75, 77–79, 88, 129, 139–149, 153, 172, 204, 258, 289, *see Glossary*
in bacteria, X
in ecosystems, 265

in plants, 143
management, 148
strong link, 2, 79, 86, 96, 99, 191, 198, 299
as stabilizer, 100
detrimental effect, 259
erosion of, 216
in business, 207
in democracy, 191
in ecoweb, 271
in immune system, 161
in industry, 206
in markets, 245
in muscle network, 166
in neural network, 169
in social network, 196, 197
predictability, 102
via slander, 210
stronglinker, 172–178, 196, 197, 255, 279
suburban sprawl, 261
suburbs, 236
success via weak links, 217
suicide bomber, 216
superconductivity, 86, 90
Superman, 225, 280
supermodular game, 297
supermodularity, 282
superorganism, 267
supersocial network, 243
superstability, 118
symbiosis, 34, 38, 105, 148, 265, 290
synapse, 167, 169
synchrony, 79–90, 138, 225, 228
human, 90
in insect networks, 182
in muscle network, 164, 165
in neural network, 167, 170, 278
in social network, 188
nested, 279
optimal level, 87, 96, 287
partial, 88
postmodern, 214
weak-link-induced, 86
syntalansis, 82, 83, 88, *see Glossary*
syphilis, 81
system logic, 273, 274

taboo, 215
tacit knowledge, 199, 260
teachers, 105
technological
change, 248, 252, 258
network, 16, 301
television, 212, 261
tension, 59, 66, 89, 133
economic, 244
terrorism, 70, 74, 207–209
September 11, 72, 205
textual network, 97
theocracy, 251
thermodynamics, 55, 112
thrift, 176
thrifty phenotype, 173, 254
thunder, 62
tick quake, 62, 171
time series, 25
tinkering, 91, 166, 240
tolerance, 193, 198, 217, 279
Tolstoy, L.N., 228, 230
top network, *see* network, top
topological phase transition, 20, 39, 74–79, 138, 140, 147, 158, 159, 182, 183, 188, 278, *see Glossary*
in firms, 203
in history, 250
in textual network, 228
totem, 215
Tourette syndrome, 63, 171
town planning, 236
Toyota, 202, 204
trade network, 72
tradition, 191, 197
traffic, 237
flow, 52
transcendent links, 257
transcription, 55
transcriptional network, 44, 127

tree, 22, 267
trophic cascade, 266, 268
trust, 198, 199, 260, 308, 309
  between nations, 249
  erosion of, 261
tulip speculation, 243
tuna, 183
Turks, 193

unemployment, 79
universality, 23
Universe, 27, 40
  as fractal, 306
  weak links, 306
unpredictability, 32

vaccination, 161
van der Waals force, 123
Venice, 234
venous network, 162
VIP club, 39, 178, 195
virtual world, 214, 215
virus, 265
viscosity, 61
volcanic activity, 61

war, 42, 248, 249
  civil, 88, 251, 258
water, 11, 44, 113, 120–122, 137
  as weak link, 122
Watts model, 71
waves in stadiums, 88, 188, 280
weak link, 2, 12, 38, 43–45, 55, 65, 66, 73, 74, 79, 86, 122, 134, 137, 139, 156, 198, *see Glossary*
  as buffer, 145
  as stabilizer, 95–109, 148, 193, 199, 217, 256, 267, 268, 287, 291, 305
  between nations, 249
  cost, 102
  definition, 3, 101, 134, 281
  energy, 117, 118
  in animal community, 184
  in complex equilibrium, 298–301
  in complex network, 292
  in consciousness, 169
  in democracy, 251
  in ecosystem, 263–267
  in firms, 205
  in immune system, 161
  in insect networks, 182
  in language network, 220, 222
  in learning, 169
  in markets, 245, 247
  in modern society, 215
  in muscle network, 164, 166
  in neural network, 168
  in psychological network, 178
  in social network, 197
  in software networks, 239
  in stability landscape, 286–288
  in towns, 237
  in Universe, 306
  intermodular, 133
  intersegmental, 192, 193
  optimal level, 100, 149, 200
  summary, 276–278
  therapy, 151
  transience, 102
  via ambiguity, 223
  via gossip, 210
  via grandmothers, 185
  via water, 122, 137
  via women, 195
weaklinker, 172–178, 196, 225, 279, 300, 301
wealth distribution, 16, 18, 27, 247
weight lifter, 164, 165
white noise, 30, 49–51
Winfree model, 86
wisdom, 307
witch-burning, 258
women, 100, 186, 195, 196, 217, 279
  in leadership, 205
  under stress, 259
woo quake, 64
wooing, 278
words, 97, 110, 219, 220

world economy, 33, 42, 154, 202, 243–247
  as top network, 258
World Wide Web, 16, 19, 70, 154

xenobiotics, 269

yeast, 16, 39, 44, 101, 132–134, 265, 304
  evolution, 147
  prion, 135, 136

Zipf law, 16, 219, 235

Printing: Krips bv, Meppel, The Netherlands
Binding: Stürtz, Würzburg, Germany